NUMERICAL METHODS AND THEIR APPLICATIONS TO LINEAR ALGEBRA

NUMERICAL METHODS AND THEIR APPLICATIONS TO LINEAR ALGEBRA

Edited by:

Olga Moreira

ARCLER
PRESS

www.arclerpress.com

Numerical Methods and their applications to Linear Algebra

Olga Moreira

Arcler Press
2010 Winston Park Drive,
2nd Floor
Oakville, ON L6H 5R7
Canada
www.arclerpress.com
Tel: 001-289-291-7705
001-905-616-2116
Fax: 001-289-291-7601
Email: orders@arclereducation.com

ISBN: 978-1-77361-557-8 (Hardcover)

DECLARATION

ABOUT THE EDITOR

Olga Moreira obtained her Ph.D. in Astrophysics from the University of Liege (Belgium) in 2010, her BSc. in Physics and Applied Mathematics from the University of Porto (Portugal). Her post-graduate travels and international collaborations with the European Space Agency (ESA) and European Southern Observatory (ESO) led to great personal and professional growth as a scientist. Currently, she is working as an independent researcher, technical writer, and editor in the fields of Mathematics, Physics, Astronomy and Astrophysics.

TABLE OF CONTENTS

SECTION III

SECTION IV

LIST OF CONTRIBUTORS

Ondrej Sluciak
TU Wien, Institute of Telecommunications, Gusshausstrasse 25/E389, 1040 Vienna, Austria

Hana Straková
University of Vienna, Faculty of Computer Science, Theory and Applications of Algorithms, Währingerstrasse 29, 1090 Vienna, Austria

Markus Rupp
TU Wien, Institute of Telecommunications, Gusshausstrasse 25/E389, 1040 Vienna, Austria

Wilfried Gansterer
University of Vienna, Faculty of Computer Science, Theory and Applications of Algorithms, Währingerstrasse 29, 1090 Vienna, Austria

A. H. Refahi Sheikhani
Department of Applied Mathematics, Faculty of Mathematical Sciences, Lahijan Branch, Islamic Azad University, Lahijan, Iran

S. Kordrostami
Department of Applied Mathematics, Faculty of Mathematical Sciences, Lahijan Branch, Islamic Azad University, Lahijan, Iran

Mahdi Tohidian
Department of Electrical Engineering, Amirkabir University of Technology, Tehran, Iran

Hamidreza Amindavar
Department of Electrical Engineering, Amirkabir University of Technology, Tehran, Iran

Ali M Reza
Electrical Engineering and Computer Science, University of Wisconsin-Milwaukee, Milwaukee, WI 53201–0784, USA

Chun-Yueh Chiang
Center for General Education, National Formosa University, Huwei 632, Taiwan

Hung-Yuan Fan
Department of Mathematics, National Taiwan Normal University, Taipei 116, Taiwan

Matthew M Lin
Department of Mathematics, National Chung Cheng University, Chia-Yi 621, Taiwan

Hsin-An Chen
Department of Mathematics, National Chung Cheng University, Chia-Yi 621, Taiwan

Kanae Akaiwa
Graduate School of Informatics, Kyoto University, Yoshida-Hommachi, Sakyo-ku, Kyoto 606-8501, Japan

Masashi Iwasaki
Department of Informatics and Environmental Sciences, Kyoto Prefectural University, 1-5 Nakaragi-cho, Shimogamo, Sakyo-ku, Kyoto 606-8522, Japan

Koichi Kondo
Graduate School of Science and Engineering, Doshisha University, 1-3 Tatara Miyakodani, Kyoto 610-0394, Japan

Yoshimasa Nakamura
Graduate School of Informatics, Kyoto University, Yoshida-Hommachi, Sakyo-ku, Kyoto 606-8501, Japan

Hongyu Qin
Wenhua College, Wuhan 430074, China

Zhiyong Wang
School of Mathematical Sciences, University of Electronic Science and Technology of China, Sichuan 611731, China

Fumin Zhu
College of Economics, Shenzhen University, Shenzhen 518060, China

Jinming Wen
Department of Electrical and Computer Engineering, University of Toronto, Toronto, M5S3G4, Canada

Murat Osmanoglu
Department of Mathematical Engineering, Chemical and Metallurgical Faculty, Yildiz Technical University, Esenler 34210, Istanbul, Turkey

Mustafa Bayram
Department of Mathematical Engineering, Chemical and Metallurgical Faculty, Yildiz Technical University, Esenler 34210, Istanbul, Turkey

Zoran Nikolic
DSP Emerging End Equipment, Texas Instruments Inc., 12203 SW Freeway, MS722, Stafford, TX 77477, USA

Ha Thai Nguyen
Coordinated Science Laboratory, Department of Electrical and Computer Engineering, University of Illinois at Urbana-Champaign, 1308 West Main Street, Urbana, IL 61801, USA

Gene Frantz
Application Specific Products, Texas Instruments Inc., 12203 SW Freeway, MS701, Stafford, TX 77477, USA

Alfredo Remon
Department of Engineering and Computer Sciences, University Jaime I, Castellon, Spain

Sergio Sanchez
Hyperspectral Computing Laboratory (HyperComp), Department of Technology of Computers and Communications, University of Extremadura, Caceres, Spain.

Sergio Bernabe
Hyperspectral Computing Laboratory (HyperComp), Department of Technology of Computers and Communications, University of Extremadura, Caceres, Spain.

Enrique S Quintana-Ort´ı
Department of Engineering and Computer Sciences, University Jaime I, Castellon, Spain

Antonio Plaza
Hyperspectral Computing Laboratory (HyperComp), Department of Technology of Computers and Communications, University of Extremadura, Caceres, Spain.

Masahiro Yukawa
Mathematical Neuroscience Laboratory, BSI, RIKEN, 2-1 Hirosawa, Wako, Saitama 351-0198, Japan

Isao Yamada
Department of Communications and Integrated Systems, Tokyo Institute of Technology, Meguro-ku, Tokyo 152-8552, Japan

Martin Bromberger
Max-Planck-Institut für Informatik, Saarland Informatics Campus E1 4, 66123 Saarbrücken, Germany
Graduate School of Computer Science, Saarland Informatics Campus E1 3, 66123 Saarbrücken, Germany

Christoph Weidenbach
Max-Planck-Institut für Informatik, Saarland Informatics Campus E1 4, 66123 Saarbrücken, Germany

LIST OF ABBREVIATIONS

ANC	Abundance Non-Negativity Constraint
APA	Affine Projection Algorithm
APSM	Adaptive Projected Subgradient Method
ARMA	Autoregressive Moving Average
AVIRIS	Airborne Visible Infra-Red Imaging Spectrometer
BLAS	Basic Linear Algebra Subprograms
BVs	Boundary Values
CDMA	Code-Division Multiple Access
DAG	Directed Acyclic Graph
DARE	Discrete-Time Algebraic Riccati Equation
DCT	Discrete Cosine Transform
DFT	Discrete Fourier transform
DSPs	Digital Signal Processors
EVD	Eigenvalue Decomposition
FPGAs	Field Programmable Gate Arrays
FWL	Fractional Word Length
GPUs	Graphics Processing Units
GSVD	Generalized Singular-Value Decomposition
IBVP	Initial Boundary Value Problem
IVs	Initial Values
IWL	Integer Word Length
LNAF	LMSNewton Adaptive Filter
LPDAEs	Linear Partial Differential Algebraic Equations
LQ	Linear Quadratic
MAC	Multiply and Accumulate
MIMO	Multiple-Input Multiple-Output
NCLS	Non-Negative Constrained Least Squares
NLMS	Normalized Least Mean Square

OFDM	Orthogonal Frequency Division Multiplexing
PCT	Principal Component Transform
QE	Quantifier Elimination
QNAF	Quasi-Newton Adaptive Filter
SARE	Stochastic Algebraic Riccati Equations
SIEP	Structured Inverse Eigenvalue Problem
SMT	Satisfiability Modulo Theories
SNR	Signal-To-Noise Ratio
SVD	Singular Value Decomposition
TDAF	Transform-Domain Adaptive Filter
V-APSM	Variable-Metric Adaptive Projected Subgradient Method
VD	Virtual Dimensionality
VLIW	Very Long Instruction Word
WL	Word Length
WSN	Wireless Sensor Network

PREFACE

This edited book, *"Numerical Methods and their applications to Linear Algebra"*, is a collection of contemporary open access articles that highlight new numerical methods and computational techniques for solving of linear algebra problems. Comprised of 11 chapters, this book consists of five major sections:

- Section 1, Chapters 1 to 3, focuses on the matrix and eigenvalue decompositions. It begins by presenting an algorithm for QR decomposition based on the Gram-Schmidt orthogonalization, followed by several iterative methods for computing the singular-value decomposition of a large sparse matrix and eigenvalue decomposition of para-Hermitian matrices.
- Section 2, Chapters 4 and 5, focuses on the matrix perturbation analysis, the nonlinear matrix differences, the inverse eigenvalue problem and the construction of a tridiagonal matrix with specified multiple eigenvalues.
- Section 3, Chapters 6 to 7, focus on the stability analysis and numerical methods for solving differential algebraic equations.
- Section 4, Chapters 8 to 9, presents several examples of numerical algebra techniques applied to engineering, scientific and real-world problems. Several iterative methods based on Newton iteration, Poisson integration model, an iterative Krylov subspace solver are also featured in this section.
- Section 5, Chapters 10 to 11, introduces the reader to the development of a variable-metric adaptive projected subgradient method, new techniques for linear arithmetic based on the linear cube transformation.

The intended audience of this edited book will mainly consist of graduate or advanced undergraduate students in engineering, science, and mathematics. The contents of this volume will also be of particular interest to researching academics in engineering and science whom to update their knowledge of modern numerical linear algebra methodologies.

SECTION I

CHAPTER 1

DISTRIBUTED GRAM-SCHMIDT ORTHOGONALIZATION WITH SIMULTANEOUS ELEMENTS REFINEMENT

Ondrej Sluciak[1], Hana Straková[2], Markus Rupp[1] and Wilfried Gansterer[2]

[1]TU Wien, Institute of Telecommunications, Gusshausstrasse 25/E389, 1040 Vienna, Austria

[2]University of Vienna, Faculty of Computer Science, Theory and Applications of Algorithms, Währingerstrasse 29, 1090 Vienna, Austria.

ABSTRACT

We present a novel distributed QR factorization algorithm for orthogonalizing a set of vectors in a decentralized wireless sensor network. The algorithm is

Citation: Slučiak, O., Straková, H., Rupp, M., & Gansterer, W. (2016). Distributed Gram-Schmidt orthogonalization with simultaneous elements refinement. EURASIP journal on advances in signal processing, 2016(1), 25.(13 pages).

based on the classical Gram-Schmidt orthogonalization with all projections and inner products reformulated in a recursive manner. In contrast to existing distributed orthogonalization algorithms, all elements of the resulting matrices **Q** and **R** are computed simultaneously and refined iteratively after each transmission. Thus, the algorithm allows a trade-off between run time and accuracy. Moreover, the number of transmitted messages is considerably smaller in comparison to state-of-the-art algorithms. We thoroughly study its numerical properties and performance from various aspects. We also investigate the algorithm's robustness to link failures and provide a comparison with existing distributed QR factorization algorithms in terms of communication cost and memory requirements.

Keywords: Distributed processing, Gram-Schmidt orthogonalization, QR factorization

INTRODUCTION

Orthogonalizing a set of vectors is a well-known problem in linear algebra. Representing the set of vectors by a matrix $A \in R^{n \times m}$, with $n \geq m$, several orthogonalization methods are possible. One example is the so-called *reduced QR factorization* (matrix decomposition), **A=Q R**, with a matrix $Q \in R^{n \times m}$ having orthonormal columns, and an upper triangular matrix $R \in R^{m \times m}$ containing the coefficients of the basis transformation [1]. In the signal processing area, QR factorization is used widely in many applications, e. g., when solving linear least squares problems or decorrelation [2, 3, 4]. In adaptive filtering, a decorrelation method is typically used as a pre-step for increasing the learning rate of the adaptive algorithm [5], ([6], p. 351), ([7], p. 700).

From an algorithmic point of view, there are many methods for computing QR factorization with different numerical properties. A standard approach is the *Gram-Schmidt orthogonalization algorithm*, which computes a set of orthonormal vectors spanning the same space as the given set of vectors. Other methods include Householder reflections or Givens rotations, which are not considered in this paper.

Optimization of QR factorization algorithms for a specific target hardware has been addressed in the literature several times (e.g., [8, 9]). Parallel algorithms for computing QR factorization, which are applicable for reliable systems with fixed, regular, and globally known topology, have been investigated extensively (e.g., [10, 11, 12, 13]).

Besides parallel algorithms, there are two other potential approaches for computation across a distributed network. In the standard—centralized—approach, the data are collected from all nodes and the computation is performed at a fusion center. Another approach is to consider distributed algorithms for fully *decentralized* networks without any fusion center where all nodes have the same functionality and each of them communicates only with its neighbors. Such an approach is typical for sensor-actuator networks or autonomous swarms of robotic networks [14]. Nevertheless, the investigation of *distributed* QR factorization algorithms designed for loosely coupled distributed systems with independently operating distributed memory nodes and with possibly unreliable communication links has only started recently [3, 15, 16]. In the following, we focus on algorithms for such decentralized networks.

Motivation

The main goal of this paper is to present a novel distributed QR factorization algorithm—DS-CGS—which is based on the classical Gram-Schmidt orthogonalization. The algorithm does not require any fusion center and assumes only local communication between neighboring nodes without any global knowledge about the topology. In contrast to existing distributed approaches, the DS-CGS algorithm computes the approximations of all elements of the new orthonormal basis *simultaneously* and as the algorithm proceeds, the values at *all* nodes are refined iteratively, approximating the exact values of **Q** and **R**. Therefore, it can deliver an estimate of the full matrix result *at any moment* of the computation. As we will show, this approach is, among others, superior to existing methods in terms of the number of transmitted messages in the network.

In Section 2, we briefly recall the concept of a consensus algorithm which we use later in the distributed orthogonalization algorithm. In Section 3, we review the basics of the QR decomposition and existing distributed methods. In Section 4, we describe the proposed distributed Gram-Schmidt orthogonalization algorithm with simultaneous refinements of all elements (DS-CGS). We experimentally compare DS-CGS with other distributed approaches in Section 5 where we also investigate the properties of DS-CGS from many different viewpoints. Section 6 concludes the paper.

Notation and terminology

In what follows, we use k as the node index, N^k denotes the set of neighbors of node k, N denotes the (known) number of nodes in the network, E the set of edges (links) of the network, d_k the kth node degree (dk=|Nk|), $\overline{d}$ the average node degree of the network, and t a discrete time (iteration) index.

We will describe the behavior of the distributed algorithm from a network (global) point of view with the corresponding vector/matrix notation. For example, the (column) vector of all ones denoted by **1**, corresponds to all nodes having value 1. In general, we denote the number of rows of a matrix by n and the number of columns by m. Element-wise division of two vectors is denoted as $\mathbf{z} = \frac{\mathbf{x}}{\mathbf{y}} \equiv \frac{x_i}{y_i}, \forall i$, element-wise multiplication of two vectors as $\mathbf{z}=\mathbf{x}\circ\mathbf{y}\equiv x_i y_i, \forall i$ and of two matrices as $\mathbf{Z}=\mathbf{X}\circ\mathbf{Y}$. The operation X⊛Y is defined as follows: Having two matrices $\mathbf{X}=(\mathbf{x}_1,\mathbf{x}_2,\ldots,\mathbf{x}_m)$ and $\mathbf{Y}=(\mathbf{y}_1,\mathbf{y}_2,\ldots,\mathbf{y}_m)$, the resulting matrix Z=X⊛Y is a stacked matrix of all matrices $\mathbf{Z}_i$ such that Zi=(x1,x2,…,xi)∘ $(\underbrace{(1,1,\ldots,1)}_{i}$ ⊗yi+1) denotes the Kronecker product; $i = 1,2,\ldots,m-1$), i.e., $\mathbf{Z} = (\underbrace{\mathbf{x}_1\circ\mathbf{y}_2}_{\mathrm{Z}_1}, \underbrace{\mathbf{x}_1\circ\mathbf{y}_3, \mathbf{x}_2\circ\mathbf{y}_3}_{\mathrm{Z}_2}, \ldots, \underbrace{\mathbf{x}_{m-2}\circ\mathbf{y}_m, \mathbf{x}_{m-1}\circ\mathbf{y}_m}_{\mathrm{Z}_{m-1}})$ thus creating a big matrix containing combinations of column vectors: $\mathbf{Z} \in \mathbb{R}^{n\times \frac{m^2-m}{2}}$. This later corresponds in our algorithm to the off-diagonal elements of the matrix **R**. Also note that all variables with the "hat" symbol, e.g., $\hat{u}$ (t) represent variables that are computed locally at nodes, while variables with the "tilde" symbol, e.g., $\tilde{u}$ (t), are updated based on the information from neighbors.

AVERAGE CONSENSUS ALGORITHM

We model a wireless sensor network (WSN) by synchronously working nodes which broadcast their data into their neighborhood within a radius ρ (so-called geometric topology). The WSN is considered to be static, connected, and with error-free transmissions (except for Section 5.4ahead). Although the practicality of synchronicity can be argued [17, 18], we note that it is not an unrealizable assumption [19].

In the following, we briefly review the classical consensus algorithm for computing the *average* of values distributed in a network. Note that the algorithm can be easily adapted to computing a *sum* by multiplying the final average value (arithmetic mean) by the total number of nodes N.

The distributed average consensus algorithm computes an estimate of the global *average* of distributed initial data $\mathbf{x}(0)$ at each node k of a WSN. In every iteration t, each node updates its estimate using the weighted data received from its neighbors, i.e.,

$$x_k(t) = [\mathbf{W}]_{kk}\, x_k(t-1) + \sum_{k' \in \mathcal{N}_k} [\mathbf{W}]_{kk'}\, x_{k'}(t-1)$$

or from a global (network) point of view

$$\mathrm{x(t)=Wx(t-1)}. \tag{1}$$

The selection of the *weight matrix* $\mathbf{W}$, representing the connections in a strongly connected network, crucially influences the convergence of the average consensus algorithm [20, 21, 22]. The main condition for the algorithm to converge is that the largest eigenvalue of $\mathbf{W}$ is equal to 1, i.e., $\lambda_{\max} = 1$, with multiplicity one, and that each row of $\mathbf{W}$ sums up to 1. It can then be directly shown [20] that the value $x_k(t)$ at each node converges to a common global value, e.g., average of the initial values.

If not stated otherwise, we use the so-called Metropolis weights [22] for matrix $\mathbf{W}$, i.e.,

$$[\mathbf{W}]_{ij} = \begin{cases} \frac{1}{1+\max\{d_i, d_j\}} & \text{if } (i,j) \in \mathcal{E}, \\ 1 - \sum_{i' \in \mathcal{N}_i} [\mathbf{W}]_{ii'} & \text{if } i = j, \\ 0 & \text{otherwise.} \end{cases} \tag{2}$$

These weights guarantee that the consensus algorithm converges to the average of the initial values.

QR FACTORIZATION

As mentioned in Section 1, there exist many algorithms for computing the QR factorization with different properties [1, 23]. In this paper we utilize the QR decomposition based on the *classical* Gram-Schmidt orthogonalization method (in ℓ^2 space).

Centralized classical Gram-Schmidt orthogonalization

Given matrix $\mathrm{A} = (\mathrm{a}_1, \mathrm{a}_2, \ldots, \mathrm{a}_m) \in \mathrm{R}^{n \times m}$, $n \geq m$, classical Gram-Schmidt orthogonalization (CGS) computes a matrix $\mathrm{Q} \in \mathrm{R}^{n \times m}$ with orthonormal columns and an upper-triangular matrix $\mathrm{R} \in \mathrm{R}^{m \times m}$, such that $\mathbf{A}=\mathbf{Q}\,\mathbf{R}$. Denoting

$$\mathbf{Q} = (\mathbf{q}_1 \quad \mathbf{q}_2 \quad \dots \quad \mathbf{q}_m)$$

$$\mathbf{R} = \begin{pmatrix} \langle \mathbf{q}_1, \mathbf{a}_1 \rangle & \langle \mathbf{q}_1, \mathbf{a}_2 \rangle & \dots & \langle \mathbf{q}_1, \mathbf{a}_m \rangle \\ 0 & \langle \mathbf{q}_2, \mathbf{a}_2 \rangle & \langle \mathbf{q}_2, \mathbf{a}_3 \rangle & \dots \\ \vdots & & \ddots & \dots \\ 0 & \dots & 0 & \langle \mathbf{q}_m, \mathbf{a}_m \rangle \end{pmatrix}, \tag{3}$$

we have

$$\mathbf{q}_i = \frac{\mathbf{u}_i}{\|\mathbf{u}_i\|_2}, i = 1, 2, \dots, m, \tag{4}$$

and

$$\mathbf{u}_i = \mathbf{a}_i - \sum_{j=1}^{i-1} \frac{\langle \mathbf{q}_j, \mathbf{a}_i \rangle}{\langle \mathbf{q}_j, \mathbf{q}_j \rangle} \mathbf{q}_j, \quad i = 1, 2, \dots, m, \tag{5}$$

where $\|\mathbf{u}\|_2 = \sqrt{\sum_{i=1}^{n} u_i^2}$ and $\langle \mathbf{q}, \mathbf{a} \rangle = \sum_{i=1}^{n} q_i a_i$

It is known that the algorithm is numerically sensitive depending on the singular values (condition number) of matrix **A** as well as it can produce vectors $\mathbf{q}_i$ far from orthogonal when the matrix **A** is close to being rank deficient even in a floating-point precision [23]. Numerical stability can be improved by other methods, e.g., modified Gram-Schmidt method, Householder transformations, or Givens rotations [1, 23].

Existing Distributed Methods

Assuming that each node k stores its local values u_k^2 and $q_k a_k$, it is then straightforward to redefine the CGS in a distributed way, suitable for a WSN, by following the definition of the ℓ^2norm, i.e., $\|\mathbf{u}\|_2^2 = u_1^2 + u_2^2 + \dots + u_n^2$ (cf. (4)), and inner products, $\langle \mathbf{q,a} \rangle = q_1 a_1 + q_2 a_2 + \dots + q_n a_n$ (cf. (5)). The summations can then be computed using any distributed aggregation algorithm, e.g., average consensus [20][1] (see Section 2), and asynchronous gossiping algorithms [24], using only communication with the neighbors.

Nevertheless, to our knowledge, all existing distributed algorithms for orthogonalizing a set of vectors are based on the gossip-based *push-sum algorithm* [16, 24]. Specifically in [3], authors used a distributed CGS based on gossiping for solving a distributed least squares problem and in [15], a gossip-based distributed algorithm for *modified* Gram-Schmidt orthogonalization (MGS) was designed and analyzed. The authors also provided a quantitative comparison to existing parallel algorithms for QR factorization. A slight modification of the latter algorithm was introduced in [25], which we use for comparison in this paper. We denote the two Gossip-based distributed Gram-Schmidt orthogonalization algorithms as G-CGS [3] and G-MGS [25], respectively.

Since the classical Gram-Schmidt orthogonalization computes each column of the matrix **Q** from the previous column recursively, i.e., to know vector $\mathbf{q}_2$, we need to compute the norm of $\mathbf{u}_2$ which depends on vector $\mathbf{q}_1$, the existing distributed algorithms always need to wait for convergence of one column before proceeding with the next column. This may be a big disadvantage in WSNs as it requires a lot of transmissions. Also, if the algorithm fails at some moment, e.g., due to transmission errors, the matrices **Q** and **R** are incomplete and unusable for further application.

In contrast, the distributed algorithm proposed in this paper overcomes these disadvantages and computes approximations of all elements of the matrices **Q** and **R** simultaneously. All the norms and inner products are refined iteratively which leads to a significant decrease of transmitted messages, and also the algorithm brings an intermediate approximation of the whole matrices **Q** and **R** at any time instance.

DISTRIBUTED CLASSICAL GRAM-SCHMIDT WITH SIMULTANEOUS ELEMENTS REFINEMENT

As mentioned in Section 3.2, the Gram-Schmidt orthogonalization method can be computed in a distributed way using any distributed aggregation algorithm. We refer to CGS based on the average consensus (see Section 2) as AC-CGS. AC-CGS as well as G-CGS [3] and G-MGS [25] have the following substantial drawback.

In all Gram-Schmidt orthogonalization methods, the computation of the norms $\|\mathbf{u}_i\|$ and the inner products $\langle \mathbf{q}_j, \mathbf{a}_i \rangle$, $\langle \mathbf{q}_j, \mathbf{q}_j \rangle$, occurring in the matrices **Q** and **R**, depends on the norms and inner products computed from the previous columns of the input matrix **A**. Therefore, each node k must *wait* until the estimates of the previous norms $\|\mathbf{u}_j\|$ ($j < i$) have achieved an acceptable accuracy before processing the next norm $\|\mathbf{u}_i\|$ (a "cascading" approach; see [15]). The same holds also for computing the inner products. We here present a novel approach overcoming this drawback.

Rewriting Eqs. (4) and (5) by a recursion, we obtain

$$\hat{\mathbf{q}}_i(t) = \frac{\hat{\mathbf{u}}_i(t)}{\sqrt{N\tilde{\mathbf{u}}_i(t-1)}}, \qquad i = 1, 2, \ldots, m, \tag{6}$$

$$\hat{\mathbf{u}}_i(t) = \mathbf{a}_i - \mathbf{p}_i(t), \qquad i = 1, 2, \ldots, m, \tag{7}$$

where $\tilde{u}_i(t)$ is the approximation of $1/N \|\mathbf{u}_i\|_2^2 1$ at time t and

$$\mathbf{p}_i(t) = \sum_{j=1}^{i-1} \frac{\tilde{\mathbf{p}}^{(2)}_{j+(i-1)(i-2)/2}(t-1) \circ \hat{\mathbf{q}}_j(t-1)}{\tilde{\mathbf{q}}_j(t-1)},$$

with $\tilde{\mathbf{p}}^{(2)}_{j+(i-1)(i-2)/2}(t)$ being an approximation of the off-diagonal inner products $1/N\ \langle \mathbf{q}_j, \mathbf{a}_i \rangle\ \mathbf{1}$ ($\forall j<i$) of matrix **R** (cf. (3)) and $\tilde{q}_j(t)$ an approximation of $1/N\ \langle \mathbf{q}_j, \mathbf{q}_j \rangle\ \mathbf{1}$ at time t. Similarly, we define $\tilde{p}^{(1)}_i(t)$ to be an approximation of $1/N\ \langle \mathbf{q}_i, \mathbf{a}_i \rangle\ \mathbf{1}$. As we show later, $\tilde{u}_i(t)$, $\tilde{q}_j(t)$, $\tilde{p}^{(1)}_i(t)$, and $\tilde{\mathbf{p}}^{(2)}_{j+(i-1)(i-2)/2}(t)$ converge to $1/N\|u_i\|^2_2 1$, $1/N\ \langle \mathbf{q}_j, \mathbf{q}_j \rangle\ \mathbf{1}$, $1/N\ \langle \mathbf{q}_i, \mathbf{a}_i \rangle\ \mathbf{1}$, and $1/N\ \langle \mathbf{q}_j, \mathbf{a}_i \rangle\ \mathbf{1}$, respectively.

Similarly to the state-of-the-art methods (see Section 3.2), we further assume that the matrices $A \in \mathbb{R}^{n\times m} A \in \mathbb{R}^{n\times m}$ and $Q \in \mathbb{R}^{n\times m} Q \in \mathbb{R}^{n\times m}$ are distributed over the network row-wise, meaning that each node stores at least one row of the matrix **A** and corresponding rows of the matrix **Q** and each node stores the whole matrix **R**. In case $n>N$, more rows must be stored at the node and each node must sum the data locally before broadcasting to neighbors. Obviously, the data distribution over the network influences the speed of convergence of the algorithm, as can be seen also in the simulations ahead (see Section 5).

Notation $\mathbf{A}_k, \mathbf{Q}_k(t)$ here represent the rows of the matrices **A** and **Q** at a given node k at time t. If more rows are stored in one node, $\mathbf{A}_k$ and $\mathbf{Q}_k(t)$ are matrices, otherwise they are row vectors. Matrix $\mathbf{R}^{(k)}(t)$ represents the whole matrix **R** at node k at time t.

From a *global* (network) point of view, the algorithm is defined in Algorithm 1.

Algorithm 1: DISTRIBUTED GRAM-SCHMIDT ORTHOGONALIZATION WITH SIMULTANEOUS REFINEMENT (DS–CGS)

• Input matrix $A = (a_1, a_2, \dots, a_m) \in R^{n\times m}$ with $n \geq m$ is distributed row-wise across N nodes. If $n > N$, some nodes store more than one row. Each node computes the rows of Q corresponding to the stored rows of A and an estimate of the whole matrix R.
indices: $k = 1, 2, \dots, N$ (nodes); $i = 1, 2, \dots, m$ (columns).

1. Initialization ($t = 0$):

$$\tilde{\mathbf{U}}(0) = \mathbf{A} \circ \mathbf{A}, \quad \hat{\mathbf{U}}(0) = \mathbf{A},$$
$$\tilde{\mathbf{Q}}(0) = \mathbf{A} \circ \mathbf{A}, \quad \hat{\mathbf{Q}}(0) = \mathbf{A},$$
$$\tilde{\mathbf{P}}^{(1)}(0) = \mathbf{A} \circ \mathbf{A}, \ \tilde{\mathbf{P}}^{(2)}(0) = \mathbf{A} \circledast \mathbf{A}$$

2. Repeat for $t = 1, 2, \dots$

(a) Compute locally at each node k

$$\mathbf{p}_i(t) = \sum_{j=1}^{i-1} \frac{\tilde{\mathbf{p}}^{(2)}_{j+(i-1)(i-2)/2}(t-1) \circ \hat{\mathbf{q}}_j(t-1)}{\tilde{\mathbf{q}}_j(t-1)}$$
$$\hat{\mathbf{u}}_i(t) = \mathbf{a}_i - \mathbf{p}_i(t)$$
$$\hat{\mathbf{q}}_i(t) = \frac{\hat{\mathbf{u}}_i(t)}{\sqrt{N\tilde{\mathbf{u}}_i(t-1)}}$$

(b) At each node k store

$$\mathbf{Q}_k(t) = (\hat{q}_{k,1}(t), \hat{q}_{k,2}(t), \dots, \hat{q}_{k,m}(t)), \quad \text{and}$$

$$\mathbf{R}^{(k)}(t) = N \begin{pmatrix} \tilde{p}^{(1)}_{k,1}(t) & \tilde{p}^{(2)}_{k,1}(t) & \dots & \tilde{p}^{(2)}_{k,(m^2-3m+4)/2}(t) \\ 0 & \tilde{p}^{(1)}_{k,2}(t) & \tilde{p}^{(2)}_{k,3}(t) & \dots \\ \vdots & & \ddots & \dots \\ 0 & \dots & 0 & \tilde{p}^{(1)}_{k,m}(t) \end{pmatrix}$$

(c) Aggregate data

$$\underbrace{\begin{matrix}\Psi^{(1)} \\ \Psi^{(2)} \\ \Psi^{(3)} \\ \Psi^{(4)}\end{matrix}}_{\Psi(t)} \begin{matrix} = \\ = \\ = \\ = \end{matrix} \underbrace{\begin{matrix}\tilde{\mathbf{U}}(t-1) \\ \tilde{\mathbf{Q}}(t-1) \\ \tilde{\mathbf{P}}^{(1)}(t-1) \\ \tilde{\mathbf{P}}^{(2)}(t-1)\end{matrix}}_{= \mathbf{X}(t-1)} \begin{matrix} + \\ + \\ + \\ + \end{matrix} \underbrace{\begin{matrix}\hat{\mathbf{U}}(t) \circ \hat{\mathbf{U}}(t) - \hat{\mathbf{U}}(t-1) \circ \hat{\mathbf{U}}(t-1) \\ \hat{\mathbf{Q}}(t) \circ \hat{\mathbf{Q}}(t) - \hat{\mathbf{Q}}(t-1) \circ \hat{\mathbf{Q}}(t-1) \\ \hat{\mathbf{Q}}(t) \circ \mathbf{A} - \hat{\mathbf{Q}}(t-1) \circ \mathbf{A} \\ \hat{\mathbf{Q}}(t) \circledast \mathbf{A} - \hat{\mathbf{Q}}(t-1) \circledast \mathbf{A}\end{matrix}}_{+ \ \Delta\mathbf{S}(t)}$$

(d) If $n > N$, sum $\mathbf{\Psi}(t)$ locally at the nodes and broadcast $\mathbf{\Psi}(t) = \left(\Psi^{(1)}, \Psi^{(2)}, \Psi^{(3)}, \Psi^{(4)}\right)$, ($\mathbf{\Psi}(t) \in \mathbb{R}^{N\times \frac{m^2+5m}{2}}$) to the neighbors, i.e.,

$$\underbrace{\left(\tilde{\mathbf{U}}(t), \tilde{\mathbf{Q}}(t), \tilde{\mathbf{P}}^{(1)}(t), \tilde{\mathbf{P}}^{(2)}(t)\right)}_{\mathbf{X}(t)} = \mathbf{W}\mathbf{\Psi}(t).$$

Proof of convergence of DS-CGS.

For the first column, vector i=1, $\hat{u}_1(t)=a_1$, and thus the convergence results of the average consensus, see Section 2, apply, i.e., as $t\to\infty$, the nodes will monotonically reach the common values, i.e., $\tilde{u}_1(t)=1/N\|a_1\|_2^2 1$ and thus also, $\hat{\mathbf{q}}_1(t) = \frac{\mathbf{a}_1}{\|\mathbf{a}_1\|_2^2}$, $\tilde{\mathbf{q}}_1(t) = 1/N\mathbf{1}$, $\tilde{\mathbf{p}}_1^{(1)}(t) = 1/N\|\mathbf{a}_1\|_2^2\mathbf{1}$, and $\tilde{\mathbf{p}}_1^{(2)}(t) = 1/N\langle\mathbf{a}_1, \mathbf{a}_2\rangle\mathbf{1}$.

Furthermore, for all columns $i>1$, all the elements depend only on the first column ($i=1$), e.g., Eq. (7), $\hat{u}_2(t)=a_2-\frac{\tilde{\mathbf{p}}_1^{(2)}(t-1)\circ\hat{\mathbf{q}}_1(t-1)}{\tilde{\mathbf{q}}_1(t-1)}$ from Eq. (6) $\hat{\mathbf{q}}_1(t) = \frac{\hat{\mathbf{u}}_1(t)}{\sqrt{N\tilde{\mathbf{u}}_1(t-1)}}$). Thus, eventually, $\hat{u}_2(t)$ will converge to $\mathbf{u}_2$ (Eq. (5)) and similarly will do all norms and inner products (Eqs. (4) and (5)) of matrix **Q** and **R**.

Intuitively, we can see that as $\tilde{u}_1(t)$ converges to its steady state, all other variables converge, with some "delay," to their steady states as well. We may say that as the first column converges, it "drags" other elements to their steady states. In the worst case, the consequent (following) column starts to converge only when the previous column is fully converged. This behavior differs from the known methods where we have to wait for $\tilde{u}\,1(t)$ $\tilde{u}\,1(t)$ to be converged before computing other terms.

Note that instead of knowing the number of nodes N and using it as a normalization constant, we could transmit an additional weight vector $\omega(t)\in\mathbb{R}^{N\times1}$, i.e., $\Psi^{(0)}(t)=\boldsymbol{\omega}(t)$ and $\boldsymbol{\Psi}(t)=(\Psi^{(0)}(t), \Psi^{(1)}(t), \Psi^{(2)}(t), \Psi^{(3)}(t), \Psi^{(4)}(t))$, such that $\boldsymbol{\omega}(0)=(1,0,\ldots,0)^{\top}$ and Eq. (6) would change only slightly[2], i.e.,

$$\hat{\mathbf{q}}_i(t) = \frac{\hat{\mathbf{u}}_i(t)}{\sqrt{\frac{\mathbf{1}}{\boldsymbol{\omega}(t)} \circ \tilde{\mathbf{u}}_i(t-1)}}.$$

We note that the normalization constant N (or $\boldsymbol{\omega}(t)$, respectively) affects only[3] the *orthonormality* (columns remain orthogonal but not normalized) of the columns of the matrix $\mathbf{Q}(t)$, and in case only orthogonality is sufficient, as in [26], we can omit this constant. We can, thus, overcome the necessity of knowing the number of the nodes or reduce the number of transmitted data in the network, respectively.

Relation to Dynamic Consensus Algorithm

The dynamic consensus algorithm is a distributed algorithm which is able to track the average of a time-varying input signal. There exist many variations of the algorithm, e.g., [27, 28, 29, 30, 31, 32, 33]. Comparing the proposed DS-CGS algorithm with a dynamic consensus algorithm from [30, 32], we observe an interesting resemblance.

Formulating DS-CGS from a global point of view, i.e.,

$X(t)=W[X(t-1)+\Delta S(t)]$,

we observe that it is a variant of the dynamic consensus algorithm with an "input signal" $\mathbf{S}(t)$. However, the "input signal" $\mathbf{S}(t)$ in our case is very complicated as it depends on $\mathbf{X}(t-1)$ and $\mathbf{S}(t-1)$ and cannot be considered as an independent signal as it is usually considered in dynamic consensus algorithms. Therefore, it is difficult to analyze the properties of this input signal and convergence conditions of DS-CGS based on the dynamic consensus algorithm. It is also beyond the scope and focus of this paper to analyze this algorithm in general. Nevertheless, some analysis of this type of dynamic consensus algorithm, for a general input signal, together with the bounds on convergence speed, has been conducted in [34].

PERFORMANCE OF DS-CGS

In our simulations, we consider a connected WSN with $N = 30$ nodes. We explore the behavior of DS-CGS for various topologies: *fully connected* (each node is connected to every other node), *regular* (each node has the same degree d), and *geometric* (each (randomly deployed) node is connected to all nodes within some radius ρ—a WSN model). If not stated otherwise, the randomly generated input matrix $A \in R^{300\times 100}$ has uniformly distributed elements from the interval [0,1] and a low condition number $\kappa(\mathbf{A})=35.7$. In Section 5.3.2, we, however, investigate the influence of various input matrices with different condition numbers on the algorithm's performance.

Also, except for the Sections 5.3.1 and 5.4, for the consensus weight matrix we use the metropolis weights (Eq. (2)).

The confidence intervals were computed from the several instantiations using a bootstrap method [35].

Orthogonality and Factorization Error

As performance metrics in the simulations, we use the following:

- *Relative factorization error*— $\frac{\|\mathbf{A}-\mathbf{Q}(t)\mathbf{R}^{(k)}(t)\|_2}{\|\mathbf{A}\|_2}$ —which measures the accuracy of the QR factorization at node k,
- *Orthogonality error*— $\|\mathbf{I}-\mathbf{Q}(t)^{\top}\mathbf{Q}(t)\|_2$ —which measures the orthogonality of the matrix $\mathbf{Q}(t)$ (see step 2 of the algorithm).

Note that both errors are calculated from the network (global) perspective and as depicted, they are not known locally at the nodes, since only $\mathbf{R}^{(k)}(t)$ is local at each node, whereas $\mathbf{Q}(t)$ is distributed row-wise across the nodes

($\mathbf{Q}_k(t)$). From now on, we simplify the notation by dropping the index t in $\mathbf{Q}(t)$ and $\mathbf{R}^{(k)}(t)$. The simulation results for a geometric topology with an average node degree $\bar{d}$ = 8.533 are depicted in Figure 1. Since both errors behave almost identically (compare Figure 1 a, b) and since each node k can compute a *local factorization error* $\|\mathbf{A}_k - \mathbf{Q}_k \mathbf{R}^{(k)}\|_2 / \|\mathbf{A}_k\|_2$ from its local data, we conjecture that such local error evaluation can be used also as a local stopping criterion in practice. Note that this fact was used in [26] for estimating a network size.

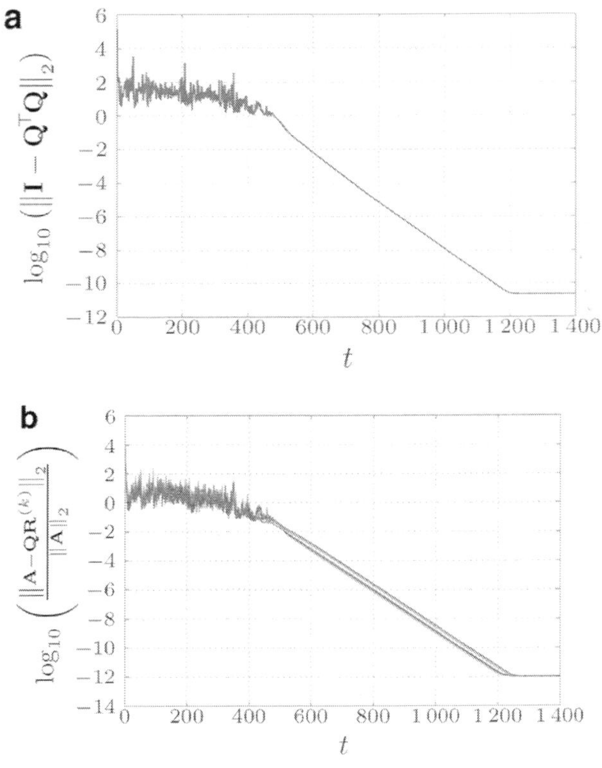

Figure 1: Example of orthogonality (**a**) and factorization error (**b**) for each node k for a geometric topology with d=8.533. N=30, k=1, 2,...,30.

Note that the error at the beginning stage in Figure 1 is caused by the disagreement and not converged norms and inner products across the nodes, i.e., the values of $\tilde{u}$ (t), $\tilde{Q}$(t), $\tilde{P}$(1)(t), and $\tilde{P}$ (2)(t). We also observe that the error floor[4] is highly influenced by the network topology, weights of matrix **W**, and condition number of input matrix **A**. We investigate these properties in Section 5.3.

Initial Data Distribution

If $n>N$, some nodes store more than one row of **A**. Thus, before doing distributed summation (broadcasting to neighbors), every node has to locally sum the values of its local rows.

Simulations show that the convergence behavior of DS-CGS strongly depends on the distribution of the rows across the network (see Figure 2). We investigate the following cases: (1) each node stores ten rows of **A** ("uniform"); (2) 271 rows are stored in the node with the lowest degree, the other 29 rows in the remaining 29 nodes; and (3) 271 rows are stored in the node with the highest degree, the rest in the remaining 29 nodes.

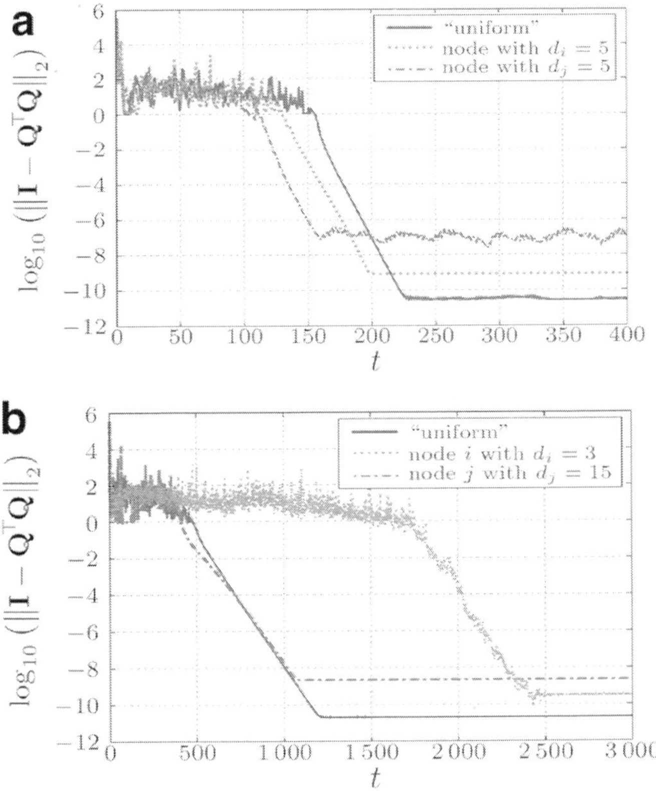

Figure 2: Convergence for networks with different topology and initial data distribution: either all nodes store the same amount of data ("uniform") or most of the data is stored in one node (with minimum or maximum degree) (**a** - Regular topology with $\bar{d}$ =5; **b** - Geometric topology with $\bar{d}$ =5). In case of the regular topology (**a**), the nodes *i,j* are picked randomly.

We observe that not only the initial distribution of the data influences the convergence behavior but also the topology of the underlying network. In the case of a regular topology (Figure 2 a), the influence of the distribution is small and relatively weak in terms of convergence time but stronger in terms of the final accuracy achieved. We recognize that the difference between the nodes comes only from the variance of the values in input matrix **A**. On the other hand, in case of a highly irregular geometric topology (see Figure 2 b), where the node with most neighbors stores most of the data, the algorithm converges much faster than in the case when most of the data are stored in a node with only few neighbors.

We further observe that in the "uniform" case, the algorithm behaves slightly differently for different distributions of the rows (although still having ten rows in each node). In Figure 3, we show results for six different placements of the data across the nodes for three different topologies, where we depict the mean value and the corresponding confidence intervals of the simulated orthogonality error. As we can observe, in case of the fully connected topology, the data distribution is of no importance, since all the nodes exchange data in every step with all other nodes. In case of the geometric topology, however, the convergence of the algorithm is influenced by the distribution of data, even if every node contains the same number of rows (ten rows in each node). This can be recognized by bigger confidence intervals of the orthogonality error. Nevertheless, the speed of convergence for all cases is bigger than the case when most data is stored in the "sparsest" node (cf. Figure 2 b). In case of the regular topology, the difference is small only due to numerical accuracy of the mixing parameters.

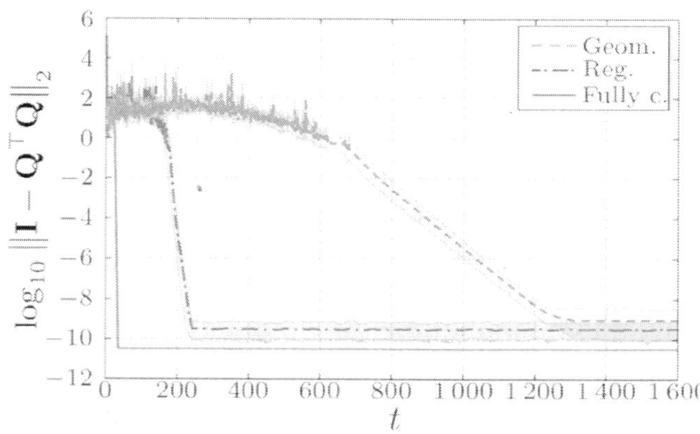

Figure 3: "Uniform" distribution for different topologies. (*Boldface line* is the mean value across six different uniform data distributions. *Shaded areas* are 95 % confidence intervals).

Numerical Sensitivity

As mentioned in Section 3.1, the classical Gram-Schmidt orthogonalization possesses some undesirable numerical properties [1, 23]. In comparison to *centralized* algorithms, numerical stability of DS-CGS is furthermore influenced by the precision of the mixing weight matrix **W**, the network topology, and properties of input matrix **A**, i.e., its condition number (see Figure 5ahead) and the distribution of the numbers in the rows of the matrix (see Figs. 2 and 3). In this section, we provide simulation results showing these dependencies.

Weights

As mentioned in Section 2, matrix **W** can be selected in many ways. Mainly, the selection of the weights influences the speed of convergence. Unlike previous simulations, where we used the metropolis weights (see Eq. (2)), here we selected constant weights for matrix **W** [20], i.e.,

$$[\mathbf{W}]_{ij} = \begin{cases} \frac{c}{N} & \text{if } (i,j) \in \mathcal{E}, \\ 1 - \frac{c}{N} d_i & \text{if } i = j, \\ 0 & \text{otherwise}, \end{cases} \tag{8}$$

where $c \in (0,1]$. Such weights, in general, lead to slower convergence. However, we can also see in Figure 4 that the weights influence not only the speed of convergence but also the numerical accuracy of the algorithm (different error floors).

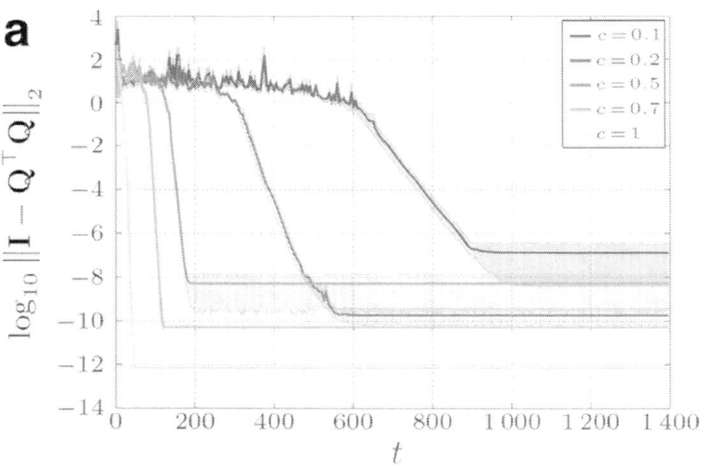

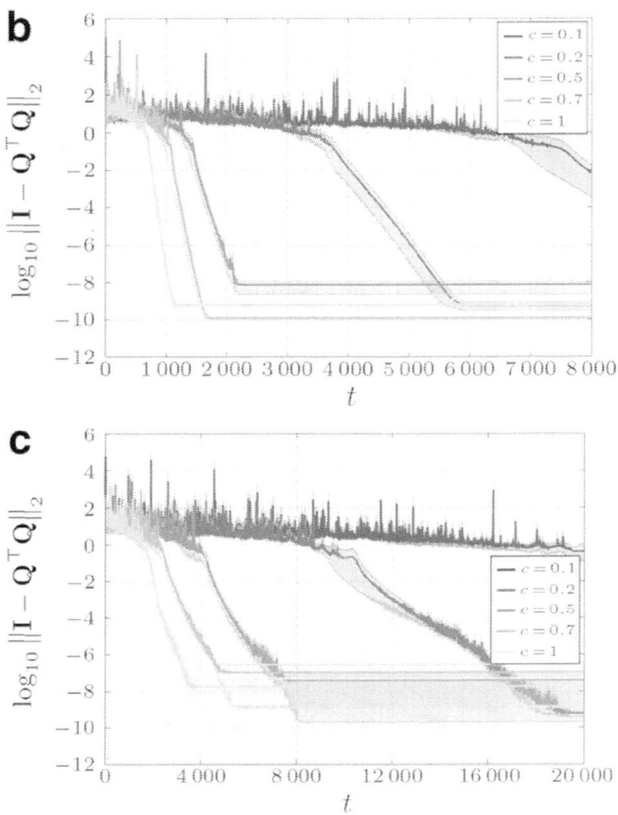

Figure 4: Influence of different constant weights c (Eq. (8)) on the algorithm's accuracy and convergence speed for three different topologies (**a** - Fully connected topology; **b** - Regular topology; **c** - Geometric topology) averaged over ten different input matrices (**a**–**c**). (*Shaded areas* are 95 % confidence intervals).

Condition Numbers

It is well known that the classical Gram-Schmidt orthogonalization is numerically unstable [23]. In cases when input matrix **A** is ill-conditioned (high condition number) or rank-deficient (matrix contains linear dependent columns), the computed vectors **Q** can be far from orthogonal even when computed with high precision.

In this section, we study the influence of the condition number of input matrix **A** on the accuracy of the orthogonality. The condition number is defined with respect to inversion as the ratio of the largest and smallest singular value. In comparison to classical (centralized) Gram-Schmidt orthogonalization, we observe (Figure 5 a) that the DS-CGS algorithm behaves similarly, although it reaches neither the accuracy of AC-CGS nor of the centralized algorithm (even in the fully connected network). We observe in all of the simulations that the orthogonality error in the first phase can reach very high values (due to divisions by numbers close to zero), which may influence the numerical accuracy in the final phase.

We further observe that the algorithm requires matrix **A** to be very well-conditioned even for the fully connected network. Unlike other methods, the factorization error in case of DS-CGS has the same characteristics as the orthogonality error and is also influenced by the condition number of the input matrix, see Figure 5 b. Although, as we noted in Section 5.1, orthogonality and factorization error of DS-CGS behave almost identically, the dependence of condition number $\kappa(\mathbf{A})$ on the factorization error would need a further investigation.

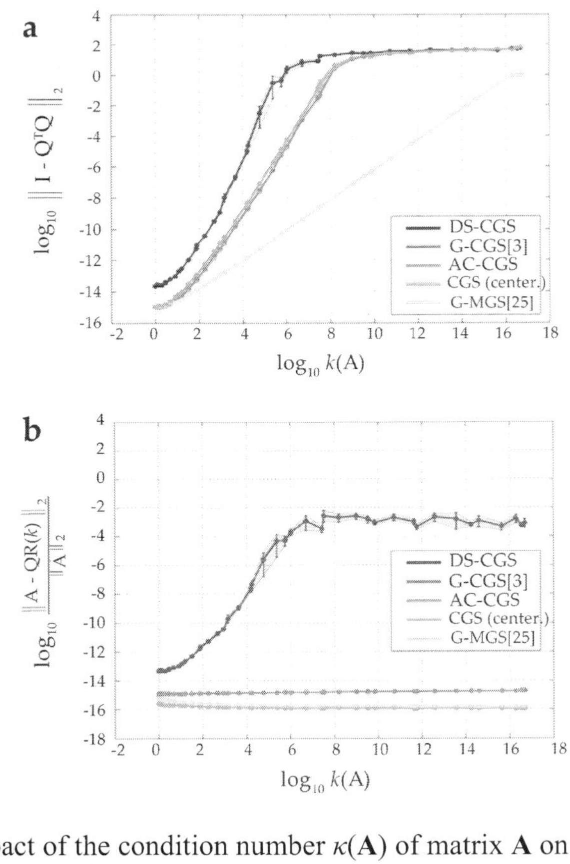

Figure 5: Impact of the condition number $\kappa(\mathbf{A})$ of matrix **A** on the orthogonality (**a**) and factorization error (**b**). Averaged over ten matrices for each condition

number. Fully connected network. (*Both axes* are in logarithmic scale. *Shaded areas* are 95 % confidence intervals).

Figure 5 also shows that G-MGS is the most robust method in comparison to the others. This is caused by the usage of the *modified* Gram-Schmidt orthogonalization instead of the classical one.

Mixing Precision

Another factor influencing the algorithm's performance is the numerical precision of the mixing weights **W**. Here, we simulate the case of a geometric topology with the Metropolis weights model, where the weights are of given precision—characterized by the number of variable decimal digits (4, 8, 16, 32, "Infinite").[5]

If we compare Figure 6 with Figure 7, we find that the numerical precision of the mixing weights have bigger influence in cases when the input matrix is worse conditioned. In Figs. 8 and 9, we can see the difference between orthogonality errors for various precisions. We observe that for the matrix **A** with higher condition number, the higher mixing precision has bigger impact on the result.

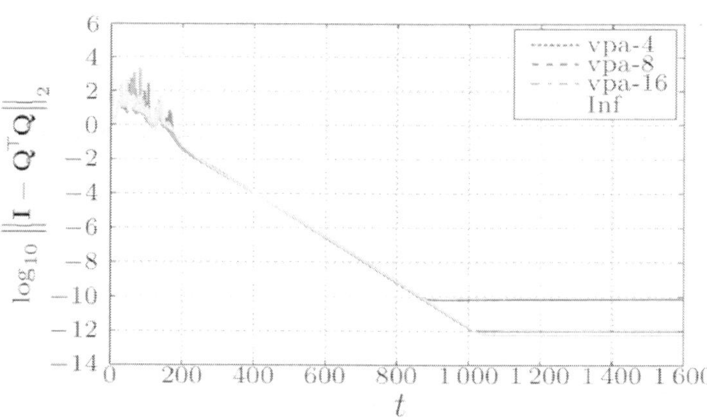

Figure 6: Influence of the numerical precision of the mixing weights on the orthogonality error of DS-CGS. Geometric topology, matrix **A** with low condition number (κ(**A**)=1.04).

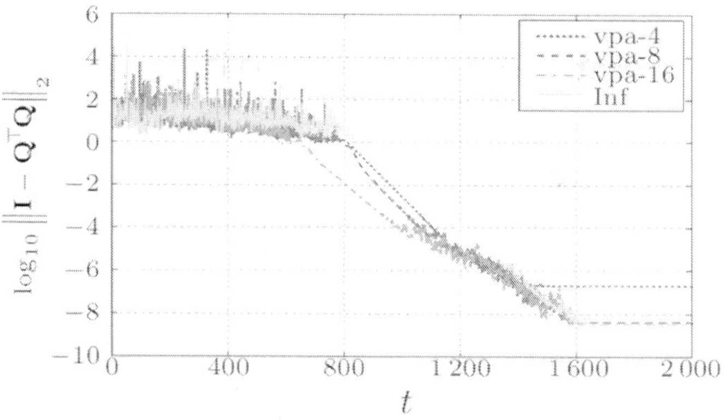

Figure 7: Influence of the numerical precision of the mixing weights on the orthogonality error of DS-CGS. Geometric topology, matrix **A** with higher condition number ($\kappa(\mathbf{A})$=76.33).

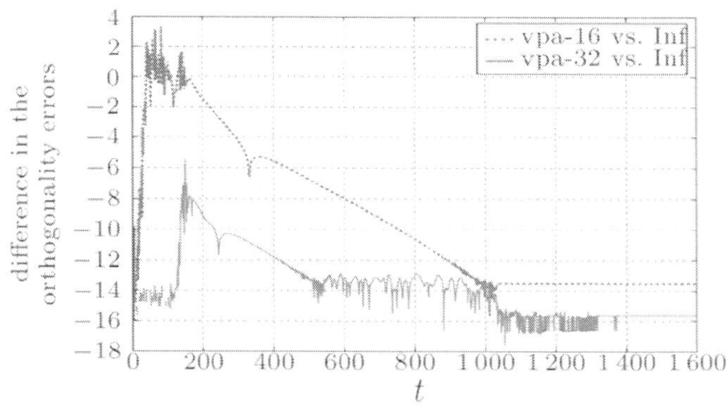

Figure 8: Difference in the orthogonality error $\left(\log_{10}\left|\|\mathbf{I} - \mathbf{Q}^\top\mathbf{Q}\|_2^{(\text{vpa}-i)} - \|\mathbf{I} - \mathbf{Q}^\top\mathbf{Q}\|_2^{(\text{Inf})}\right|\right)$ for the case of 16 and 32 decimal digits versus "infinite" precision (converted to double).

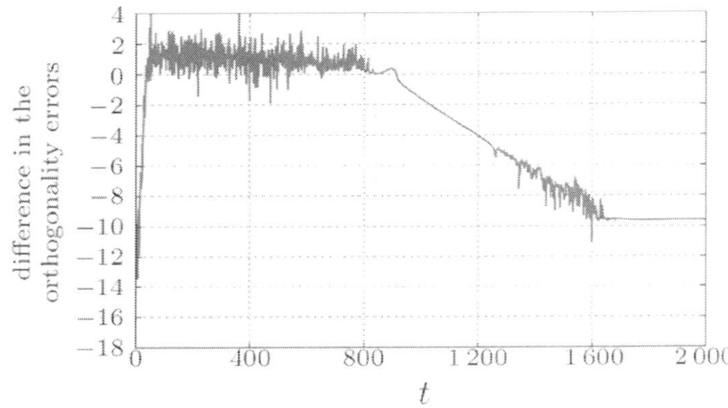

Figure 9: Difference in the orthogonality error $\left(\log_{10}\left|\left\|\mathbf{I}-\mathbf{Q}^{\top}\mathbf{Q}\right\|_2^{(\text{vpa}-16)} - \left\|\mathbf{I}-\mathbf{Q}^{\top}\mathbf{Q}\right\|_2^{(\text{Inf})}\right|\right)$ for the case of 16 decimal digits versus "infinite" precision (converted to double). Note that in comparison to Figure 8, the difference between "infinite" and more than 16 digits is below the machine precision (exact same results).

As we find in Figure 6, the error floor moves with the mixing precision. However, we must note that even for the "infinite" mixing precision the orthogonality error stalls at an accuracy ($\sim 10^{-12}$) lower than the used machine precision—taking into account also the conversion to double precision. From the simulations, we conclude that this is caused by high numerical dynamic range in the first phases of the algorithm as well as by the errors created by the misagreement among the nodes during the transient phase of the algorithm.

Robustness to Link Failures

In case of distributed algorithms, it is of big importance that the algorithm is robust against network failures. Typical failures in WSN are message losses or link failures, which occur due to many reasons, e.g., channel fading, congestions, message collisions, moving nodes, or dynamic topology.

We model link failures as a temporary drop-out of a bidirectional connection between two nodes, meaning that no message can be transmitted between the nodes. In every time step, we randomly remove some percentage of links in the network. As a weight model, we picked the constant weights model, Eq. (8), due to its property that every node can compute at each time step the weights *locally* based only on the number of received messages (d_i). Thus, no global knowledge is required. However, the nodes must still work synchronously.[6]

From Figure 10, we conclude that the algorithm is very robust and even if we drop in every time step, a big percentage (up to 60 %) of the links, the algorithm still achieves some accuracy (at least 10^{-2}; Figure 10 c).

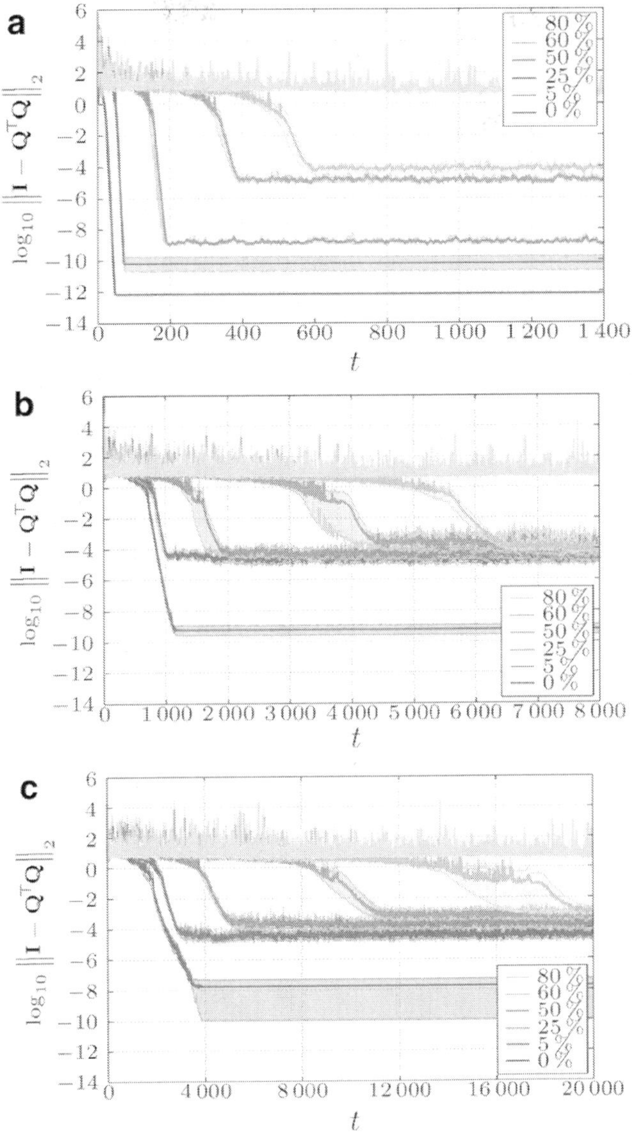

Figure 10: Robustness to link failures for different percentages of failed links at every time step (**a** - Fully connected; **b** - Regular topology; **c** - Geometric topology). Constant weight model with c=1, i.e., the fastest option (see Figure 4). (*Shaded areas* are 95 % confidence intervals).

It is worth noting that moving nodes and dynamic network topology can be modeled in the same way. We therefore argue that the algorithm is robust also to such scenarios (assuming that synchronicity is guaranteed).

Performance Comparison with Existing Algorithms

We compare our new DS-CGS algorithm with AC-CGS, G-CGS, and G-MGS introduced in Section 3.2. Although all approaches have iterative aspects, the cost per iteration strongly differs for each algorithm. Thus, instead of providing a comparison in terms of number of iterations to converge, we compare the communication cost needed for achieving a certain accuracy of the result. We investigate the total number of messages sent as well as the total amount of data (real numbers) exchanged.

Simulation results for various topologies are shown in Figs. 11 and 12. The gossip-based approaches exchange, in general, less data (Figure 12), but since their message size is much smaller than in DS-CGS, the total number of messages sent is higher (Figure 11).

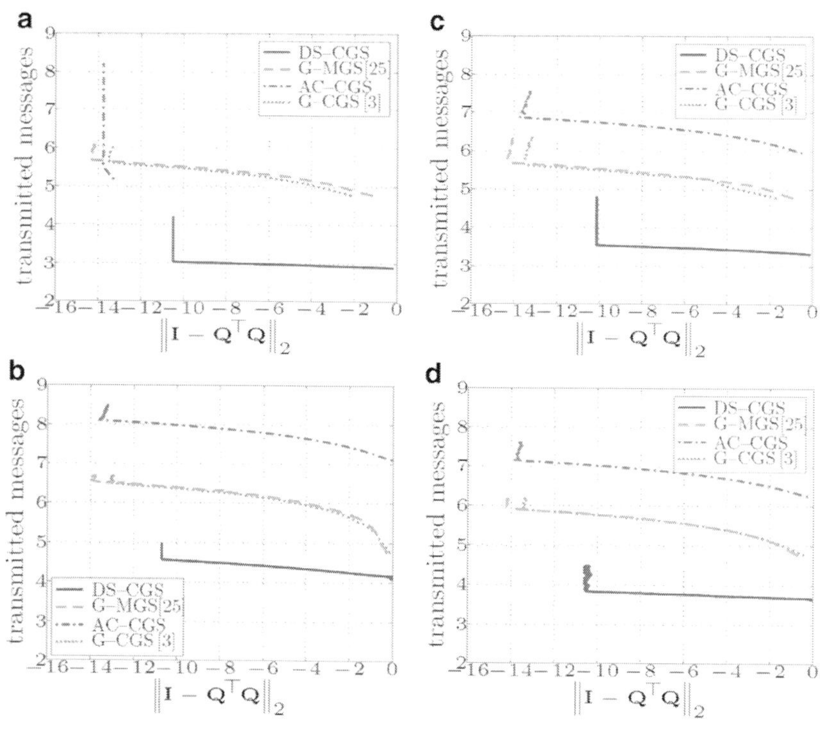

Figure 11: Total number of transmitted messages in the network vs. orthogonality error (*both axes*are in *logarithmic* scale log10) (**a** - Fully connected topol-

ogy; **b** - Geometric topology with $\overline{d}$ =8.53 ; **c** - Geometric topology with $\overline{d}$ =24.46; **d** - Regular topology with $\overline{d}$ =5).

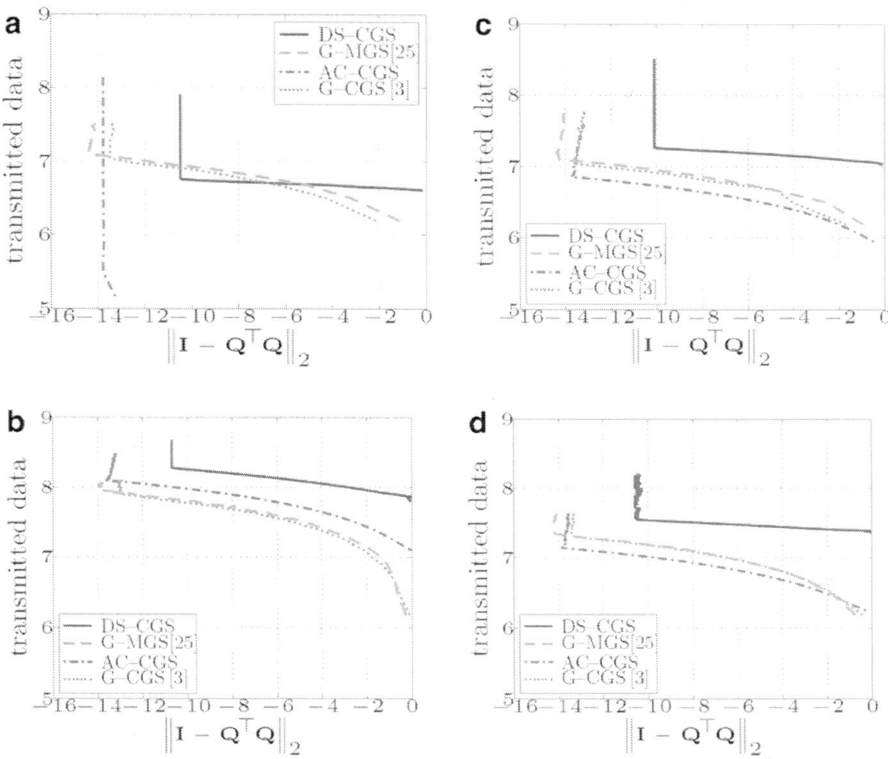

Figure 12: Total number of transmitted real numbers (data) in the network vs. orthogonality error (*both axes* are in *logarithmic* scale log10) (**a** - Fully connected topology; **b** - Geometric topology with $\overline{d}$ =8.53; **c** - Geometric topology with $\overline{d}$ =24.46; **d** - Regular topology with $\overline{d}$ =5).

Because the message size of AC-CGS is even smaller than in the gossip-based approaches, it sends the highest number of messages. Since the energy consumption in a WSN is mostly influenced by the number of transmissions [36, 37], it is better to transmit as few messages as possible (with any payload size); therefore, DS-CGS is the most suitable method for a WSN scenario. However, we notice that in many cases, DS-CGS does not achieve the same final accuracy of the result as the other methods.

Note that in fully connected networks, AC-CGS delivers a highly accurate result from the beginning, because within the first iterations, all nodes exchange the required information with all other nodes.

In Table 1, we summarize the total communication cost and local memory requirements of the algorithms. However, due to different parameters, it is difficult to rank the approaches in a general case. The requirements depend especially on the topology of the underlying network, the number of iterations $I^{(s)}$ and $I^{(d)}$ required for convergence in "static" and "dynamic" consensus-based algorithms or the number of rounds R needed for convergence of push-sum in the gossip-based approaches. For example, in a *fully connected* network $R=O(\log N)$ [24], $I^{(s)}=1$. Thus, AC-CGS requires $O(m^2 N)$ messages sent as well as data exchanged, whereas gossip-based approaches need $O(mN \log N)$ messages and $O(m^2 N \log N)$ data. Note that G-CGS and G-MGS have theoretically identical communication cost; however, G-MGS is numerically more stable (see Figure 5) and achieves a higher final accuracy (see Figs. 11 and 12). In case of DS-CGS and a fully connected network, we can interpret DS-CGS in the worst case as m consequent static consensus algorithms (one for each column); thus, $I^{(d)}=O(m)$, and the number of transmitted messages is $O(mN)$ and data $O(m^3 N)$. Nevertheless, theoretical convergence bounds of DS-CGS (on $I^{(d)}$) remain an open research question.

Table 1: Comparison of various distributed QR factorization algorithms

	Total number of	**Total amount of**	**Local memory**
	sent messages	**data (real numbers)**	**requirements**
			per node
DS-CGS	$N \cdot I^{(d)}$	N·I(d)·m2+5m2N·I(d)·m2+5m2	$O(mn/N + m^2)$
AC-CGS	N·I(s)·(m+1) m2N·I(s)·(m+1)m2	N·I(s)·(m+1)m2N·I(s)·(m+1)m2	$O(mn/N + m^2)$
G-CGS	$N \cdot R \cdot (2m-1)$	N·R·m2+5m−22N·R·m2+5m−22	$O(nm/N)$
G-MGS	$N \cdot R \cdot (2m-1)$	N·R·m2+5m−22N·R·m2+5m−22	$O(nm/N)$

$I^{(d)}$ denotes the number of iterations of "dynamic" consensus, $I^{(s)}$ the number of iterations of "static" consensus, R the number of rounds per push-sum, N the number of nodes, m the number of columns of the input matrix

CONCLUSIONS

We presented a novel distributed algorithm for computing QR decomposition and provided an analysis of its properties. In contrast to existing methods, which compute the columns of the resulting matrix **Q** consecutively, our method iteratively refines all elements at once. Thus, in any moment, the algorithm can deliver an estimate of both matrices **Q** and **R**. The algorithm dramatically outperforms known distributed orthogonalization algorithms in terms of transmitted messages, which makes it suitable for energy-constrained WSNs. Based on our empirical observation, we argue that the evaluation of the local factorization error at each node might lead to a suitable stopping criterion for the algorithm. We also provided a thorough study of its numerical properties, analyzing the influence of the precision of the mixing weights and condition numbers of the input matrix. We furthermore analyzed the robustness of the algorithm to link failures and showed that the algorithm is capable to reach a certain accuracy even for a high percentage of link failures.

The biggest drawback of the algorithm is the necessity to have synchronously working nodes. This leads to poor robustness when the messages are sent (or lost) asynchronously. As we showed, since the algorithm originates from the classical Gram-Schmidt orthogonalization, also the numerical sensitivity of the algorithm is a big issue and needs to be addressed in the future. The optimization of the weights and design of algorithm in such way that it avoids a big dynamic numerical range, especially in the first phases, is also of interest.

An alternative approach, not considered here, which could be worth of future research, would be to find a distributed algorithm as an optimization problem, e.g., mins.t. **Q** $^{\top}$ **Q**=**I**||**A**−**Q** **R**||. In literature, there exist many distributed optimization methods, e.g., [38, 39], which could lead to even superior algorithms, with even faster convergence and smaller error floors.

Last but not least, theoretical bounds of DS-CGS for the convergence time and rate remain an open issue. A first application of the algorithm has already been proposed in [26]. Also, since the proposed algorithm is not restricted to the usage in wireless sensor networks only, a transfer of the proposed algorithm onto so-called network-on-chip platforms [40] could possibly lead to further new interesting and practical applications as well.

ENDNOTES

[1]Knowing n, $\|\mathbf{u}\|_2^2 = n \lim_{t\to\infty} \mathbf{W}^t(\mathbf{u} \circ \mathbf{u}) = \sum_{i=1}^{n} u_i^2$.

[2] $\lim_{t\to\infty} \omega(t) = 1/N\mathbf{1}$.

[3]Not considering numerical properties.

[4]Error level at which the algorithm stalls at given computational precision.
[5]The simulations were performed in Matlab R2011b 64-bit using the Symbolic Math Toolbox with variable precision arithmetic. "Infinite" precision denotes weights represented as an exact ratio of two numbers. The depicted result after "infinite" precision multiplication was converted to double precision.
[6]If there is a link, nodes see each other and immediately exchange messages. From a mathematical point of view, this implies that weight matrix **W** will be doubly stochastic [1] in every time step.

APPENDIX: LOCAL ALGORITHM

For a better clarity, we here reformulate DS-CGS algorithm from the point of view of an individual node *i* (local point of view). Note that input matrix **A** is stored row-wise in the nodes, and for simplicity, we show here the case when the number of rows of matrix $A \in R^{n\times m}$ is equal to the number of nodes in the network. For a formulation from the network (global) point of view and arbitrary size of matrix **A**, see Section 4.

1. Initialization (t = 0). Node i stores the following vectors.

$$\hat{\mathbf{u}}_i(0) = (a_{i,1},\ a_{i,2},\ \ldots\ a_{i,m})$$

$$\hat{\mathbf{q}}_i(0) = (a_{i,1},\ a_{i,2},\ \ldots\ a_{i,m})$$

$$\tilde{\mathbf{u}}_i(0) = (a_{i,1}^2,\ a_{i,2}^2,\ \ldots\ a_{i,m}^2)$$

$$\tilde{\mathbf{q}}_i(0) = (a_{i,1}^2,\ a_{i,2}^2,\ \ldots\ a_{i,m}^2)$$

$$\tilde{\mathbf{p}}_i^{(1)}(0) = (a_{i,1}^2,\ a_{i,2}^2,\ \ldots\ a_{i,m}^2)$$

$$\tilde{\mathbf{p}}_i^{(2)}(0) = (a_{i,1}a_{i,2},\ a_{i,1}a_{i,3},\ a_{i,2}a_{i,3},\ \ldots\ a_{i,m-1}a_{i,m})$$

2. Repeat for t = 1, 2, ...

(a) Compute vectors locally.

$$\mathbf{p}_i(t) = \Big(0,\ \tilde{p}_{i1}^{(2)}(t-1)\frac{\hat{q}_{i,1}(t-1)}{\tilde{q}_{i,1}(t-1)}\ ,$$

$$\tilde{p}_{i,2}^{(2)}(t-1)\frac{\hat{q}_{i,1}(t-1)}{\tilde{q}_{i,1}(t-1)} + \tilde{p}_{i,3}^{(2)}(t-1)\frac{\hat{q}_{i,2}(t-1)}{\tilde{q}_{i,2}(t-1)},\ \ldots,$$

$$\sum_{j=1}^{m-1} \tilde{p}_{i,j+(m-1)(m-2)/2}^{(2)}(t-1)\frac{\hat{q}_{i,j}(t-1)}{\tilde{q}_{i,j}(t-1)} \Big)$$

$$\hat{\mathbf{u}}_i(t) = \big(a_{i,1} - p_{i,1}(t),\ a_{i,2} - p_{i,2}(t),\ \ldots,\ a_{i,m} - p_{i,m}(t)\big)$$

$$\hat{\mathbf{q}}_i(t) = \left(\frac{\hat{u}_{i,1}(t)}{\sqrt{N\tilde{u}_{i,1}(t-1)}},\ \frac{\hat{u}_{i,2}(t)}{\sqrt{N\tilde{u}_{i,2}(t-1)}},\ \ldots,\ \frac{\hat{u}_{i,m}(t)}{\sqrt{N\tilde{u}_{i,m}(t-1)}} \right)$$

(b) Store the local part of the resulting matrix Q and the whole matrix R, i.e.,

$$\hat{\mathbf{q}}_i(t) = \left(\frac{\hat{u}_{i,1}(t)}{\sqrt{N\tilde{u}_{i,1}(t-1)}}, \frac{\hat{u}_{i,2}(t)}{\sqrt{N\tilde{u}_{i,2}(t-1)}}, \ldots, \frac{\hat{u}_{i,m}(t)}{\sqrt{N\tilde{u}_{i,m}(t-1)}} \right)$$

$$\mathbf{R}^{(i)}(t) = N \begin{pmatrix} \tilde{p}^{(1)}_{i,1}(t) & \tilde{p}^{(2)}_{i,1}(t) & \ldots & \tilde{p}^{(2)}_{i,(m^2-3m+4)/2}(t) \\ 0 & \tilde{p}^{(1)}_{i,2}(t) & \tilde{p}^{(2)}_{i,3}(t) & \ldots \\ \vdots & & \ddots & \ldots \\ 0 & \ldots & 0 & \tilde{p}^{(1)}_{i,m}(t) \end{pmatrix}$$

(c) Aggregate the following data into one message:

$$\psi^{(1)}_i = \big(\tilde{u}_{i,1}(t-1) + \hat{u}^2_{i,1}(t) - \hat{u}^2_{i,1}(t-1), \ \ldots, \ \tilde{u}_{i,m}(t-1) + \hat{u}^2_{i,m}(t) - \hat{u}^2_{i,m}(t-1)\big)$$

$$\psi^{(2)}_i = \big(\tilde{q}_{i,1}(t-1) + \hat{q}^2_{i,1}(t) - \hat{q}^2_{i,1}(t-1), \ \ldots, \ \tilde{q}_{i,m}(t-1) + \hat{q}^2_{i,m}(t) - \hat{q}^2_{i,m}(t-1)\big)$$

$$\psi^{(3)}_i = \Big(\tilde{p}^{(1)}_{i,1}(t-1) + a_{i,1}\hat{q}_{i,1}(t) - a_{i,1}\hat{q}_{i,1}(t-1), \ \ldots, \ \tilde{p}^{(1)}_{i,m}(t-1) + a_{i,m}\hat{q}_{i,m}(t) - a_{i,m}\hat{q}_{i,m}(t-1)\Big)$$

$$\psi^{(4)}_i = \Big(\tilde{p}^{(2)}_{i,1}(t-1) + \hat{q}_{i,1}(t)a_{i,2} - \hat{q}_{i,1}(t-1)a_{i,2}, \ \tilde{p}^{(2)}_{i,2}(t-1) + \hat{q}_{i,1}(t)a_{i,3} - \hat{q}_{i,1}(t-1)a_{i,3}, \ \ldots, \ \tilde{p}^{(2)}_{i,(m^2-m)/2}(t-1) + \hat{q}_{i,m-1}(t)a_{i,m} - \hat{q}_{i,m-1}(t-1)a_{i,m}\Big)$$

(d) Broadcast the message containing the vectors $\left\{\psi^{(1)}_i, \psi^{(2)}_i, \psi^{(3)}_i, \psi^{(4)}_i\right\}$ to the neighbors and update the own local data $\left\{\tilde{\mathbf{u}}_i(t), \tilde{\mathbf{q}}_i(t), \tilde{\mathbf{p}}^{(1)}_i(t), \tilde{\mathbf{p}}^{(2)}_i(t)\right\}$ from received data.

ACKNOWLEDGEMENTS

This work was supported by the Austrian Science Fund (FWF) under project grants S10608-N13 and S10611-N13 within the National Research Network SISE. Preliminary parts of this work were previously published at the 46th Asilomar Conf. Sig., Syst., Comp., Pacific Grove, CA, USA, Nov. 2012 [32].

REFERENCES

1. GH Golub, CF Van Loan, *Matrix Computations*, 3rd Ed. (Johns Hopkins Univ. Press, Baltimore, USA, 1996).
2. JM Lees, RS Crosson, in *Spatial Statistics and Imaging*, 20, ed. by A Possolo. Bayesian ART versus conjugate gradient methods in tomographic seismic imaging: an application at Mount St. Helens, Washington (IMS Lecture Noted-Monograph SeriesHayward, CA, 1991), pp. 186–208.
3. C Dumard, E Riegler, in *Int. Conf. on Telecom. ICT '09*. Distributed sphere decoding (IEEEMarrakech, 2009), pp. 172–177.
4. G Tauböck, M Hampejs, P Svac, G Matz, F Hlawatsch, K Gröchenig, Low-complexity ICI/ISI equalization in doubly dispersive multicarrier systems using a decision-feedback LSQR algorithm. IEEE Trans. Signal Process.**59**(5), 2432–2436 (2011).
5. E Hänsler, G Schmidt, *Acoustic Echo and Noise Control* (Wiley, Chichester, New York, Brisabne, Toronto, Singapore, 2004).
6. PSR Diniz, *Adaptive Filtering—Algorithms and Practical Implementation* (Springer, US, 2008).
7. AH Sayed, *Adaptation, Learning, and Optimization over Networks*, vol. 7 (Foundations and Trends in Machine Learning, Boston-Delft, 2014).
8. K-J Cho, Y-N Xu, J-G Chung, in *IEEE Workshop on Signal Processing Systems*. Hardware efficient QR decomposition for GDFE (IEEEShanghai, China, 2007), pp. 412–417.
9. X Wang, M Leeser, A truly two-dimensional systolic array FPGA implementation of QR decomposition. ACM Trans. Embed. Comput. Syst.**9**(1), 3–1317 (2009).
10. A Buttari, J Langou, J Kurzak, J Dongarra, in *Proc. of the 7th International Conference on Parallel Processing and Applied Mathematics*. Parallel tiled QR factorization for multicore architectures (SpringerBerlin, Heidelberg, 2008), pp. 639–648.
11. J Demmel, L Grigori, MF Hoemmen, J Langou, Communication-optimal parallel and sequential QR and LU factorizations (2008). Technical report, no. UCB/EECS-2008-89, EECS Department, University of California, Berkeley.
12. F Song, H Ltaief, B Hadri, J Dongarra, in *International Conference for High Performance Computing, Networking, Storage and Analysis*.

Scalable tile communication-avoiding QR factorization on multicore cluster systems (IEEE Computer SocietyWashington, DC, USA, 2010), pp. 1–11.

13. M Shabany, D Patel, PG Gulak, A low-latency low-power QR-decomposition ASIC implementation in 0.13 μm CMOS. IEEE Trans. Circ. Syst. I. **60**(2), 327–340 (2013).

14. A Nayak, I Stojmenović, *Wireless Sensor and Actuator Networks: Algorithms and Protocols for Scalable Coordination and Data Communication* (Wiley, Hoboken, NJ, 2010).

15. H Straková, WN Gansterer, T Zemen, in *Proc. of the 9th International Conference on Parallel Processing and Applied Mathematics, Part I. Lecture Notes in Computer Science*, 7203. Distributed QR factorization based on randomized algorithms (Springer Berlin HeidelbergBerlin, Heidelberg, 2012), pp. 235–244.

16. H Straková, Truly distributed approaches to orthogonalization and orthogonal iteration on the basis of gossip algorithms (2013). PhD thesis, University of Vienna.

17. Slučiak, M Rupp, in *Proc. of the 36th IEEE International Conference on Acoustics, Speech and Signal Processing (ICASSP)*. Reaching consensus in asynchronous WSNs: algebraic approach (Prague, 2011), pp. 3300–3303. Chap. Acoustics, Speech and Signal Processing (ICASSP), 2011.

18. Slučiak, M Rupp, in *Proc. of Statistical Sig. Proc. Workshop (SSP)*. Almost sure convergence of consensus algorithms by relaxed projection mappings (IEEEAnn Arbor, MI, USA, 2012), pp. 632–635.

19. F Sivrikaya, B Yener, Time synchronization in sensor networks: a survey. IEEE Netw. Mag. Special Issues Ad Hoc Netw. Data Commun. Topol. Control. **18**(4), 45–50 (2004).

20. R Olfati-Saber, JA Fax, RM Murray, Consensus and cooperation in networked multi-agent systems. Proc. IEEE. **95**(1), 215–233 (2007).

21. L Xiao, S Boyd, Fast linear iterations for distributed averaging. Syst. Control Lett.**53:**, 65–78 (2004).

22. L Xiao, S Boyd, S Lall, in *Proc. ACM/IEEE IPSN–05*. A scheme for robust distributed sensor fusion based on average consensus (IEEELos Angeles, USA, 2005), pp. 63–70.

23. LN Trefethen, D Bau III, *Numerical Linear Algebra* (SIAM: Society for Industrial and Applied Mathematics, Philadelphia, 1997).

24. D Kempe, A Dobra, J Gehrke, in *Foundations of Computer Science, 2003. Proceedings. 44th Annual IEEE Symposium on*. Gossip-based computation of aggregate information, (2003), pp. 482–491. ISSN:0272-5428, doi:10.1109/SFCS.2003.1238221.
25. H Straková, WN Gansterer, in *21st Euromicro Int. Conf. on Parallel, Distributed, and Network-Based Processing (PDP)*. A distributed eigensolver for loosely coupled networks (IEEEBelfast, UK, 2013), pp. 51–57.
26. Slučiak, M Rupp, Network size estimation using distributed orthogonalization. IEEE Sig. Proc. Lett.**20**(4), 347–350 (2013).
27. P Braca, S Marano, V Matta, in *Proc. Int. Conf. Inf. Fusion (FUSION 2008)*. Running consensus in wireless sensor networks (IEEECologne, Germany, 2008), pp. 152–157.
28. W Ren, in *Proc. of the 2007 American Control Conference*. Consensus seeking in multi-vehicle systems with a time-varying reference state (IEEENew York, NY, 2007), pp. 717–722.
29. V Schwarz, C Novak, G Matz, in *Proc. 43rd Asilomar Conf. on Sig., Syst., Comp*. Broadcast-based dynamic consensus propagation in wireless sensor networks (IEEEPacific Grove, CA, 2009), pp. 255–259.
30. M Zhu, S Martínez, Discrete-time dynamic average consensus. Automatica. **46**(2), 322–329 (2010).
31. Slučiak, O Hlinka, M Rupp, F Hlawatsch, PM Djurić, in *Rec. of the 45th Asilomar Conf. on Signals, Systems, and Computers*. Sequential likelihood consensus and its application to distributed particle filtering with reduced communications and latency (IEEEPacific Grove, CA, 2011), pp. 1766–1770.
32. Slučiak, H Straková, M Rupp, WN Gansterer, in *Rec. of the 46th Asilomar Conf. on Signals, Systems, and Computers*. Distributed Gram-Schmidt orthogonalization based on dynamic consensus (IEEEPacific Grove, CA, 2012), pp. 1207–1211.
33. P Braca, S Marano, V Matta, AH Sayed, in *Proc. of the 39th IEEE International Conference on Acoustics, Speech and Signal Processing (ICASSP)*. Large deviations analysis of adaptive distributed detection (IEEEFlorence, Italy, 2014), pp. 6153–6157.
34. Slučiak, Convergence analysis of distributed consensus algorithms (2013). PhD thesis, TU Vienna.

35. B Efron, RJ Tibshirani, *An Introduction to the Bootstrap* (Chapman & Hall/CRC Monographs on Statistics & Applied Probability 57, London, UK, 1994).

36. P Rost, G Fettweis, in *GLOBECOM Workshops, 2010 IEEE*. On the transmission-computation-energy tradeoff in wireless and fixed networks (IEEEMiami, FL, 2010), pp. 1394–1399.

37. R Shorey, A Ananda, MC Chan, WT Ooi, *Mobile, Wireless, and Sensor Networks: Technology, Applications, and Future Directions* (Wiley, Hoboken, NJ, 2006).

38. B Johansson, On distributed optimization in networked systems (2008). PhD thesis, KTH, Stockholm.

39. I Matei, JS Baras, Performance evaluation of the consensus-based distributed subgradient method under random communication topologies. IEEE J. Sel. Top. Signal Process.**5**(4), 754–771 (2011).

40. L Benini, GD Micheli, Networks on chips: a new SoC paradigm. IEEE Comput.**35**(1), 70–78 (2002).

CHAPTER 2

NEW ITERATIVE METHODS FOR GENERALIZED SINGULAR-VALUE PROBLEMS

A. H. Refahi Sheikhani, S. Kordrostami

Department of Applied Mathematics, Faculty of Mathematical Sciences, Lahijan Branch, Islamic Azad University, Lahijan, Iran

ABSTRACT

This paper presents two new iterative methods to compute generalized singular values and vectors of a large sparse matrix. To reach acceleration in the convergence process, we have used a different inner product instead of the common one, Euclidean one. Furthermore, at each restart, a different inner product has been chosen by the researchers. A number of numerical experiments illustrate the performance of the above-mentioned methods.

Citation (APA): Sheikhani, A. R., & Kordrostami, S. (2017). New iterative methods for generalized singular-value problems. Mathematical Sciences, 11(4), 257-265. (9 pages).

Keywords: Generalized singular value, Krylov subspace, Iterative Sparse

INTRODUCTION

There are a number of applications for generalized singular-value decomposition (GSVD) in the literature including the computation of the Kronecker form of the matrix pencil A−λB [5], solving linear matrix equations [1], weighted least squares [2], and linear discriminant analysis [6] to name but a few. In a number of applications like the generalized total least squares problem, the matrices A and B are large and sparse, so in such cases, only a few of the generalized singular vectors corresponding to the smallest or largest generalized singular values are needed. There is a kind of close connection between the GSVD problem and two different generalized eigenvalue problems. In fact, there are many efficient numerical methods to solve generalized eigenvalue problems [8, 9, 10, 11]. In this paper, we will examine the Jacobi–Davidson-type subspace method which is related to the Jacobi–Davidson for the SVD [5], which in turn is inspired by the Jacobi–Davidson method to solve the eigenvalue problem [4]. The main step in Jacobi–Davidson-type method for the (GSVD) is solving the correction equations in an exact manner requiring the solution of linear systems of original size at each iteration. In general, these systems are considered as large, sparse, and nonsymmetrical. For this matter, we use the weighted Krylov subspace process to solve the correction equations in an exact manner, and we show that our proposed method has the feature of asymptotic quadratic convergence.

The paper is organized as follows. In "Preparations", we will remind the readers of basic definitions of the generalized singular-value decomposition problems and their elementary properties. "A new iterative method for GSVD" introduces our new numerical methods to solve generalized eigenvalue problems together with an analysis of the convergence of these methods. Several numerical examples are presented in "Numerical experiments". Finally, the conclusions are given in the last section.

PREPARATIONS

Definition 2.1

Supposes that $A \in R^{m\times n}$ and $B \in R^{p\times n}$. The generalized singular values of the pair (A,B) are presented as

$$\sum(A,B) = \{\sigma | \sigma \geq 0,\ \det(A^T A - \sigma^2 B^T B) = 0\}.$$

Definition 2.2

A generalized singular value is called simple if $\sigma_i \neq \sigma_j$, for all $i \neq j$.

Theorem 2.3

Suppose $A \in R^{m\times n}$, $B \in R^{p\times n}$, *and* $m \geq n$. *Here, taking the previous theorem into consideration, we see that there are orthogonal matrices* $U_{m\times m}$, $V_{p\times p}$ *and a nonsingular matrix* $X^{n\times n}$, *such that*

$$U^T AX = \sum\nolimits_1 = \mathrm{diag}(\alpha_1, \ldots, \alpha_n) \quad \alpha_i \geq 0,$$
$$V^T BX = \sum\nolimits_2 = \mathrm{diag}(\beta_1, \ldots, \beta_n) \quad \beta_i \geq 0, \tag{1}$$

where $q=\min\{p,n\}$, $r=\mathrm{rank}(B)$, *and* $\beta_1 \geq \cdots \geq \beta_r > \beta_{r+1} = \cdots = \beta_n = 0$. *If* $\alpha_j = 0$ *for any* j, $r+1 \leq j \leq n$, *then* $\sum(A,B) = \{\sigma | \sigma \geq 0\}$. *Otherwise,* $\sum(A,B) = \left\{\frac{\alpha_i}{\beta_i} | i = 1, \ldots, r\right\}$

.

Proof

Refer to [3].

Theorem 2.4

Let $A \in R^{n\times n}$, $B \in R^{n\times n}$ *have the GSVD:*

$$U^T AX = \sum\nolimits_1 = \mathrm{diag}(\alpha_i), \quad V^T BX = \sum\nolimits_2 = \mathrm{diag}(\beta_i)$$

furthermore, consider it as nonsingular. Here, then, the matrix pencil

$$\begin{pmatrix} 0 & A \\ A^T & 0 \end{pmatrix} - \lambda \begin{pmatrix} I & 0 \\ 0 & B^T B \end{pmatrix} \tag{2}$$

has eigenvalues $\lambda_j = \pm\alpha_j/\beta_j$, $j=1,\ldots,n$ *which corresponds to the eigenvectors:*

$$\begin{pmatrix} u_j \\ \pm x_j/\beta_j \end{pmatrix}, \quad j = 1, \ldots, n \tag{3}$$

where u_j *is the ith column of* U *and* x_j *is the ith column of* X.

Proof

Refer to [3].

Let D be a diagonal matrix, that is, $D=diag(d_1,d_2,\ldots,d_n)$. If u and v are two vectors of R^n, we define the D-scalar product of $(u,v)_D=v^TDu$. which is well defined if and only if the matrix D is positively definite or to say di > 0,i=1,…,n. The norm associated with this inner product is the D-norm $\|\cdot\|D$ which is defined as $\|u\|_D = \sqrt{(u,u)_D} = \sqrt{u^TDu}\ \forall u \in R^n$.

As assumption B has full rank, $(x,y)_{(BTB)-1}:=y^T(B^TB)^{-1}x$ is an inner product, and due to this, the corresponding norm satisfies $\|x\|^2_{(BTB)-1}:=(x,x)_{(BTB)-1}$. Inspired by the equality $\|Z\|^2_F=trace(Z^TZ)$ for a real matrix Z, we define the $(B^TB)^{-1}$-Frobenius norm of Z by

$$\|Z\|^2_{(B^TB)^{-1},F} = \text{trace}(Z^T(B^TB)^{-1}Z). \tag{4}$$

A NEW ITERATIVE METHOD FOR GSVD

We will advance different extraction methods here which are often more appropriate for small generalized singular values than the standard one from "A new iterative method for GSVD". Before dealing with these new methods, we should refer to our main idea which is developed considering Krylov subspace methods.

Theorem 3.1

Assume that (σ,u,v) *is a generalized singular triple:* $Aw=\sigma u$ *and* $A^Tu=\sigma B^TBw$, *where* σ *is a simple nontrivial generalized singular value, and* $\|u\|=\|Bw\|=1$, *and suppose that the correction equations*

$$P = \begin{pmatrix} I - \tilde{u}\tilde{u}^T & 0 \\ 0 & I - B^TB\tilde{w}\tilde{w}^T \end{pmatrix}, \tag{5}$$

are solved exactly in every step. Provided that the initial vectors $(\tilde{u},\tilde{w})$ *are close enough to* (u,w) *the sequence of approximations* $(\tilde{u},\tilde{w})$ *converges quadratically to* (u,w).

Proof

Refer to [4].

Lemma 3.2

Having in mind the Theorem 3.1, *now suppose that* mm *steps of the weighted Arnoldi process*[7] *have been performed on the following matrix:*

$$\begin{pmatrix} I - uu^{\mathrm{T}} & 0 \\ 0 & I - B^{\mathrm{T}}Bww^{\mathrm{T}} \end{pmatrix} \begin{pmatrix} -\theta I & A \\ A^{\mathrm{T}} & -\theta B^{\mathrm{T}}B \end{pmatrix}. \tag{6}$$

Furthermore, consider the matrix $\widetilde{H}_m$ *as the Hessenberg matrix, whose non-zero entries are the scalars* $\tilde{h}_{i,j}$, *constructed by the Weighted Arnoldi process. Here, we notice that the basis* $\widetilde{V}_m = [\tilde{v}_1, \ldots, \tilde{v}_m]$ *constructed by this algorithm is* D-*orthonormal and we have*

$$\widetilde{V}_m^{\mathrm{T}} D \widetilde{V}_m = I_m, \tag{7}$$

$$\begin{pmatrix} I - uu^{\mathrm{T}} & 0 \\ 0 & I - B^{\mathrm{T}}Bww^{\mathrm{T}} \end{pmatrix} \begin{pmatrix} -\theta I & A \\ A^{\mathrm{T}} & -\theta B^{\mathrm{T}}B \end{pmatrix} \widetilde{V}_m$$

(I $$= \widetilde{V}_{m+1} \begin{pmatrix} \widetilde{H}_m \\ h_{m+1,m} e_m^{\mathrm{T}} \end{pmatrix}. \tag{8}$$

Proof

See [4].

We know that similar to Krylov methods, the *m*th (m≥1) iterate $x_m=[s_m,t_m]^t$ of the weighted-FOM and weighted-GMRES methods belong to the affine Krylov subspace:

$$\begin{pmatrix} s_0 \\ t_0 \end{pmatrix} + \mathcal{K}_m\left(\begin{pmatrix} I - uu^{\mathrm{T}} & 0 \\ 0 & I - B^{\mathrm{T}}Bww^{\mathrm{T}} \end{pmatrix} \times \begin{pmatrix} -\theta I & A \\ A^{\mathrm{T}} & -\theta B^{\mathrm{T}}B \end{pmatrix}, \begin{pmatrix} r_0^{(s)} \\ r_0^{(t)} \end{pmatrix}\right). \tag{9}$$

Now, it is the time to prove our main theorem.

Theorem 3.3

Considering Theorem 3.1, mm *steps of the weighted Arnoldi process have been run on* (7). *Here, the iterate* $x_m=[s_m,t_m]^t$ *is the exact solution of the correction equation:*

$$P\begin{pmatrix} -\theta I & A \\ A^{\mathrm{T}} & -\theta B^{\mathrm{T}}B \end{pmatrix}\begin{pmatrix} s \\ t \end{pmatrix} = -r, \quad s \perp \tilde{u}, \quad t \perp \tilde{w}. \tag{10}$$

Proof

The iterate x^{WF}_m of the weighted-FOM method is selected, because its residual is D-orthonormal or

$$\begin{pmatrix} r_m^{(s)} \\ r_m^{(t)} \end{pmatrix}^{\mathrm{WF}} \perp_D \mathcal{K}_m\left(\begin{pmatrix} I-uu^{\mathrm{T}} & 0 \\ 0 & I-B^{\mathrm{T}}Bww^{\mathrm{T}} \end{pmatrix} \times \begin{pmatrix} -\theta I & A \\ A^{\mathrm{T}} & -\theta B^{\mathrm{T}}B \end{pmatrix}, \begin{pmatrix} r_0^{(s)} \\ r_0^{(t)} \end{pmatrix}\right). \tag{11}$$

The iterate x^{WG}_m of the weighted-GMRES method is selected to lessen the residual D-norm in (9). Here, we notice that it is the solution of the least squares problem:

$$\text{minimize}_{[s,t]'\in(4.4)}\left\| \begin{pmatrix} A\tilde{w}-\theta\tilde{u} \\ A^{\mathrm{T}}\tilde{u}-\theta B^{\mathrm{T}}B\tilde{w} \end{pmatrix} - P\begin{pmatrix} -\theta I & A \\ A^{\mathrm{T}} & -\theta B^{\mathrm{T}}B \end{pmatrix}\begin{pmatrix} s \\ t \end{pmatrix}\right\|_D. \tag{12}$$

In these methods, we use the D-inner product and the D-norm to calculate the solution in the affine subspace (9) and we create a D-orthonormal basis of the Krylov subspace:

$$\mathcal{K}_m\left(\begin{pmatrix} I-uu^{\mathrm{T}} & 0 \\ 0 & I-B^{\mathrm{T}}Bww^{\mathrm{T}} \end{pmatrix}\begin{pmatrix} -\theta I & A \\ A^{\mathrm{T}} & -\theta B^{\mathrm{T}}B \end{pmatrix}, \begin{pmatrix} r_0^{(s)} \\ r_0^{(t)} \end{pmatrix}\right). \tag{13}$$

by the weighted Arnoldi process. An iterate x_m of these two methods can be transcribed as

$$\begin{pmatrix} s_m \\ t_m \end{pmatrix} = \begin{pmatrix} s_0 \\ t_0 \end{pmatrix} + \widetilde{V}_m\begin{pmatrix} y_m^{(s)} \\ y_m^{(t)} \end{pmatrix},$$

where $y_m \in R^m$.

Therefore, the matching residual $r_m = \left[r_m^{(s)}, r_m^{(t)}\right]^t$ satisfies

$$\begin{aligned}
\begin{pmatrix} r_m^{(s)} \\ r_m^{(t)} \end{pmatrix} &= \begin{pmatrix} A\tilde{w}-\theta\tilde{u} \\ A^{\mathrm{T}}\tilde{u}-\theta B^{\mathrm{T}}B\tilde{w} \end{pmatrix} - \begin{pmatrix} I-uu^{\mathrm{T}} & 0 \\ 0 & I-B^{\mathrm{T}}Bww^{T} \end{pmatrix}\begin{pmatrix} -\theta I & A \\ A^{\mathrm{T}} & -\theta B^{\mathrm{T}}B \end{pmatrix}\begin{pmatrix} s_m \\ t_m \end{pmatrix} \\
&= \begin{pmatrix} A\tilde{w}-\theta\tilde{u} \\ A^{\mathrm{T}}\tilde{u}-\theta B^{T}B\tilde{w} \end{pmatrix} - P\begin{pmatrix} -\theta I & A \\ A^{\mathrm{T}} & -\theta B^{\mathrm{T}}B \end{pmatrix}\left(\begin{pmatrix} s_0 \\ t_0 \end{pmatrix} + \widetilde{V}_m\begin{pmatrix} y_m^{(s)} \\ y_m^{(t)} \end{pmatrix}\right), \\
&= \begin{pmatrix} r_0^{(s)} \\ r_0^{(t)} \end{pmatrix} - P\begin{pmatrix} -\theta I & A \\ A^{\mathrm{T}} & -\theta B^{\mathrm{T}}B \end{pmatrix}\widetilde{V}_m\begin{pmatrix} y_m^{(s)} \\ y_m^{(t)} \end{pmatrix}, \\
&= \widetilde{V}_{m+1}\left(\beta e_1 - \begin{pmatrix} \widetilde{H}_m \\ h_{m+1,m}e_m^{\mathrm{T}} \end{pmatrix}\begin{pmatrix} y_m^{(s)} \\ y_m^{(t)} \end{pmatrix}\right),
\end{aligned}$$

where $\beta = \|r_0\|_D$, $r_0 = \left[r_0^{(s)}, r_0^{(t)}\right]^t$, and e_1 is the first vector of the canonical basis. At this point, the weighted-FOM method entails finding the vector

$y_m^{\mathrm{WF}} = \left[y_m^{(s)}, y_m^{(t)} \right]^t$ solution of the problem:

$\widetilde{V}_m^{\mathrm{T}} D \widetilde{V}_{m+1} (\beta e_1 - \widetilde{H}_m y_m^{\mathrm{WF}}) = 0,$

which is equal to solve

$$\widetilde{H}_m y_m^{\mathrm{WF}} = \beta e_1. \tag{14}$$

To the extent that the weighted-GMRES method is considered, the matrix $\widetilde{V}_{m+1}$ is D-orthonormal, so we have

$\|r_m\|_D^2 = \left\| \widetilde{V}_{m+1} (\beta e_1 - \widetilde{H}_m y_m) \right\|_D^2 = \left\| \beta e_1 - \widetilde{H}_m y_m \right\|_2^2,$

and problem (12) is condensed to find the vector y_m^{WG} solution of the minimization problem:

$$\text{minimize}_{y \in R^m} \left\| \beta e_1 - \widetilde{H}_m y \right\|_2. \tag{15}$$

We can reach the solution of (14) and (15) with the use of the **QR** decomposition of the matrix $\widetilde{H}_m$, as for the FOM and GMRES algorithms. When m is equal to the degree of the minimal polynomial of

$$\begin{pmatrix} I - uu^{\mathrm{T}} & 0 \\ 0 & I - B^{\mathrm{T}} B w w^{\mathrm{T}} \end{pmatrix} \begin{pmatrix} -\theta I & A \\ A^{\mathrm{T}} & -\theta B^{\mathrm{T}} B \end{pmatrix}$$

for $r_0 = \left[r_0^{(s)}, r_0^{(t)} \right]^t$, the Krylov subspace (13) will be invariant. Therefore, the iterate $x_m = [s_m, t_m]^t$ gained by both methods is the exact solution of the correction Eq. (10).■

It is time to write the main algorithm in this paper now. The following algorithm applies FOM, GMRES, weighted-FOM, and weighted-GMRES processes to solve the correction Eq. (10) and as a final point to solve the generalized singular-value decomposition problem. They are represented as F-JDGSVD, G-JDGSVD, WF-JDGSVD, and WG-JDGSVD.

Algorithm 3.1 F-JDGSVD, G-JDGSVD, WF-JDGSVD and WG-JDGSVD

Input: *Starting vectors* u_1 *and* w_1, *and a tolerance* ε. *parameter* m, *and initial vector* x_0

Output: *An approximate triple* (θ, u, w) *for the largest generalized singular triple satisfying*

$$\left\| \begin{pmatrix} Aw - \theta u \\ A^T u - \theta B^T B w \end{pmatrix} \right\| \leq \varepsilon$$

1. $s = u_1, t = w_1, U_0 = [\ ], W_0 = [\ ]$

 for $k = 1, 2, \ldots$ **do**

2. $U_k = RGS\,(U_{k-1}, s)$

   $W_k = RGS_{B^T B}\,(W_{k-1}, t)$

3. *Compute the kth column of* AW_k, $A^T U_k$, *and* $B^T B W_k$

   Compute the kth row and column of $H_k = U_k^T A W_k$

4. *Compute the approximately largest generalized singular triple* (θ, c, d), *of the projected system using the standard extraction technique.*

5. $u = U_k c$, $w = W_k d$

6. $r = \begin{pmatrix} Aw - \theta u \\ A^T u - \theta B^T B w \end{pmatrix}$

7. *Stop if* $\|r\| \le \varepsilon$

8. *Compute*

$$\mathrm{A} = \begin{pmatrix} I - uu^T & 0 \\ 0 & I - B^T B w w^T \end{pmatrix} \begin{pmatrix} -\theta I & A \\ A^T & -\theta B^T B \end{pmatrix}$$

9. *for solving the system* $\mathrm{A}\begin{pmatrix} s \\ t \end{pmatrix} = -r$

   compute $r_0 = b - Ax_0$

   for $l = 1, 2, \ldots$ **do**

   for solving the system by FOM or GMRES methods:

   Compute $\beta = \|r_0\|_2$ *and* $v_1 = r_0/\beta$.

   Construct the orthonormal basis V_m *by the Arnoldi process, starting with* v_1

   Form the approximate solution:

   FOM: *Solve the system* $H_m y_m = \beta e_1$ *by the QR factorization of* H_m *and set*

   $x_m = x_0 + V_m y_m$, $r_m = b - Ax_m$.

GMRES: *Compute* $y_m = \operatorname{argmin}_{y \in R^m} \|\beta e_1 - H_m y\|_2$, *by the QR factorization of* H_m *and set*

$x_m = x_0 + V_m y_m$, $r_m = b - Ax_m$

for solving the system by WFOM or WGMRES methods:

Choose the vector d *such as* $\|d\|_2 = \sqrt{n}$, *and set* $D = diag(d)$

Compute $\tilde{\beta} = \|r_0\|_D$ *and* $\tilde{v}_1 = r_0/\tilde{\beta}$.

Construct the D*-orthonormal basis* V_m *by the weighted Arnoldi process, starting with* $\tilde{v}_1$

Form the approximate solution:

WFOM: *Solve the system* $H_m y_m = \tilde{\beta} e_1$ *by the QR factorization of* H_m *and set*

$x_m = x_0 + V_m y_m$, $r_m = b - Ax_m$.

WGMRES: *compute* $y_m = \operatorname{argmin}_{y \in R^m} \left\| \tilde{\beta} e_1 - \overline{H}_m y \right\|_2$, *by the QR factorization of* $\overline{H}_m$,

set $x_m = x_0 + V_m y_m$, $r_m = b - Ax_m$.

If $\|r_m\|_2 \le \varepsilon_1$ *stop else set* $x_0 = x_m$, $r_0 = r_m$.

As Algorithm 3.1 displays, there are two loops in this algorithm. One of them computes the largest generalized singular value called the outer iteration, and the other called the inner iteration solves the system of linear equation at each iteration. Numerical tests indicate that there is a significant relation between parameter mm and the norm of residual vector and the computational time.

Convergence

We will now demonstrate that the method we have proposed has asymptotically quadratic convergence to generalized singular values when the correction equations are solved in an exact manner and tend toward linear convergence when they are solved with a sufficiently small residual reduction.

Theorem 3.4

Having in mind Theorem 3.3, *suppose that* mm *steps of the weighted Arnoldi process have been performed on* (6) *and* $x_m = [s_m, t_m]^t$ *is the exact solution of the correction Eq.* (10). *Provided that he initial vectors* $(\tilde{u}, \tilde{w})$ *are close enough to* (u,w) , *the sequence of approximations* $(\tilde{u}, \tilde{w})$ *converges quadratically to* (u,w).

Proof

Suppose

$$A = \begin{pmatrix} 0 & A \\ A^T & 0 \end{pmatrix}, \quad B = \begin{pmatrix} I & 0 \\ 0 & B^T B \end{pmatrix}$$

and P are like what you have seen in (5). Let $[s_m, t_m]^T$ with $s_m \perp \tilde{u}$ and $t_m \perp \tilde{w}$ be the exact solution to the correction equation:

$$P(\mathrm{A}-\theta\mathrm{B})\begin{pmatrix} s_m \\ t_m \end{pmatrix} = -r. \tag{16}$$

Besides, let $\alpha u = \tilde{u} + s,\ \ s \perp \tilde{u}$, and ,$\beta w = \tilde{w} + t,\ \ t \perp \tilde{w}$ for certain scalars α and β, satisfy (15); note that these decompositions are possible meanwhile $u^T \tilde{u} \neq 0$ and $w^T \tilde{w} \neq 0$ because of the assumption that the vectors ($\tilde{u}$, $\tilde{w}$) are close to (u,w). Projecting (16) yields

$$P(\mathrm{A}-\theta\mathrm{B})\begin{pmatrix} s \\ t \end{pmatrix} = -r + P\begin{pmatrix} (\mu_1-\theta)s \\ (\mu_2-\theta)B^T Bt \end{pmatrix}. \tag{17}$$

Subtracting (16) from (17) gives

$$P(\mathrm{A}-\theta\mathrm{B})\begin{pmatrix} s-s_m \\ t-t_m \end{pmatrix} = P\begin{pmatrix} (\mu_1-\theta)s \\ (\mu_2-\theta)B^T Bt \end{pmatrix}.$$

Thus for ($\tilde{u}$, $\tilde{w}$) close enough to (u,w), P(A−θB) is a bijection from $\tilde{u}^{\perp} \times \tilde{w}^{\perp}$ onto itself. Together with

$$\mu_1 = \tilde{u}^T A(\tilde{w}+t) = \theta + O(\|t\|),$$

$$\mu_2 = (\tilde{w}+t)^T A^T(\tilde{u}+s)/\|B(\tilde{w}+t)\|^2 = \theta + O(\|s\| + \|t\|),$$

this implies asymptotic quadratic convergence:

$$\left\|\begin{pmatrix} \alpha u - (\tilde{u}+s_m) \\ \beta w - (\tilde{w}+t_m) \end{pmatrix}\right\| = \left\|\begin{pmatrix} s-s_m \\ t-t_m \end{pmatrix}\right\| = O\left(\left\|\begin{pmatrix} s \\ t \end{pmatrix}\right\|^2\right).$$

NUMERICAL EXPERIMENTS

In this section, we look for the largest generalized singular value, using the following default options of the proposed method:

Maximum dimension of search spaces	30
Maximum iterations to solve correction equation	10
Fix target until ‖r‖≤ε	0.01
Initial search spaces	Random

Example 4.1

The matrix pair (A,B) is constructed, such that that they are similar to experiments as [7]. We choose two diagonal matrices of dimension n=1000. For j=1,2,…,1000

$C = \mathrm{diag}(c_j), \quad c_j = (n - j + 1)/2n, \quad S = \sqrt{1 - C^2},$

$D = \mathrm{diag}(d_j), \quad d_j = \lceil j/250 \rceil + r_j$

where the r_j uniformly distributed on the interval (0,1) and $\lceil\cdot\rceil$ denotes the ceil function. We take

$A = Q_1 C D Q_2, \quad B = Q_1 S D Q_2$

where Q_1 and Q_2 are two random orthogonal matrices. The estimated condition numbers of A and B are 4.4e2 and 5.7e0, respectively (Table 1).

Table 1: Implementation of Algorithm 3.1 for (A,B) with different values of m

m	**F-JDGSVD**			**G-JDGSVD**			**WF-JDGSVD**			**WG-GSVD**		
	σ_{max}	$\|r\|_2$	**Time**	σ_{max}	$\|r\|_2$	**Time**	σ_{max}	$\|r\|_2$	**Time**	σ_{max}	$\|r\|_2$	**Time**
4	0.5766	0.0084	23.95	0.5767	0.0062	28.35	0.5773	9.22e−5	31.13	0.5770	8.88e−6	22.08
6	0.5773	0.0052	19.82	0.5770	0.0043	23.32	0.5772	4.82e−5	28.76	0.5768	4.01e−6	17.51
8	0.5773	0.0023	16.10	0.5771	0.0028	19.30	0.5773	7.92e−6	23.66	0.5772	1.00e−6	14.70
10	0.5772	0.0058	14.85	0.5772	0.0014	16.31	0.5773	2.81e−6	17.99	0.5772	9.94e−7	12.04

We can see that by increasing the value of mm, the number of outer and inner iterations decreases. Therefore, the consuming time also decreases. But not that if m is very large, the number of iterations increases because of loosing the orthogonality property. This example is given to show the improvement brought by the weighted methods WF-JDGSVD and WG-JDGSVD is simultaneously on the relative error and on the computational time (Figure 1).

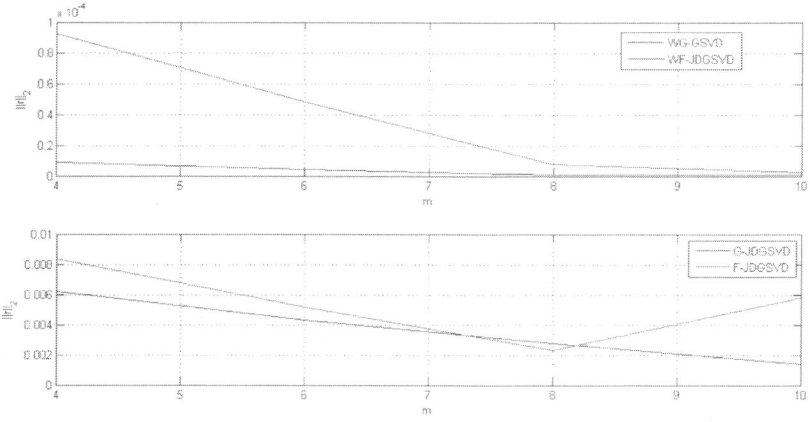

Figure 1: *Errors plot* created by F-JDGSVD, G-JDGSVD, WF-JDGSVD, and WG-GSVD.

From figure one, we can see that the suggested method WG-JDGSVD is more accurate form the other methods.

Example 4.2

In this experiment, we take A=CD and B=SD of various dimension n=400,800,1000,1200.

This example is given to show the performance of two new methods on the large sparse problems. In this test, we have difficulties in computing the largest singular value for ill-conditioned matrices A and B. We note that in this experiments, due to the ill-conditioning of A and B, it turned out to be advantageous to turn of the Krylov option.

Example 4.3

Consider the matrix pair (A,B), where AA is selected from the university of Florida sparse matrix collection [8] as *lp*-ganges. This matrix arises from a linear programming problem. Its size is 1309×1706 and it has a total of Nz=6937 nonzero elements. The estimated condition number is 2.1332e4, and B is the 1309×1706 identity matrix (Tables 2, 3).

Table 2: Implementation of Algorithm 3.1 for (A,B) with various dimensions and m=6

n	F-JDGSVD		G-JDGSVD		WF-JDGSVD		WG-GSVD		κ(A)	κ(B)
	$\|r\|_2$	**Time**	$\|r\|_2$	**Time**	$\|r\|_2$	**Time**	$\|r\|_2$	**Time**		
400	8.82e−4	7.03	0.0098	6.08	2.47e−8	11.85	2.14e−9	11.78	3.5e2	3.2e0
800	0.0085	19.59	0.0063	21.89	9.19e−8	26.09	4.44e−8	22.25	3.6e2	5.6e0
1200	0.0034	27.83	0.0073	29.35	6.74e−6	41.18	5.19e−7	42.35	4.8e2	6.6e0
1600	0.0075	38.65	0.0084	35.89	1.19e−5	49.09	4.99e−5	58.17	6.0e2	8.9e0

Table 3: Implementation of Algorithm 3.1 for (A,B) with different values of mm

m	F-JDGSVD			G-JDGSVD			WF-JDGSVD			WG-GSVD		
	σ_{max}	$\|r\|_2$	**Time**	σ_{max}	$\|r\|_2$	**Time**	σ_{max}	$\|r\|_2$	**Time**	σ_{max}	$\|r\|_2$	**Time**
10	3.9889	0.0075	52.57	3.9865	0.0079	48.86	2.7297	0.00034	63.59	3.9890	0.00015	55.36
20	3.9907	0.0054	46.63	3.9889	0.0035	42.84	2.7298	0.00098	56.99	3.9890	0.00041	47.39
30	2.7298	0.0016	39.78	3.9889	0.0097	36.08	3.9907	0.00043	48.74	3.9888	0.00040	39.65
40	3.9897	0.0091	33.17	3.9888	0.0052	30.89	2.7298	0.00027	38.37	3.9887	0.00014	32.68

We should mention that, for all considered Krylov subspaces sizes, each weighted method converges in less iterations and less time than its corresponding standard method. The convergence of F-JDGSVD and G-JDGSVD is slow, and we have linear asymptotic convergence. However, the two WF-JDGSVD and WG-JDGSVD methods have quadratic asymptotic convergence, because the correction Eq. (10) is solved exactly.

Remark 4.4

From the above examples and tables, we can see that the two suggested methods are more accurate than G-JDGSVD and F-JDGSVD for the same value *m*, but its computational times are often a little longer than G-JDGSVD and F-JDGSVD. Therefore, we can use WF-JDGSVD and WG-GSVD if the computational time is less important.

Remark 4.5

The algorithm we have described finds the largest generalized singular triple. We can compute multiple generalized singular triples of the pair (A,B) using a deflation technique. Suppose that $U_f=[u_1,\ldots,u_f]$ and $W_f=[w_1,\ldots,w_f]$ contain the already found generalized singular vectors, where BW_f has orthonormal columns. We can check that the pair of deflated matrices

$$\hat{A} := (I - U_f U_f^T) A (I - W_f W_f^T B^T B)$$
$$\hat{B} := B(I - W_f W_f^T B^T B) \quad (18)$$

has the same generalized singular values and vectors as the pair (A,B) (see [3]).

Example 4.6

In generalized singular-value decomposition, if $B=I_n$, the n×n identity matrix, we get the singular value of A. SVD has important applications in image and data compression. For example, consider the following image.

This image is represented by a 1185×1917 matrix A. Which we can then decompose via the singular-value decomposition as $A=U\Sigma V^T$ where U is 1185×1185, Σ is 1185×1917, and V is 1917×1917. The matrix A, however, can also be written as a sum of rank 1 matrices $A = \sum_{j=1}^{r} \sigma_j u_j v_j^T$, where $\sigma_1 \geq \sigma_2 \geq \cdots \geq \sigma_r > 0$ are the r nonzero singular value of A. In digital image processing, any matrix A of order m×n(m≥n) generally has a large number of small singular values. Suppose there are (n−k) small singular values of A that can be neglected (Figure 2).

Figure 2: Original image.

Then, the matrix $A_k = \sigma_1 u_1 v_1^T + \sigma_2 u_2 v_2^T + \cdots + \sigma_k u_k v_k^T$ is a very good approximation of A, and such an approximation can be adequate. Even when k is chosen much less then n, the digital image corresponding to A_k can be very close to the original image. Below are the subsequent approximations using various numbers of singular values.

5 singular values

10 singular values

The observation on those examples, we found when k≤20, the images are blurry but with the increase of singular values, when their numbers are about 50, we have a good approach to the original image.

20 singular values

30 singular values

50 singular values

100 singular values

CONCLUSIONS

In this paper, we have suggested two new iterative methods, namely, WF-JDGSVD and WG-JDGSVD, for the computation of some of the generalized singular values and corresponding vectors. Various examples studied illustrate these methods. To accelerate the convergence, we applied the Krylov subspace method for solving the correction equations in large sparse problems. In our methods, we see the existence of asymptotically quadratic convergence, because the correction equations are solved exactly. In the meantime, the correction equations in F-JDGSVD and G-JDGSVD methods are solved inexactly for large sparse problems, so we have linear convergence.

As the amount of the WF-JDGSVD and WG-JDGSVD methods is not much larger than that of the F-JDGSVD and G-JDGSVD methods, and as the weighted methods need less iterations to convergence, the parallel version of the weighted methods seems very interesting. From the tables and the figures, we see that when m increases, the suggested methods are more accurate than the previous methods; moreover, by increasing the dimension of the matrix, two suggested methods are applicable; this results are supported by convergence theorem which shows the asymptotically quadratic convergence to generalized singular values.

REFERENCES

1. Betcke, T.: The generalized singular value decomposition and the method of particular solutions. SIAM. Sci. Comput. **30**, 1278–1295 (2008)
2. Hochstenbach, M.E.: Harmonic and refined extraction methods for the singular value problem, with applications in least square problems. BIT **44**, 721–754 (2004)
3. Hochstenbach, M.E.: A Jacobi–Davidson type method for the generalized singular value problem. Linear Algebra Appl. **431**, 471–487 (2009)
4. Hochstenbach, M.E., Sleijpen, G.L.C.: Two-sided and alternating Jacobi–Davidson. Linear Algebra Appl. **358**(1–3), 145–172 (2003)
5. Kagstrom, B.: The generalized singular value decomposition and the general $A - \lambda B$problem. BIT **24**, 568–583 (1984)
6. Park, C.H., Park, H.: A relationship between linear discriminant analysis and the generalized minimum squared error solution. SIAM J. Matrix Anal. Appl. **27**, 474–492 (2005)
7. Saad, Y.: Krylov subspace methods for solving large unsymmetrical linear systems. Math. Comput. **37**, 105–126 (1981)
8. Saberi Najafi, H., Refahi Sheikhani, A.H: A new restarting method in the Lanczos algorithm for generalized eigenvalue problem. Appl. Math. Comput. **184**, 421–428 (2007)
9. Saberi Najafi, H., Refahi Sheikhani, A.H.: FOM-inverse vector iteration method for computing a few smallest, (largest) eigenvalues of pair (A, B). Appl. Math. Comput. **188**, 641–647 (2007)
10. Saberi Najafi, H., Refahi Sheikhani, A.H., Akbari, M.: Weighted FOM-inverse vector iteration method for computing a few smallest (largest) eigenvalues of pair (A, B). Appl. Math. Comput. **192**, 239–246 (2007)
11. Saberi Najafi, H., Edalatpanah, S.A., Refahi Sheikhani, A.H.: Convergence analysis of modified iterative methods to solve linear systems. Mediterr. J. Math. **11**(3), 1019–1032 (2014)

CHAPTER 3

A DFT-BASED APPROXIMATE EIGENVALUE AND SINGULAR VALUE DECOMPOSITION OF POLYNOMIAL MATRICES

Mahdi Tohidian[1], Hamidreza Amindavar[1] and Ali M Reza[2]

[1]Department of Electrical Engineering, Amirkabir University of Technology, Tehran, Iran

[2]Electrical Engineering and Computer Science, University of Wisconsin-Milwaukee, Milwaukee, WI 53201–0784, USA

ABSTRACT

In this article, we address the problem of singular value decomposition of polynomial matrices and eigenvalue decomposition of para-Hermitian matrices. Discrete Fourier transform enables us to propose a new algorithm based on uniform sampling of polynomial matrices in frequency domain.

Citation (APA): Tohidian, M., Amindavar, H., & Reza, A. M. (2013). A DFT-based approximate eigenvalue and singular value decomposition of polynomial matrices. EURASIP Journal on Advances in Signal Processing, 2013(1), 93. (16 pages).

This formulation of polynomial matrix decomposition allows for controlling spectral properties of the decomposition. We set up a nonlinear quadratic minimization for phase alignment of decomposition at each frequency sample, which leads to a compact order approximation of decomposed matrices. Compact order approximation of decomposed matrices makes it suitable in filterbank and multiple-input multiple-output (MIMO) precoding applications or any application dealing with realization of polynomial matrices as transfer function of MIMO systems. Numerical examples demonstrate the versatility of the proposed algorithm provided by relaxation of paraunitary constraint, and its configurability to select different properties.

Keywords: Singular Value Decomposition, Discrete Fourier Transform, Singular Vector, Polynomial Matrix, Polynomial Matrice

INTRODUCTION

Polynomial matrices have been used for a long time for modeling and realization of multiple-input multiple-output (MIMO) systems in the context of control theory [1]. Nowadays, polynomial matrices have a wide spectrum of applications in MIMO communications [2, 3, 4, 5, 6], source separation [7], and broadband array processing [8]. They also have a dominant role in development of multirate filterbanks [9].

More recently, there have been much interest in polynomial matrix decomposition such as QR decomposition [10, 11, 12], eigenvalue decomposition (EVD) [13, 14], and singular value decomposition (SVD) [5, 11]. Lambert [15] has utilized Discrete Fourier transform (DFT) domain to change the problem of polynomial EVD to pointwise EVD. Since EVD is obtained at each frequency separately, eigenvectors are known at each frequency up to a scaling factor. Therefore, this method requires many frequency samples to avoid abrupt changes in adjacent eigenvectors.

Although, many methods of designing principle component filterbanks have been developed that are equivalent to EVD of pseudo circulant polynomial matrices [16, 17], the next pioneering work on polynomial matrix EVD is presented by McWhirter et al. [13]. They use an extension of Jacobi algorithm known as SBR2 for EVD of para-Hermitian polynomial matrices which guarantees exact paraunitarity of eigenvector matrix. Since final goal of SBR2 algorithm is to have strong decorrelation, the decomposition does not necessarily satisfy spectral majorization property. SBR2 algorithm has also been modified for QR decomposition and SVD [10, 11].

Jacobi-type algorithms are not the only proposed methods for polynomial matrix decomposition. Another iterative method for spectrally majorized EVD is presented in [14] which is based on the maximization of zeroth-order diagonal energy. Spectral majorization property of this algorithm is verified via simulation. Followed by the work of [6], a DFT-based approximation of polynomial SVD is also proposed in [18] which uses model order truncation by phase optimization.

In this article, we present polynomial EVD and SVD based on DFT formulation. It transforms the problem of polynomial matrix decomposition to the problem of, pointwise in frequency, constant matrix decomposition. At first it seems that applying inverse DFT on the decomposed matrices leads to polynomial EVD and SVD of the corresponding polynomial matrix. However, we will show later in this article that in order to have compact order decomposition, phase alignment of decomposed constant matrices in DFT domain results in polynomial matrices with considerably lower order. For this reason, a quadratic nonlinear minimization problem is set up to minimize the decomposition error for a given finite order constraint. Consequently, the required number of frequency samples and computational complexity of decomposition reduce dramatically. The algorithm provides compact order matrices as an approximation of polynomial matrix decomposition for an arbitrary polynomial order. This is suitable in MIMO communications and filterbank applications, where we deal with realization of MIMO linear time invariant systems. Moreover, formulation of polynomial EVD and SVD in DFT domain enables us to select the property of decomposition. We show that if eigenvalues (singular values) intersect at some frequencies in frequency domain, smooth decomposition, and spectrally majorized decomposition are distinct. The proposed algorithm is able to reach to either of these properties.

The remainder of this article is organized as follows. The relation between polynomial matrix decomposition and DFT matrix decomposition is formulated in Section 2. In Section 3, two important spectral properties of decomposition, namely spectral majorization and smooth decomposition, are provided using appropriate arrangement of singular values (eigenvalues) and corresponding singular vectors (eigenvectors). The equality of polynomial matrix and dft matrix decomposed matrices decompositions are guaranteed via the finite duration constraint, which is investigated in Section 4. The finite duration constraint imposes the phase angles of singular vector (eigenvector) to minimize a nonlinear quadratic function. A solution for this problem is proposed in Section 5. Section 6 presents the results of some computer simulations which are considered to demonstrate performance of the proposed decomposition algorithm.

Notation

Some notational conventions are as follows: constant values, vectors, and matrices are in regular character lower case, lower case over-arrow, and upper case, respectively. Coefficients of polynomial (scalar, vector, and matrix) are with indeterminate variable n in the square brackets. Any polynomial (scalar, vector, and matrix) is distinguished by bold character and indeterminate variable z in the parenthesis and its DFT by bold character and indeterminate variable k in the brackets.

PROBLEM FORMULATION

Denote a $p \times q$ polynomial matrix $\mathbf{A}(z)$ such that each element of $\mathbf{A}(z)$ is a polynomial. Equivalently, we can indicate this type of matrix by coefficient matrix $A[n]$,

$$\mathbf{A}(z) = \sum_{n=N_{\min}}^{N_{\max}} A[n]\, z^{-n} \tag{1}$$

where $A[n]$ is only non-zero in the interval $[N_{\min}, N_{\max}]$. Define the effective degree of $\mathbf{A}(z)$ as $N_{\max} - N_{\min}$ (or the length of $A[n]$ as $N_{\max} - N_{\min} + 1$).

The polynomial matrix multiplication of a $p \times q$ matrix $\mathbf{A}(z)$ and a $q \times t$ matrix $\mathbf{B}(z)$ is defined as

$$\begin{aligned} \mathbf{C}(z) &= \mathbf{A}(z)\mathbf{B}(z) \\ \mathbf{c}_{ij}(z) &= \textstyle\sum_{k=1}^{q} \mathbf{a}_{ik}(z)\mathbf{b}_{kj}(z). \end{aligned}$$

We can obtain the coefficient matrix of product by matrix convolution of $A[n]$ and $B[n]$, that is defined as

$$\begin{aligned} C[n] &= A[n] * B[n] \\ c_{ij}[n] &= \textstyle\sum_{k=1}^{q} a_{ik}[n] * b_{kj}[n] \end{aligned}$$

where $*$ denotes the linear convolution operator.

Denote para-conjugate of a polynomial matrix as

$$\tilde{\mathbf{A}}(z) = \mathbf{A}_*^T(z^{-1}) = \sum_{N_{\min}}^{N_{\max}} A^H[n]\, z^n.$$

in which, $*$ as a subscript denotes the complex conjugate of coefficients in the polynomial matrix $\mathbf{A}(z)$.

A matrix is said to be para-Hermitian if A˜(z)=A(z) or equivalently $A[n]=A^H[-n]$. We call a polynomial matrix paraunitary if U˜(z)U(z)=I, where **I** is a $q\times q$ identity matrix.

Thin EVD of a $p\times p$ para-Hermitian polynomial matrix $\mathbf{A}(z)$ is of the form

$$\mathbf{A}(z) = \mathbf{U}(z)\mathbf{\Lambda}(z)\tilde{\mathbf{U}}(z), \tag{2}$$

and thin SVD of a $p\times q$ arbitrary polynomial matrix is of the form,

$$\mathbf{A}(z) = \mathbf{U}(z)\mathbf{\Sigma}(z)\tilde{\mathbf{V}}(z) \tag{3}$$

where $\mathbf{U}(z)$ and $\mathbf{V}(z)$ are $p\times r$ and $q\times r$ paraunitary matrices, respectively. $\mathbf{\Lambda}(z)$ and $\mathbf{\Sigma}(z)$ represent $r\times r$ diagonal matrices where r is the rank of $\mathbf{A}(z)$.

We can equivalently write EVD of a para-Hermitian matrix and SVD of a polynomial matrix in coefficient matrix form

$$A[n] = U[n] * \Lambda[n] * U^H[-n] \tag{4}$$

$$A[n] = U[n] * \Sigma[n] * V^H[-n] \tag{5}$$

in which, $U[n]$, $V[n]$, $\Lambda[n]$, and $\Sigma[n]$ are the coefficient matrices corresponding to $\mathbf{U}(z)$, $\mathbf{V}(z)$, $\mathbf{\Lambda}(z)$, and $\mathbf{\Sigma}(z)$.

In general, EVD and SVD of a finite-order polynomial matrix are not finite order. As an example, suppose EVD of para-Hermitian polynomial matrix

$$A(z) = \begin{bmatrix} 2 & z^{-1}+1 \\ z+1 & 2 \end{bmatrix}. \tag{6}$$

Eigenvalues and eigenvectors of the polynomial matrix in (6) are neither of finite order nor rational

$$\Lambda(z)=\begin{bmatrix} 2+(z^{-1}+2+z)^{1/2} & 0 \\ 0 & 2-(z^{-1}+2+z)^{1/2} \end{bmatrix}$$

$$\mathbf{U}(z)=\frac{1}{\sqrt{2}}\begin{bmatrix} (z^{-1}+1)(z^{-1}+2+z)^{-1/2} & z^{-1} \\ 1 & (z^{-1}+1)(z^{-1}+2+z)^{-1/2} \end{bmatrix}.$$

The same results can be found for polynomial QR decomposition in [12].

We mainly explain the proposed algorithm for polynomial SVD, yet wherever it seems necessary we explain the result for both decomposition.

The decomposition in (3) can also be approximated by samples of discrete-time Fourier transform, yields a decomposition off the form

$$\mathbf{A}[k] = \mathbf{U}'[k]\,\mathbf{\Sigma}'[k]\,\mathbf{V}'^{H}[k], \quad k = 0, 1, \ldots, K-1. \tag{7}$$

Such a decomposition can be obtained by taking the K-point DFT of coefficient matrix $A[n]$,

$$\mathbf{A}[k] = \mathbf{A}(z)\big|_{z=w_K^k} = \sum_{N_{\min}}^{N_{\max}} A[n]\, w_K^{kn}, \quad k = 0, 1, \ldots, K-1, \tag{8}$$

where $w_K = \exp(-j\,2\pi/K)$.

DFT formulation plays an important role in decomposition of polynomial matrices because it replaces the problem of polynomial SVD that involves many protracted steps with K conventional SVD that are pointwise in frequency. It also enables us to control spectral properties of the decomposition. However, it causes two inherent drawbacks:

- Regardless of what is the trajectory of polynomial singular values in frequency domain, conventional SVD order singular values irrespectively of the ordering in neighboring frequency samples.
- In frequency domain, samples of polynomial singular vectors are known up to a scalar complex exponential by using the SVD at each frequency sample, which yields to discontinuous variation between neighboring frequency samples.

The first issue is directly dealt with the spectral properties of the decomposition. In Section 3, we would explain why arranging singular values in decreasing order yields to approximate spectral majorization, while smooth decomposition requires rearrangement of singular values and their corresponding singular vectors.

For the second issue, suppose conventional SVD of an arbitrary constant matrix A. If the pair $\vec{u}$ and $\vec{v}$ are the left and right singular vectors corresponding to a non-zero singular value, for an arbitrary scalar phase angle θ, the pair $e^{j\theta}\vec{u}$ and $e^{j\theta}$ are also left and right singular vectors corresponding to the same singular value. Although this non-uniqueness is trivial in conventional SVD, it plays a crucial role in polynomial SVD. When we perform SVD at each frequency of DFT matrix as in (7), these non-uniquenesses in phase exist at each frequency regardless of other frequency samples.

Denote $\vec{u}_i[k]$ and $\vec{v}_i[k]$ the i th column vector of the desired matrices $\mathbf{U}(z)$ and $\mathbf{V}(z)$. Then all the vectors of the form

$$\begin{cases} \vec{\mathbf{u}}_i'[k] = e^{j\theta_i[k]}\vec{\mathbf{u}}_i[k] \\ \vec{\mathbf{v}}_i'[k] = e^{j\theta_i[k]}\vec{\mathbf{v}}_i[k], \quad i = 1, 2, \ldots, r \\ \sigma_i'[k] = \sigma_i[k] \end{cases} \tag{9}$$

have the chance to appear as the i th column of $\mathbf{U}'[k]$ and $\mathbf{V}'[k]$, and i th diagonal element of $\mathbf{\Sigma}'[k]$, respectively. Moreover, in many applications, specially those which are related to MIMO precoding, we can relax constraints of the problem by letting singular values to be complex (see applications of polynomial SVD in [4, 18])

$$\begin{cases} \vec{\mathbf{u}}_i'[k] = e^{j\theta_i^u[k]}\vec{\mathbf{u}}_i[k] \\ \vec{\mathbf{v}}_i'[k] = e^{j\theta_i^v[k]}\vec{\mathbf{v}}_i[k] \quad , \quad i = 1, 2, \ldots, r. \\ \sigma_i'[k] = e^{j(\theta_i^v[k] - \theta_i^u[k])}\sigma_i[k] \end{cases} \tag{10}$$

Given this situation, singular values have not all their conventional meaning. For instance, the greatest singular value is conventionally 2-norm of the corresponding matrix, which is not true for complex singular values. The process of compensating singular vectors for these phases is what we call *phase alignment* and is developed in Section 4.

Based on what was mentioned above, Algorithm 1 gives the descriptive pseudo code for DFT-based SVD. Modifications of the algorithm for EVD of para-Hermitian matrices are straightforward. If at each frequency sample all singular values are in decreasing order, REARRANGE function (which is described in Algorithm 2) is only required for smooth decomposition, otherwise for spectral majorization, no further arrangement is required. For the phase alignment, first we need to compute phase angles which is indicated in the algorithm by DOGLEG function and is described in Algorithm 3.

SPECTRAL MAJORIZED DECOMPOSITION VERSUS SMOOTH DECOMPOSITION

Two of the most appealing decomposition properties are smooth decomposition [19] and spectral majorization [13]. These two objectives do not always occur at the same time, hence we should choose which one we are willing to use as our main objective.

In many filterbank applications which are dealt with principle components filterbank, spectral majorization and strong decorrelation are both required [16]. Since smooth decomposition leads to more compact decomposition, in cases that the only objective is strong decorrelation, exploiting smooth decomposition is reasonable. The DFT-based approach of polynomial matrix decomposition is capable of decomposing a matrix with either of these properties with small modification.

Algorithm 1 Approximate SVD

U[n] , [n] ,V[n] ← ASVD (A[n])

for k = 0, 1, ⋯ , K − 1

Compute A[k] from (8):

$$\mathbf{A}[k] = \sum_{N_{\min}}^{N_{\max}} A[n]\, w_K^{kn}$$

Decompose A[k] from (7):

$$\left[\mathbf{U}'[k], \mathbf{\Sigma}[k], \mathbf{V}'[k]\right] \leftarrow \mathrm{SVD}(\mathbf{A}[k])$$

end(for)

If smooth decomposition is required use Algorithm 2:

$$\left[\mathbf{U}'[k], \mathbf{\Sigma}'[k], \mathbf{V}'[k]\right] \leftarrow \mathbf{REARRANGE}\left(\mathbf{U}'[k], \mathbf{\Sigma}'[k], \mathbf{V}'[k]\right)$$

for i = 1, 2, ⋯ ,r

Compute phase angles using Algorithm 3:

$$\left[\theta_i^u[k], \theta_i^v[k]\right] \leftarrow \mathbf{DOGLEG}\left(\vec{\mathbf{u}}_i'[k], \vec{\mathbf{v}}_i'[k]\right)$$

for k = 0, 1, ... , K − 1

Phase alignment using (9) or (10)

end(for)

end(for)

for n = 0, 1, ... , M − 1

Compute decomposed polynomial matrices:

$$U[n] \leftarrow \sum_{k=0}^{K} \mathbf{U}[k]\, W^{-kn}$$

$$V[n] \leftarrow \sum_{k=0}^{K} \mathbf{V}[k]\, W^{-kn}$$

$$\Sigma[n] \leftarrow \texttt{Diagonal elements of } U^H[-n] * A[n] * V[n]$$

end(for)

Polynomial EVD of a para-Hermitian matrix is said to have spectral majorization property if [13, 16]

$$\lambda_1(e^{j\omega}) \geq \lambda_2(e^{j\omega}) \geq \cdots \geq \lambda_r(e^{j\omega}), \quad \forall \omega.$$

Note that, eigenvalues corresponding to para-Hermitian matrices are real in all frequencies.

We can extend the definition to the polynomial SVD, replacing singular values with eigenvalues in the definition, we have

$$\sigma_1(e^{j\omega}) \geq \sigma_2(e^{j\omega}) \geq \cdots \geq \sigma_r(e^{j\omega}), \quad \forall\omega.$$

If we let singular values to be complex, we can replace absolute value of singular values in the definition.

A polynomial matrix have no discontinuity in frequency domain, hence we modify definition of smooth decomposition presented in [19] to fit with our problem and avoid unnecessary discussions.

Polynomial EVD (SVD) of a matrix is said to possess smooth decomposition if eigenvectors (singular vectors) have no discontinuity in frequency domain, that is

$$\left|\frac{d}{d\omega}\mathbf{u}_{il}(e^{j\omega})\right| < \infty, \quad \forall\omega \text{ and } \begin{matrix} i = 1, 2, \ldots, r \\ l = 1, 2, \ldots, p, \end{matrix} \tag{11}$$

where $\mathbf{u}_{il}$ is the l th element of $\vec{u}_i$.

If eigenvalues (singular values) of a polynomial matrix intersect at some frequencies, the spectral majorization and smooth decomposition are not simultaneously realizable. As an example, suppose $\mathbf{A}(z)$ is a polynomial matrix with $\vec{u}_1(z)$ and $\vec{u}_2(z)$ are eigenvectors corresponding to distinct eigenvalues $\lambda_1(z)$ and $\lambda_2(z)$, respectively. Lets assume $\vec{u}_1(e^{j\omega})$ and $\vec{u}_2(e^{j\omega})$ have no discontinuity in frequency domain, and $\lambda_1(e^{j\omega})$ and $\lambda_2(e^{j\omega})$ intersect at some frequencies. Denote

$$\lambda_1'(e^{j\omega}) = \begin{cases} \lambda_1(e^{j\omega}) & \lambda_1(e^{j\omega}) \geq \lambda_2(e^{j\omega}) \\ \lambda_2(e^{j\omega}) & \lambda_1(e^{j\omega}) < \lambda_2(e^{j\omega}) \end{cases},$$

$$\lambda_2'(e^{j\omega}) = \begin{cases} \lambda_2(e^{j\omega}) & \lambda_1(e^{j\omega}) \geq \lambda_2(e^{j\omega}) \\ \lambda_1(e^{j\omega}) & \lambda_1(e^{j\omega}) < \lambda_2(e^{j\omega}) \end{cases}, \tag{12}$$

Algorithm 2 Rearrangement for smooth decomposition

$$\left[\mathbf{U}'[k], \Sigma[k], \mathbf{V}'[k]\right] \leftarrow \mathbf{REARRANGE}\left(\mathbf{U}'[k], \Sigma[k], \mathbf{V}'[k]\right)$$

for k = 1, 2, ... , K
Define S = {1, 2, ... ,r}
for i = 1, 2, ... ,r

$$\imath \leftarrow \arg\max_{j\in S} \frac{|c_{ij}^{u'}[k]| + |c_{ij}^{v'}[k]|}{2}$$

$$\vec{\mathbf{u}}_i''[k] \leftarrow \vec{\mathbf{u}}_\imath'[k]$$

$$\vec{\mathbf{v}}_i''[k] \leftarrow \vec{\mathbf{v}}_\imath'[k]$$

$$\sigma_i''[k] \leftarrow \sigma_\imath[k]$$

$$S \leftarrow S - \{\imath\}$$

end(for)

$$\mathbf{U}'[k] \leftarrow \mathbf{U}''[k]$$

$$\mathbf{V}'[k] \leftarrow \mathbf{V}''[k]$$

$$\Sigma[k] \leftarrow \Sigma''[k]$$

end(for)

and

$$\vec{\mathbf{u}}_1'(e^{j\omega}) = \begin{cases} \vec{\mathbf{u}}_1(e^{j\omega}) & \lambda_1(e^{j\omega}) \geq \lambda_2(e^{j\omega}) \\ \vec{\mathbf{u}}_2(e^{j\omega}) & \lambda_1(e^{j\omega}) < \lambda_2(e^{j\omega}) \end{cases},$$

$$\vec{\mathbf{u}}_2'(e^{j\omega}) = \begin{cases} \vec{\mathbf{u}}_2(e^{j\omega}) & \lambda_1(e^{j\omega}) \geq \lambda_2(e^{j\omega}) \\ \vec{\mathbf{u}}_1(e^{j\omega}) & \lambda_1(e^{j\omega}) < \lambda_2(e^{j\omega}) \end{cases}. \quad (13)$$

Obviously, $\vec{u}'_1(e^{j\omega})$ and $\vec{u}'_2(e^{j\omega})\vec{\mathbf{u}}_2'(e^{j\omega})$ are eigenvectors corresponding to distinct eigenvalues $\lambda'_1(e^{j\omega})$ and $\lambda'_2(e^{j\omega})$, respectively. Note that, $\lambda'_1(e^{j\omega}) \geq \lambda'_2(e^{j\omega})$ for all frequencies, which means $\lambda_1(e^{j\omega})$ and $\lambda_2(e^{j\omega})$ are spectrally majorized. However, $\vec{u}'_1(e^{j\omega})$ and $\vec{u}'_2(e^{j\omega})$ are discontinuous at intersection frequencies of $\lambda_1(e^{j\omega})$ and $\lambda_2(e^{j\omega})$, which implies that they are not smooth anymore. In this situation, although $\lambda'_1(e^{j\omega})$, $\lambda'_2(e^{j\omega})$, $\vec{u}'_1(e^{j\omega})$, and $\vec{u}'_2(e^{j\omega})$are not even analytic, we can approximate them with finite order polynomials.

If a decomposition has spectral majorization, its eigenvalues (singular values) are of decreasing order in all frequencies. Therefore, they are in decreasing order in any arbitrary frequency sample set, including DFT frequencies. Obviously the converse is only approximately true. Hence, for polynomial EVD to possess spectral majorization approximately, it suffices to arrange sampled eigenvalues (singular values) of (7) in decreasing order. Since we only justify spectral majorization at DFT frequency samples, the resulting EVD (SVD) may possess the property only approximately. Similar results can be seen in [14, 20].

To have smooth singular vectors, we propose an algorithm based on inner product of consecutive frequency samples of singular vectors. We can accumulate smoothing requirement in (11) for all r elements as

$$\left\| \frac{d}{d\omega} \vec{\mathbf{u}}_i(e^{j\omega}) \right\| < \infty, \quad \forall\omega \text{ and } \quad i = 1, 2, \ldots, r. \tag{14}$$

Let B be the upper bound of norm of derivative and $R\{\cdot\}$ be the real value of a complex value.

For an arbitrary $\Delta\omega$ we have

$$\begin{aligned} \left\| \vec{\mathbf{u}}_i(e^{j(\omega+\Delta\omega)}) - \vec{\mathbf{u}}_i(e^{j\omega}) \right\|^2 &= 2 - 2\Re\left\{ \vec{\mathbf{u}}_i^H(e^{j(\omega+\Delta\omega)}) \vec{\mathbf{u}}_i(e^{j\omega}) \right\} \\ &< (\Delta\omega B)^2 \quad \forall\omega, \end{aligned} \tag{15}$$

that is, for a smooth singular vector $R\{\vec{u}_i(e^{j(\omega+\Delta\omega)})\,\vec{u}_i(e^{j\omega})\}$ can be made to be as close to unity as desired by making $\Delta\omega$ sufficiently small. In our problem $\vec{u}_i(e^{j\omega})$ is sampled uniformly with $\Delta\omega = 2\pi/K$. Since EVD is performed at each frequency sample independently, $\vec{u}_i[k]$ and $\vec{u}_i[k+1]$ are not necessarily two consecutive frequency samples of a smooth eigenvector. Therefore, we should rearrange eigenvalues and eigenvectors to yield smooth decomposition. This can be done for each sample of eigenvector $\vec{u}_i[k]$ by seeking for the eigenvector of successor sample $\vec{u}_j[k+1]$ with the most value of $R\{\vec{u}_i^H[k]\,\vec{u}_j[k+1]\}$.

Define inner product $c_{ij}^u[k]$ as

$$c_{ij}^u[k] = \vec{\mathbf{u}}_i^H[k-1]\,\vec{\mathbf{u}}_j[k].$$

Since, $\vec{u}'_i[k]$ is a scalar phase multiplication of $\vec{u}_i[k]$, computation of $R\{c_{ij}^u[k]\}$ is not possible before phase alignment. Due to (15), for sufficiently small $\Delta\omega$, two consecutive samples of a smooth singular vector can be as close as desired and we can approximate

$$\Re\left\{c_{ij}^{u}[k]\right\} \approx |c_{ij}^{u}[k]| = |c_{ij}^{u'}[k]|,$$

which allows us to use inner product of $\vec{u}$ '[k] instead of $\vec{u}$ [k]. From (12) and (13), it can be seen that before the intersection of eigenvalues, consecutive eigenvectors which are sorted by conventional EVD in decreasing order, are from the same smooth eigenvector and so $|c_{11}^{u'}[k]|$ and $|c_{22}^{u'}[k]|$ are near unity. However, if $k-1$ and k are two frequency sample before and after the intersection, respectively, due to decreasing order of eigenvalues, smoothed eigenvectorsareswappedafterintersection.Therefore, $|c_{11}^{u'}[k]|$ and $|c_{22}^{u'}[k]|$ are some values near zero, instead $|c_{12}^{u'}[k]|$ and $|c_{21}^{u'}[k]|$ are near unity.

Algorithm 2 describes a simple rearrangement procedure to track eigenvectors (singular vectors) for smooth decomposition.

FINITE DURATION CONSTRAINT

Phase alignment is critical to have compact order decomposition. Another aspect of this fact is revealed in the coefficient's domain perspective of (7). In this domain, the multiplication is replaced by circular convolution

$$\begin{aligned} A[((n))_K] &= U'[((n))_K] \circledast \Sigma'[((n))_K] \circledast V'^H[((-n))_K] \\ &= U[((n))_K] \circledast \Sigma[((n))_K] \circledast V^H[((-n))_K] \end{aligned} \qquad (16)$$

in which $\circledast$ is the circular convolution operator and $((n))_K$ denotes n module K.

Polynomial SVD corresponds to linear convolution in the coefficients domain, however the decomposition obtained from DFT corresponds to circular convolution. Recalling from discrete-time signal processing, it is well known that we can equivalently utilize circular convolution instead of linear convolution if convoluted signals are zero-padded adequately. That is, for $x_1[n]$ and $x_2[2]$ are two signals with the length of N_1 and N_2, respectively, apply zero padding such that zero padded signals have the length N_1+N_2-1 [21]. Hence, if the last M–1 coefficients of $U[n]$, $\Sigma[n]$, and $V[n]$, are zero, the following results are hold:

$$A[((n))_K] = U[((n))_K] \circledast \Sigma[((n))_K] \circledast V^H[((-n))_K] \Rightarrow A[n] = U[n] * \Sigma[n] * V^H[-n],$$
$$U[((n))_K] \circledast U^H[((-n))_K] = \delta[((n))_K] I \Rightarrow U[n] * U^H[-n] = \delta[n] I,$$
$$V[((n))_K] \circledast V^H[((-n))_K] = \delta[((n))_K] I \Rightarrow V[n] * V^H[-n] = \delta[n] I. \quad (17)$$

Therefore, the problem is to obtain the phase set $\{\{\vec{\theta}_i[k]\}$ and correcting the singular vectors using (9). The phase set $\{\vec{\theta}_i[k]\}$ should be such that the resulting coefficients satisfy (17).

Without loss of generality, let $U[n]$ and $V[n]$ be causal, i.e., $U[n]=V[n]=0$ for $n<0$. $U[n]$ and $V[n]$ (which are supposed to be of length M) should be multi-sequence zero-padded at least with $(M-1)$ zeros.

$$U[n] = \frac{1}{K}\sum_{k=0}^{K-1} \mathbf{U}[k]\, w_K^{-kn} = \mathbf{0} \quad (18)$$

for $n=M, M+1, \ldots, K-1$, in which $K \geq 2M-1$. If these conditions are satisfied, circular convolution can be used instead of linear convolution.

Since the available matrix of singular vectors at each frequency is $\mathbf{U}'[k]$, inserting (9) in (18) for each singular vector separately leads to

$$\sum_{k=0}^{K-1} \mathbf{u}_i'[k]\, e^{-j\theta_i^u[k]} w_K^{-kn} = \mathbf{0} \quad (19)$$

for $n=M, M+1, \ldots, K-1$.

Without loss of generality, let $\theta_i[0]=0$. In a more compact form we can express these $(K-M)$-folded equations in matrix form

$$\mathcal{F}_M(\vec{\mathbf{u}}_i')\vec{\mathbf{x}}(\vec{\theta}_i^u) = -\vec{f}_M(\vec{\mathbf{u}}_i') \qquad i = 1, 2, \ldots, r \quad (20)$$

in which $\vec{\mathbf{x}}(\vec{\theta}_i^u) = [\exp(j\theta_i^u[1]), \ \exp(j\theta_i^u[2]), \ldots, \exp(j\theta_i^u[K-1])]^T$, $\vec{f}_M(\vec{\mathbf{u}}_i') = [\vec{\mathbf{u}}_i'^T[0], \vec{\mathbf{u}}_i'^T[0], \ldots, \vec{\mathbf{u}}_i'^T[0]]^T$ is a $p(K-M)\times 1$ vector, and

$$\mathcal{F}_M(\vec{\mathbf{u}}_i') = \begin{bmatrix} \vec{\mathbf{u}}_i'[1]w_K^{-M} & \vec{\mathbf{u}}_i'[2]w_K^{-2M} & \cdots & \vec{\mathbf{u}}_i'[K-1]w_K^{-(K-1)M} \\ \vec{\mathbf{u}}_i'[1]w_K^{-(M+1)} & \vec{\mathbf{u}}_i'[2]w_K^{-2(M+1)} & \cdots & \vec{\mathbf{u}}_i'[K-1]w_K^{-(K-1)(M+1)} \\ \vdots & \vdots & & \vdots \\ \vec{\mathbf{u}}_i'[1]w_K^{-(K-1)} & \vec{\mathbf{u}}_i'[2]w_K^{-2(K-1)} & \cdots & \vec{\mathbf{u}}_i'[K-1]w_K^{-(K-1)^2} \end{bmatrix}.$$

For polynomial EVD, Equation (20) is enough, however, for polynomial SVD we have two options. To approximate SVD with approximately positive singular values, we must augment $F_M(\vec{u}_i')$ and $\vec{f}_M(\vec{u}_i')$ with similar defined matrix and vector for $\mathbf{v}\, i'[k]$

$$\mathcal{F}_M(\vec{\mathbf{u}}_i', \vec{\mathbf{v}}_i') = \begin{bmatrix} \mathcal{F}_M(\vec{\mathbf{u}}_i') \\ \mathcal{F}_M(\vec{\mathbf{v}}_i') \end{bmatrix} \text{ and } \vec{f}_M(\vec{\mathbf{u}}_i', \vec{\mathbf{v}}_i') = \begin{bmatrix} \vec{f}_M(\vec{\mathbf{u}}_i') \\ \vec{f}_M(\vec{\mathbf{v}}_i') \end{bmatrix},$$

then solve

$$\mathcal{F}_M(\vec{\mathbf{u}}_i', \vec{\mathbf{v}}_i')\vec{\mathbf{x}}(\vec{\theta}_i) = -\vec{f}_M(\vec{\mathbf{u}}_i', \vec{\mathbf{v}}_i') \quad i = 1, 2, \ldots, r. \tag{21}$$

An additional degree of freedom is obtained by letting singular values to be complex. However, an straightforward solution which yield to singular values and singular vectors of order M is complicated. Instead, we impose the finite duration constraint only two singular vectors

$$\begin{cases} \mathcal{F}_M(\vec{\mathbf{u}}_i')\vec{\mathbf{x}}(\vec{\theta}_i^v) = -\vec{f}_M(\vec{\mathbf{u}}_i') \\ \mathcal{F}_M(\vec{\mathbf{v}}_i')\vec{\mathbf{x}}(\vec{\theta}_i^v) = -\vec{f}_M(\vec{\mathbf{v}}_i') \end{cases} \quad i = 1, 2, \ldots, r. \tag{22}$$

If $K \geq 2M-1$, then the last $M-1$ coefficients of resulting polynomial vectors are zero. Therefore, according to (17), $\mathbf{U}(z)$ and $\mathbf{V}(z)$ are paraunitary. On the other hand, if $K \geq 2M+N_{max}-N_{min}-1$, circular convolution relation of coefficient

$$\Sigma[\,((n))_K] = U^H[\,((-n))_K] \circledast A[\,((n))_K] \circledast V[\,((n))_K]$$

results in the linear convolution $\Sigma[n] = U^H[-n] * A[n] * V[n]$. This guarantee that $\mathbf{\Sigma}(z)$ is a diagonal polynomial matrix of order $2M+N_{max}-N_{min}-2$. Obviously, if $\mathbf{U}(z)$ and $\mathbf{V}(z)$ are paraunitary and $\Sigma(z)=\tilde{U}(z)A(z)V(z)$ is a diagonal matrix, $A(z)=U(z)\Sigma(z)\tilde{V}(z)$is the polynomial SVD of $\mathbf{A}(z)$.

Once the set of phase $\{\theta_i^u[k], \theta_i^v[k]\}$ are obtained from (20), (21), or (22), phase alignment of $\vec{u}'_i[k]$ and $\vec{v}'_i[k]$ can be done using (10) and inverse DFT of $\mathbf{U}[k]$ and $\mathbf{V}[k]$ yield to coefficient matrices $U[n]$ and $V[n]$. For obtaining singular values, we have two options, we can either set $K \geq 2M-1$ and phase align $\boldsymbol{\sigma}\ i'[k]$ using (10). After inverse DFT of $\boldsymbol{\Sigma}[k]$, we should truncate $\Sigma[n]$ to have duration M. Another option which yields to more accurate results is by calculating $\vec{U}$ (z)A(z)V(z) and replacing off-diagonal elements with zero.

Next, we provide a minimization approach to determine the unknown set $\{\theta_i[k]\}$.

GRADIENT DESCENT SOLUTION

In general, there may exist no phase vector $\vec{\theta}$ which satisfies (20). Even when there exists a phase vector that satisfies the finite duration constraint, the solution is not straightforward. For these reasons, we can view (20) as a minimization problem [6]. We use energy of the highest order coefficients (the coefficients that we equate to zero in (18)) as the objective to the minimization problem

$$J(\vec{\theta}_i) = \left\| \mathcal{F}_M(\vec{\mathbf{u}}'_i)\vec{\mathbf{x}}(\vec{\theta}_i^u) + \vec{f}_M(\vec{\mathbf{u}}'_i) \right\|^2, \quad i = 1, 2, \ldots, r. \tag{23}$$

An alternative minimization technique as a solution for this phase optimization problem is proposed in [6], which we describe it in this section.

Throughout this section, we focus on solving $\vec{\theta}_i$=argminJ($\vec{\theta}$ i) as a least square solution for a single singular vector $\vec{u}_i[k]$, so we drop the subscript "i" from the quantity $\vec{\theta}$ i and use F and $\vec{f}$, instead of FM($\vec{u}'_i$) and $\vec{f}$ M($\vec{u}'_i$) to simplify the notation. The objective J($\vec{\theta}$) is intentionally presented as a function of $\vec{\theta}$ to emphasize the fact that our problem is classified as an unconstrained optimization.

We exploit the trusted region strategy for the problem (23). By utilizing the information about the gradient vector and Hessian matrix in each step, trusted region strategy constructs a model function m_k which have a similar behavior close to the current point $\vec{\theta}$ (k). The model m_k is usually defined as the second-order Taylor series expansion (or its approximation) of J($\vec{\theta}$ + $\vec{\varphi}$) around $\vec{\theta}$, that is

$$m_k(\vec{\varphi}) = J(\vec{\theta}) + \vec{\varphi}^T \nabla J(\vec{\theta}) + \vec{\varphi}^T \nabla^2 J(\vec{\theta})\vec{\varphi},$$

where ∇J($\vec{\theta}$) and ∇^2J($\vec{\theta}$) are the gradient vector and the Hessian matrix corresponding to J($\vec{\theta}$), respectively. The model m_k is designed to be a good approximation of J($\vec{\theta}$) near the current point and is not trustworthy on regions far from the current point. Consequently, the restriction in minimization of m_k on a region around $\vec{\theta}$ (k) is crucial, that is

$$\vec{\varphi} = \arg\min_{\vec{\varphi}} m_k(\vec{\varphi}) \quad \|\vec{\varphi}\| < R. \tag{24}$$

where R is the trusted region radius.

The decision about shrinking of the trusted region is determined by comparing the actual reduction inobjective function and predicted reduction. Given a step $\vec{\varphi}$, the ratio

$$\rho = \frac{J(\vec{\theta}) - J(\vec{\theta} + \vec{\varphi})}{m_k(0) - m_k(\vec{\varphi})} \tag{25}$$

is used as a criterion to indicate if the trusted region is small enough.

Among methods which approximate the solution of the constrained minimization (24) dogleg procedure is the only one which leads to analytical approximation. It also promises to achieve at least as much reduction in m_k as is possible by Cauchy point (the minimizer of m_k along the steepest descent direction −∇J($\vec{\theta}$), subject to the trusted region) [22]. However, this procedure requires Hessian matrix (or an approximation of it) to be positive definite.

Hessian Matrix Modification

The gradient vector and Hessian matrix corresponding to J($\vec{\theta}$) are as follows

$$\begin{aligned} \vec{g}(\vec{\theta}) &= 2\Im\left\{\mathbf{X}(\vec{\theta})\mathcal{F}^H(\mathcal{F}\vec{\mathbf{x}}(\vec{\theta}) + \vec{f})\right\}, \\ H(\vec{\theta}) &= 2\Re\left\{\mathbf{X}(\vec{\theta})\mathcal{F}^H\mathcal{F}\mathbf{X}(\vec{\theta})^H\right\} \\ &\quad -2\Re\left\{\text{diag}\left(\mathbf{X}(\vec{\theta})\mathcal{F}^H(\mathcal{F}\vec{\mathbf{x}}(\vec{\theta}) + \vec{f})\right)\right\}, \end{aligned} \tag{26}$$

where $X(\vec{\theta})$ is a diagonal matrix with the k th diagonal element $\exp(-j\,\theta[k])$ and $k=0, 1, \ldots, K-1$.

In general, Hessian matrix in (26) does not promise to be always positive definite. Therefore, we should modify Hessian matrix to yield a positive definite approximation.

We provide a simple modification which brings some desirable features by omitting the second term from the Hessian matrix and diagonal loading to guarantee positive definiteness

$$H(\vec{\theta}) \approx 2\Re\left\{\mathbf{X}(\vec{\theta})\mathcal{F}^H\mathcal{F}\mathbf{X}(\vec{\theta})^H\right\} + \alpha I. \qquad (27)$$

The term $2R\{X(\vec{\theta})F^H F X(\vec{\theta})^H\}$ is positive semi-definite and in many situations, it is much more significant than the second term of Hessian matrix in (26). Hence, with diagonal loading αI (I is with conformable size and α is very small), the modified Hessian matrix guarantees (27) to be positive definite and provides the desired properties in contrast to use the exact Hessian matrix.

Dogleg Method

Dogleg method starts with the unconstrained minimization of (24)

$$\phi^h = -H^{-1}\vec{g} \qquad (28)$$

When the trusted region radius is so large that $\|\phi^H\| \leq R$, it is the exact solution of (24) and we select it as the dogleg method answer. On the other hand, for small R the solution of (24) is $-R\vec{g}/\|\vec{g}\|$. For intermediate values of R, the optimal solution lies on a curved trajectory between these two points [22].

Algorithm 3 Trusted region dogleg algorithm

Obtain $\vec{\theta}_0$ from (33)
Given $R_0 < \bar{R}$
for $i = 0, 1, 2, \cdots$
  Obtain $\vec{\phi}_h$ from (28)
  if $\left\|\vec{\phi}^h\right\| < R_0$
    $\vec{\phi}_i \leftarrow \vec{\phi}^u$
  else
    Obtain $\vec{\phi}^g$ from (29)
    if $\left\|\vec{\phi}^g\right\| > R_i$

   else
     Obtain $\vec{\phi}^g$ from (29)
     if $\left\|\vec{\phi}^g\right\| > R_i$
       $\vec{\phi}_i \leftarrow R_i \frac{\vec{\phi}^g}{\|\vec{\phi}^g\|}$
     else
       $\vec{\phi}^d \leftarrow \vec{\phi}^h - \vec{\phi}^g$
       Solve $\left\|\vec{\phi}^d\right\| \tau^2 + 2(\vec{\phi}^g)^H \vec{\phi}^d \tau + \left\|\vec{\phi}^g\right\| - R_i^2 = 0$
       Select the root $\tau > 0$
       $\vec{\phi}_i \leftarrow \vec{\phi}^g + \tau \vec{\phi}^h$

  Evaluate ρ_i from (25)
  if $\rho_i < \frac{1}{4}$
    $R_{i+1} \leftarrow 0.25 R_i$
  elseif $\rho > \frac{3}{4}$ and $\|\vec{\phi}_i\| = R_i$
    $R_{i+1} \leftarrow min(2R_i, \bar{R})$
  else
    $R_{i+1} \leftarrow R_i$
  if $\rho_i > 0$
    $\theta_{i+1} \leftarrow \theta_i + \phi_i$
  else
    $\theta_{i+1} \leftarrow \theta_i$
end(for)

The dogleg method approximates this trajectory by a path consisting of two line segments. The first line segment starts from the origin to the unconstrained minimized point along the steepest descent direction

$$\phi^g = -\frac{\vec{g}^T\vec{g}}{\vec{g}^T H\vec{g}}\vec{g}. \tag{29}$$

The second line segment starts from ϕ^g to ϕ^h. These two line segments form an approximate trajectory which its intersection with the sphere $\|\phi\|=R$ is the approximate solution of (24) when $\|\phi^h\|>R$.

Alternating Minimization

Another solution of (23) is provided by converting the problem of multivariate minimization to a sequence of single-variate minimization problem via alternating minimization [6]. In each iteration, a series of single-variate minimization is performed, while other parameters are held unchanged. Each Iteration consists of $k-1$ steps, which at each step one parameter $\theta[k]$ is updated. Suppose we are at step k of i th iteration. At this step $k-1$ first parameters were updated in the current iteration, and $K-k-2$ last parameters were updated from the previous iteration. These parameters are held fixed, while $\theta[k]$ is minimized at the current step,

θi[k]=argminθ[k]J(θi[1],...,θi[k−1],θ[k],θi−1[k+1],...,θi−1[K−1]).

$$\theta_i[k] = \arg\min_{\theta[k]} J\left(\theta_i[1],\ldots,\theta_i[k-1],\theta[k],\theta_{i-1}[k+1],\ldots,\theta_{i-1}[K-1]\right). \tag{30}$$

The cost function is guaranteed to be non-incremental at each step; however, this method is also converges to a local minima which highly depend on the initial guess of the algorithm. For solving (30) it is suffices to make the k th element of gradient vector in (26) equal to zero. Suppose the calculation are performed for phase alignment of

$\vec{\mathbf{u}}'[k] \quad k = 0, 1, \ldots, K-1,$

$$\begin{aligned}\frac{\partial J}{\partial\theta[k]} &= \Im\left\{e^{-j\theta[k]}t_i[k]\right\}\\ &= 2\,|t_i[k]|^2 \sin(\angle t_i[k] - \theta[k]) = 0,\end{aligned} \tag{31}$$

where $\angle t_u^{(i)}[k]$ is the phase angle of $t_u^{(i)}[k]$ and

$$t_i[k] = \sum_{l=0}^{k-1} e^{j\theta_i[l]} \vec{\mathbf{u}}'^H[k]\, \vec{\mathbf{u}}'[l] \frac{w_K^{(k-l)M} - 1}{1 - w_K^{(k-l)}}$$
$$+ \sum_{l=k+1}^{K-1} e^{j\theta_{i-1}[l]} \vec{\mathbf{u}}'^H[k]\, \vec{\mathbf{u}}'[l] \frac{w_K^{(k-l)M} - 1}{1 - w_K^{(k-l)}}.$$

Fortunately, Equation (31) has a closed form solution

$$\theta_i[k] = \begin{cases} \angle t_i[k] \\ \angle t_i[k] + \pi \end{cases}. \tag{32}$$

However, only the second case of (32) has positive second partial deviation. Therefore, the global minima of (30) is

$$\theta_i[k] = \angle t_i[k] + \pi.$$

Initial Guess

All algorithms of unconstrained minimization require to be supplied by a starting point, which we denoted by $\vec{\theta}_0$. To avoid getting stuck in local minima, we should select a good initial guess. This can be accomplished by minimizing a different but similar cost function denoted by J′($\vec{\theta}$)

$$J'(\vec{\theta}) = \left\| \vec{\mathbf{x}}(\vec{\theta}_i) - \mathcal{F}^{\dagger}\vec{f} \right\|^2$$

in which $\dagger$ represents pseudo inverse.

Solving J′($\vec{\theta}$) yields to a simple initial guess

$$\vec{\theta}_0 = \angle \mathcal{F}^{\dagger}\vec{f}. \tag{33}$$

Based on what have been mentioned in this section, a pseudo-code description of the trusted region dogleg algorithm is given by Algorithm 3. In this algorithm, we start with the initial guess of (33) and a trusted region radius upper bound $\bar{R}$. Then we continue the trusted region minimization procedure as described in this section.

SIMULATION RESULTS

In this section, we present some examples to demonstrate the performance of the proposed algorithm. For the first example, our algorithm is applied to a polynomial matrix example from [11]

$$\mathbf{A}(z) = \begin{bmatrix} 2 & 0 & 2z^{+1} \\ z^{+1} & 1 & 0 \\ 0 & z^{-1} & 1 \end{bmatrix}. \tag{34}$$

Frequency behavior of singular values can be seen in Figure 1. There is no intersection of singular values, so the setup of the algorithm either for spectral majorization or frequency smoothness leads to identical decomposition.

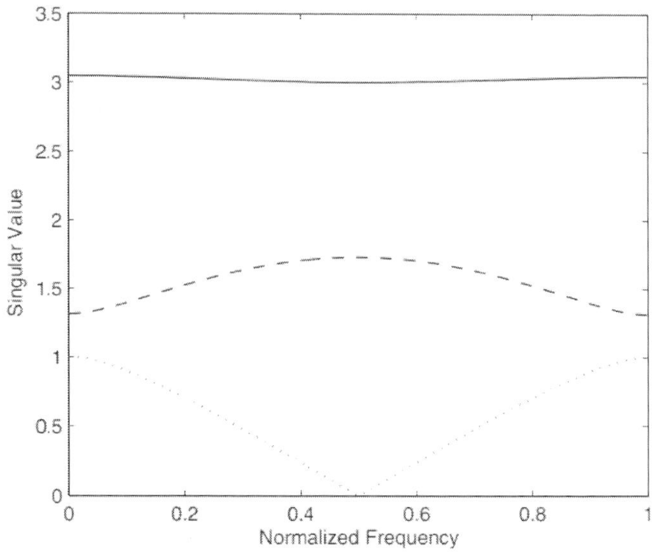

Figure 1: Singular values versus frequency.

For having approximately positive singular values, we use (21). Define the average energy of highest order coefficients for the pair of polynomial singular vectors $\vec{u}_i$ and $\vec{v}_i$ as $E_i^{u,v}=J(\vec{\theta}_i)/(K-M)$ (we expect energy of highest order coefficients to be zero or at least minimized). A plot of E_i versus iteration for each pair of singular vectors is depicted in Figure 2. The decomposition length is M=9 (order is 8) and we use $K=2M+(N_{max}-N_{min})=20$ number of DFT points.

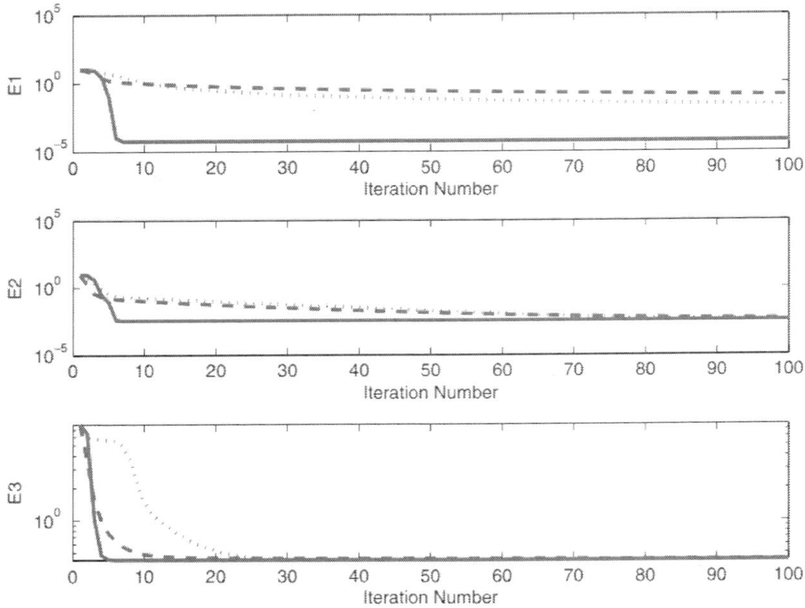

Figure 2: Average highest order coefficients energy E_i versus iteration number for a decomposition with approximately positive singular values. Dotted line: Cauchy points. Dashed line: Alternative minimization. Solid Line: proposed algorithm.

As it is seen, the use of dogleg method with approximate Hessian matrix leads to a fast convergence in contrast with using alternative minimization and Cauchy-point (which is always selected along the gradient direction). Of course we should consider that due to matrix inversion, computational complexity of Dogleg method is $O(K^3)$ while computational complexity of alternative minimization and Cauchy point is $O(K^2)$.

The final value of average highest order coefficient for three pair of singular vectors are 5.54×10^{-5}, 3.5×10^{-3}, and 0.43, respectively. The first singular vector satisfies finite duration constraint almost exactly. The second singular vector fairly satisfies this constraint. However, highest order coefficients of last singular vector, possess considerable amount of energy, that seems to cause decomposition error.

Denote the relative error of the decomposition as

$$E^A = \frac{\left\| \mathbf{A}(z) - \mathbf{U}(z)\Sigma(z)\tilde{\mathbf{V}}(z) \right\|_F}{\|\mathbf{A}(z)\|_F}$$

in which $\|\cdot\|_F$ is the extension of Frobenius norm for polynomial matrices and is defined by

$$\|\mathbf{A}(z)\|_F = \sqrt{\sum_n \|A[n]\|_F^2}.$$

Since in our optimization procedure we only seek for finite duration approximation, $\mathbf{U}(z)$ and $\mathbf{V}(z)$ are only approximately paraunitary. Therefore, we also define relative error of paraunitarity as

$$E^U = \frac{\left\| \tilde{\mathbf{U}}(z)\mathbf{U}(z) - \mathbf{I} \right\|_F}{r}.$$

An upper bound for E^U can be obtained as

$$E^U \leq 2\sqrt{\frac{K-M}{K}\sum_{i=1}^{r} E_i^{(u)}(1 - E_i^{(u)})} + \sqrt{\frac{K-M}{K}\sum_{i=1}^{r}(E_i^{(u)})^2},$$

which means as average energy on $K-M$ highest order goes to zero, E^U diminishes.

The relative error of this decomposition is $E^A = 1.18 \times 10^{-2}$ while the error of $\mathbf{U}(z)$ and $\mathbf{V}(z)$ are $E^U = 3.3 \times 10^{-2}$ and $E^V = 3.08 \times 10^{-2}$, respectively. The paraunitarity error is relatively high in contrast with decomposition error. This is due to the difference between the first two singular values and the last singular value.

A plot of relative errors E^A, E^U, and E^V for various amount of M is shown in Figure 3. The number of frequency samples is fixed at $K=2M+2(N_{max}-N_{min})$.

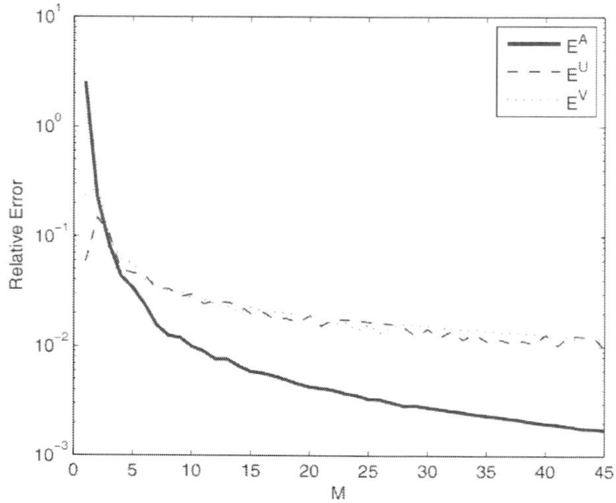

Figure 3: Relative error versus M for a decomposition with approximately positive singular values. $K=2M=2$.

The number of frequency samples K is an optional choice, however as discussed in Section 4, it should satisfy $K \geq 2M+N_{max}-N_{min}-1$. In order to demonstrate the effect of number of frequency samples on the decomposition error, a plot of relative error versus different amount of K is depicted in Figure 4. Increasing the number of frequency samples does not lead to reduction of relative error. Moreover, it increases computational burden. Therefore, a value near $2M+(N_{max}-N_{min})-1$ is a reasonable choice for the number of frequency samples.

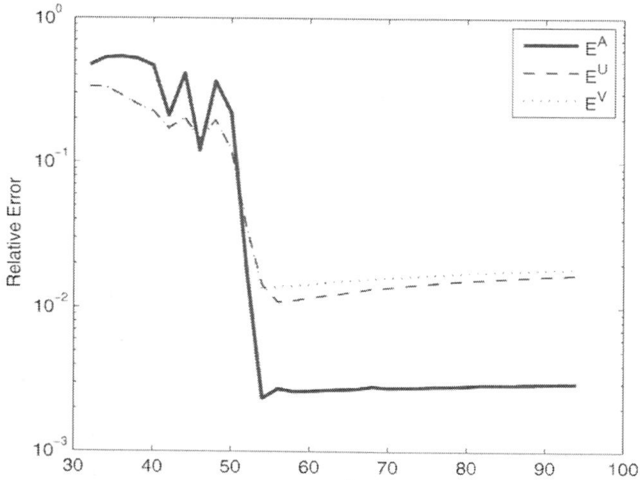

Figure 4: Relative error versus K for a decomposition with approximately positive singular values. $M=31$.

Now, lets relax the problem by allowing singular values to be complex and using (22). A plot of E^u_i and E^v_i versus iteration for each pair of singular vectors is depicted in Figure 5. The decomposition length is M=9 (order is 8) and we use $K=2M+(N_{max}-N_{min})=20$ number of DFT points.

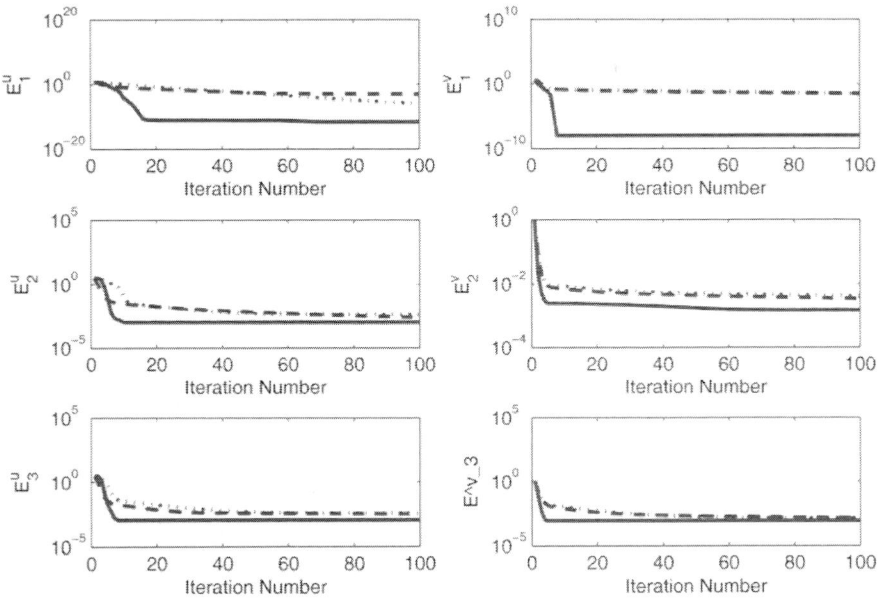

Figure 5: Average highest order coefficients energy E_i versus iteration number for a decomposition with complex singular values. Dotted line: Cauchy points. Dashed line: Alternative minimization. Solid Line: proposed algorithm.

Again Dogleg method converges very rapidly while alternative minimization and Cauchy point converge slowly. The final value of average energy for three left singular vectors are 1.23×10^{-10}, 9.7×10^{-4}, and 10^{-3}, respectively. This is while these values for right singular vectors are 1.12×10^{-10}, 1.4×10^{-3}, and 8.7^{-4}, respectively.

Note that the average energy of highest order coefficients for the third pair of singular vectors alleviate meaningfully. Figure 1 shows that the third singular value goes to zero and then returns to positive values. If we constrain singular values to be positive, a phase jump of πradian, is imposed to one of third singular vectors near the frequency which singular vector goes to zero. However, by letting singular values to be complex, the zero crossing occur which requires no discontinuity of singular vectors.

The relative error of this decomposition is $E^A=4.9\times10^{-3}$ while the error of **U**(z) and **V**(z) are $E^U=2.5\times10^{-3}$ and $E^V=3.5\times10^{-3}$, respectively.

In contrast with constraining singular values to be positive, having complex singular values decrease decomposition and paraunitarity error significantly.

Plots of relative errors E^A, E^U, and E^V for various amount of M and K are shown in Figures 6and 7, respectively. Letting singular values be complex causes significant reduction of all relative errors. As it was mentioned, Figure 7 shows that increasing K from $2M+N_{max}-N_{min}-1$ causes no improvement in relative errors while it makes additional computational burden.

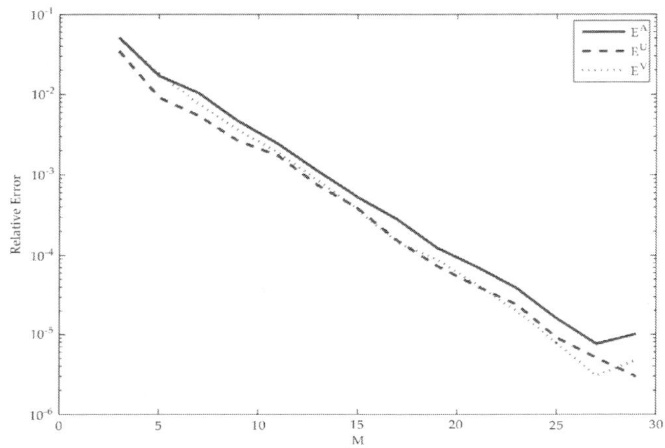

Figure 6: Relative error versus M for a decomposition with complex singular values. $K=2M+2$.

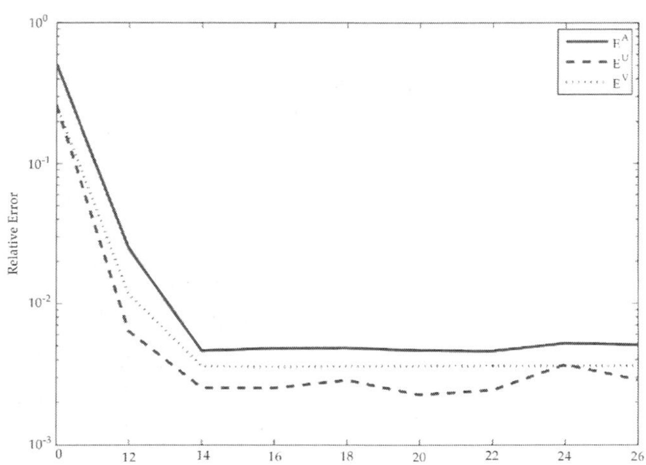

Figure 7: Relative error versus K for a decomposition with complex singular values. $M=9$.

McWhirter and coauthors [11] have reported the relative error of decomposition. Provided that paraunitary matrices **U**(z) and **V**(z) are of order 33, the relative error of their algorithm is 0.0469. This is while our algorithm only requires paraunitary matrices of order 3 for relative error of 0.035 with positive singular values and relative error of 2.45×10^{-6} with complex singular values. In addition, in the new approach, exploiting paraunitary matrices of order 33, the relative error is 0.0032 with positive singular values and 4.7×10^{-6} with complex singular values.

This large difference is not caused by iteration numbers because we compare results while all algorithms relatively converges, and with continuation of iterations trivial improvement are obtained. The main reason lies on different constraints of the solution presented in [11] in contrast to our proposed method. While they impose paraunitary constraint on $\vec{U}$ (z)A(z) V(z) to yield a diagonalized $\Sigma(z)$, we impose the finite duration constraint and obtain approximation of **U**(z) and **V**(z) with fair fitting to the decomposed matrices at each frequency samples. Therefore, we can consider this method as a finite duration polynomial regression of matrices which is obtained by uniformly sampling **U**(z) and **V**(z) on the unit circle in z-plane.

As a second example, consider EVD of the following para-Hermitian matrix

$$\mathbf{A}(z) = \begin{bmatrix} .5z^2 + 3 + .5z^{-2} & -.5z^2 + .5z^{-2} \\ .5z^2 - .5z^{-2} & -.5z^2 + 1 - .5z^{-2} \end{bmatrix}$$

The exact smooth EVD of this matrix is of finite order

$$\mathbf{U}(z) = \frac{1}{2}\begin{bmatrix} 1 + z^{-1} & 1 - z^{-1} \\ 1 - z^{-1} & 1 + z^{-1} \end{bmatrix},$$

$$\Lambda(z) = \begin{bmatrix} z^1 + 2 + z^{-1} & 0 \\ 0 & -z^1 + 2 - z^{-1} \end{bmatrix}. \tag{35}$$

Frequency behavior of eigenvalues can be seen in Figure 8. Since eigenvalues intersect at two frequencies, smooth decomposition and spectrally majorized decomposition result two distinct solutions.

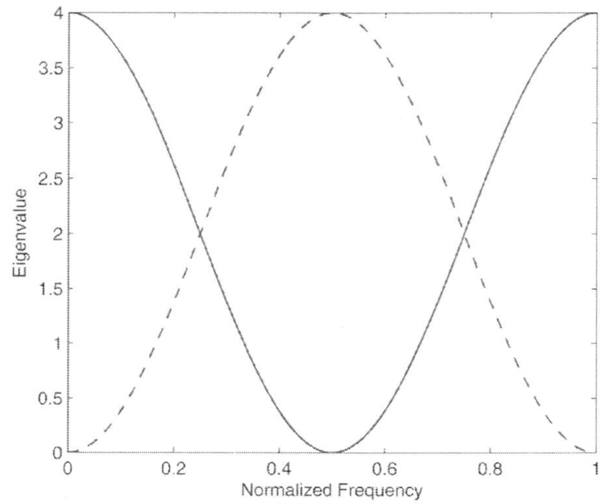

Figure 8: Eigenvalues of smooth decomposition versus frequency.

To perform smooth decomposition, we need to track and rearrange eigenvectors to avoid any discontinuity using Algorithm 2. The resulting $|c_{ij}^{u'}[k]|$ are shown in Figure 9 for k=0, 1, …, $K-1$. Using these $|c_{ij}^{u'}[k]|$ the Algorithm 2 swap first and second eigenvalues and eigenvectors for k=12:32 which results in continuity of eigenvalues and eigenvectors.

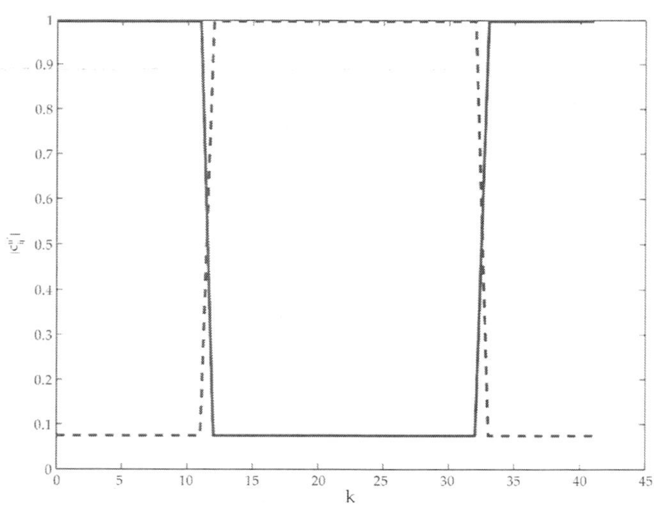

Figure 9: Rearrangement of eigenvalues and eigenvectors. K=42. Dashed Line: $c_{12}^{u'}[k]$ and $c_{21}^{u'}$. Solid Line: $c_{11}^{u'}[k]$ and $c_{22}^{u'}[k]$.

Now that all eigenvalues and eigenvectors are rearranged in DFT domain, it's time for phase alignment of eigenvectors. A plot of E_i versus iteration for M=3 and smooth decomposition is depicted in Figure 10. It is predictable that dogleg algorithm converges rapidly while the alternative minimization and Cauchy point has a long way to converge.

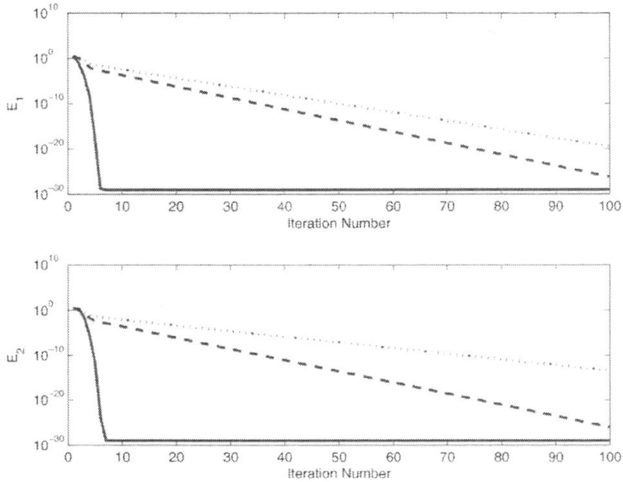

Figure 10: E_i versus iteration number corresponding to smooth decomposition. Dotted line: Cauchy points. Dashed line: Alternative minimization. Solid Line: proposed algorithm.

Since the energy of highest order coefficients of eigenvectors are trifling, using the proposed method for smooth decomposition results in very high accuracy, as seem in the figures. Relative error of smooth decomposition versus M is shown in Figure 11.

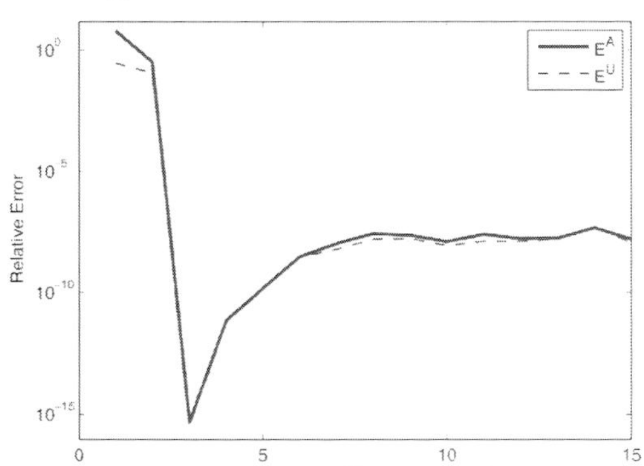

Figure 11: Relative error of smooth decomposition versus M.

While using frequency smooth EVD of (35) leads to relative error below 10^{-5} for $M \geq 3$ with a few number of iterations, Spectrally majorized EVD requires a lot more polynomial order to reach a reasonable relative error.

Unlike smooth decomposition which requires rearrangement of eigenvalues and eigenvectors, spectral majorization requires only to sort eigenvalues at each frequency sample in decreasing order. Most of conventional EVD algorithm sort eigenvalues in decreasing order, which we should only align eigenvector phases using 3. A plot of E_i versus iteration for M=20 and spectrally majorized decomposition is depicted in Figure 12.

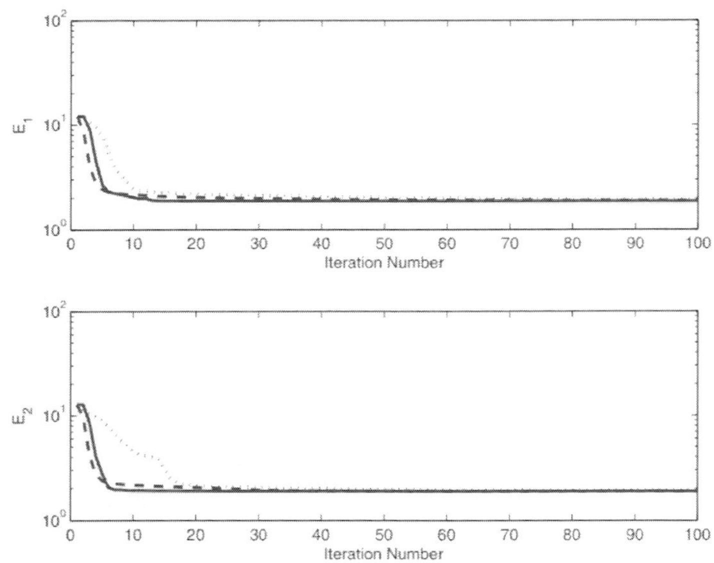

Figure 12: E_i versus iteration number corresponding to spectrally majorized decomposition. Dotted line: Cauchy points. Dashed line: Alternative minimization. Solid Line: proposed algorithm.

Due to an abrupt change in eigenvectors at the intersection frequency of eigenvalues, increasing the decomposition order leads to a slow decay of relative error. Figure 13 shows the relative error as a function of M.

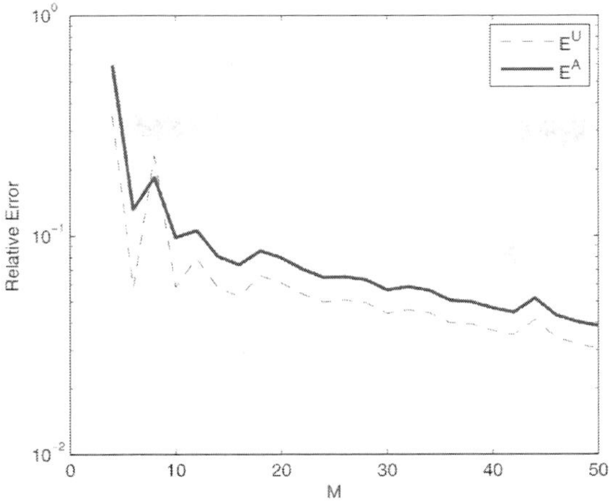

Figure 13: Relative error of spectrally majorized decomposition versus M.

To see the difference between smooth and spectrally majorized decomposition results, eigenvalues of spectrally majorized decomposition is shown in Figure 14, which is comparable with Figure 8 which corresponds to eigenvalues of smooth decomposition. Therefore, a low order polynomial is required using smooth decomposition and much higher polynomial order for spectrally majorized decomposition. Even with M=20 the decomposition have relatively high error.

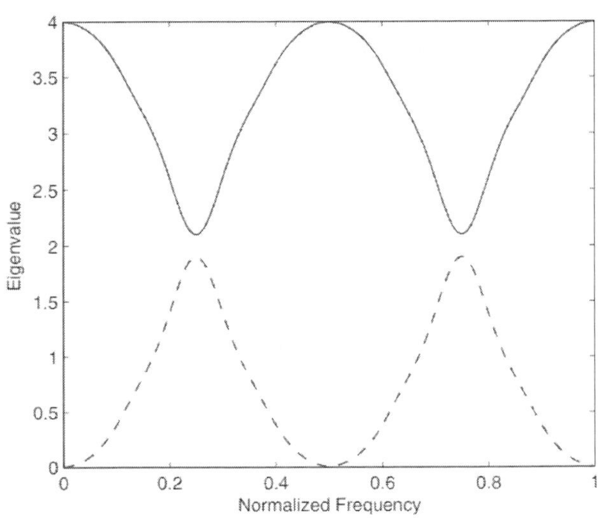

Figure 14: Eigenvalues of spectrally majorized decomposition versus frequency. M=20.

CONCLUSION

An algorithm for polynomial EVD and SVD based on DFT formulation has been presented. One of the advantages of the DFT formulation is that it enables us to control properties of decomposition. Among these properties, we introduce how to setup the decomposition to achieve spectrally majorization and frequency smoothness. We have shown, if singular values (eigenvalues) intersect at some frequency, then simultaneous achievement of spectral majorization and smooth decomposition is not possible. In this situation, setting up the decomposition to possess spectral majorization requires considerably higher order polynomial decomposition and more computational complexity. Highest order polynomial coefficients of singular vectors (eigenvectors) are utilized as square error to obtain a compact decomposition based on phase alignment of frequency samples. The algorithm has the flexibility to compute a decomposition with approximately positive singular values, and a more relaxed decomposition with complex singular values. A solution for this nonlinear quadratic problem is proposed via Newton's method. Since we apply an approximate Hessian matrix to assist the Newton optimization, a fast convergence is achieved. The algorithm capability to control the order of polynomial elements of decomposed matrices and to select properties of decomposition, make the proposed method as a good choice for filterbank and MIMO precoding applications. Finally, performance of the proposed algorithm under different conditions is demonstrated via simulations. Simulation results reveal superior decomposition accuracy in contrast with coefficient domain algorithms due to relaxation of paraunitarity.

REFERENCES

1. Kailath T: *Linear Systems*. NJ: Prentice Hall, Englewood Cliffs; 1980.
2. Tugnait J, Huang B.: Multistep linear predictors-based blind identification and equalization of multiple-input multiple-output channels. *IEEE Trans. Signal Process*2000, 48(1):569-571.
3. Fischer R: Sorted spectral factorization of matrix polynomials in MIMO communications. *IEEE Trans. Commun* 2005, 53(6):945-951. 10.1109/TCOMM.2005.849639
4. Zamiri-Jafarian H, Rajabzadeh M: A polynomial matrix SVD approach for time domain broadband beamforming in MIMO-OFDM systems. *IEEE Vehicular Technology Conference, VTC Spring 2008* 2008, 802-806.
5. Brandt R: Polynomial matrix decompositions: evaluation of algorithms with an application to wideband MIMO communications. 2010.
6. Palomar D, Lagunas M, Pascual A, Neira A: Practical implementation of jointly designed transmit-receive space-time IIR filters. *International Symposium on Signal Processing and Its Applications, ISSPA* 2001, 521-524.
7. Lambert R, Bell A: Blind separation of multiple speakers in a multipath environment. *Proceedings of the International Conference on Acoustics, Speech, and Signal Processing*1997, 423-426.
8. Redif S, McWhirter J, Baxter P, Cooper T: Robust broadband adaptive beamforming via polynomial eigenvalues. *OCEANS 2006* 2006, 1-6.
9. Vaidyanathan P: *Multirate Systems and, Filterbanks*. Prentice Hall, Englewood Cliffs; 1993.
10. Foster J, McWhirter J, Chamber J: A novel algorithm for calculating the QR decomposition for a polynomial matrix. *Proceedings of the International Conference on Acoustics, Speech, and Signal Processing* 2009, 3177-3180.
11. Foster J, Mcwhirter J, Davies M, Chambers J: An algorithm for calculating the QR and singular value decompositions of polynomial matrices. *IEEE Trans. Signal Process* 2010, 58(3):1263-1274.
12. Cescato D, Bolcskei H: QR decomposition of Laurent polynomial matrices sampled on the unit circle. *IEEE Trans. Inf. Theory* 2010, 56(9):4754-4761.

13. Mcwhirter J, Baxter P, Cooper T, Redif S: An EVD algorithm for para-Hermitian polynomial matrices. *IEEE Trans. Signal Process* 2007, 55(5):2158-2169.
14. Tkacenko A: Approximate eigenvalue decomposition of para-Hermitian systems through successive FIR paraunitary transformations. In *Proceedings of the International Conference on Acoustics, Speech, and Signal Processing*. (Dallas, Texas, USA; 2010:4074-4077.
15. Lambert R: Multichannel blind deconvolution: FIR matrix algebra and separation of multipath mixtures. 1996.
16. Vaidyanathan P: Theory of optimal orthonormal subband coders. *IEEE Trans. Signal Process* 1998, 46(4):1528-1543.
17. Tkacenko A, Vaidyanathan P: On the Spectral Factor Ambiguity of FIR Energy Compaction Filter Banks. *IEEE Trans. Signal Process* 2006, 54(1):146-160.
18. Brandt R, Bengtsson M: Wideband MIMO channel diagonalization in the time domain. *International Symposium on Personal, Indoor, and Mobile Radio Communication* 2011, 1958-1962.
19. Dieci L, Eirola T: On smooth decomposition of matrices. *SIAM J. Matrix Anal. Appl* 1999, 20(3):800-819. 10.1137/S0895479897330182
20. Redif S, McWhirter J, Weiss S: Design of FIR paraunitary filter banks for subband coding using a polynomial eigenvalue decomposition. *IEEE Trans. Signal Process* 2011, 59(11):5253-5264.
21. Oppenheim A, Schafer R, Buck J: *Discrete-Time Signal Processing*. Prentice Hall, Englewood Cliffs; 1999.
22. Nocedal J, Wright S: *Numerical Optimization*. New York: Springer; 1999.

SECTION II

CHAPTER 4

PERTURBATION ANALYSIS OF THE STOCHASTIC ALGEBRAIC RICCATI EQUATION

Chun-Yueh Chiang[1], Hung-Yuan Fan[2] , Matthew M Lin[3] and Hsin-An Chen[3]

[1]Center for General Education, National Formosa University, Huwei 632, Taiwan

[2]Department of Mathematics, National Taiwan Normal University, Taipei 116, Taiwan

[3]Department of Mathematics, National Chung Cheng University, Chia-Yi 621, Taiwan

ABSTRACT

In this paper we study a general class of stochastic algebraic Riccati equations (SARE) arising from the indefinite linear quadratic control and stochastic H_∞ problems. Using the Brouwer fixed point theorem, we provide sufficient conditions for the existence of a stabilizing solution of the perturbed SARE. We obtain a theoretical perturbation bound for measuring

Citation (APA): Chiang, C. Y., Fan, H. Y., Lin, M. M., & Chen, H. A. (2013). Perturbation analysis of the stochastic algebraic Riccati equation. Journal of Inequalities and Applications, 2013(1), 580. (22 pages).

accurately the relative error in the exact solution of the SARE. Moreover, we slightly modify the condition theory developed by Rice and provide explicit expressions of the condition number with respect to the stabilizing solution of the SARE. A numerical example is applied to illustrate the sharpness of the perturbation bound and its correspondence with the condition number.

Keywords: Brouwer fixed-point theorem, perturbation bound, stochastic algebraic Riccati equations, condition number

INTRODUCTION

In this paper we consider a general class of continuous-time stochastic algebraic Riccati equations

$$A^{\top}X+XA+C^{\top}XC-(XB+C^{\top}XD+S)(R+D^{\top}XD)^{-1}(B^{\top}X+D^{\top}XC+S^{\top})+H=0, \quad (1a)$$

$$R+D^{\top}XD\succ 0, \quad (1b)$$

where $A\in R^{n\times n}$, $C\in R^{n\times n}$, $B\in R^{n\times m}$, $D\in R^{n\times m}$, $S\in R^{n\times m}$, respectively. Moreover, $H\in R^{n\times n}$ and $R\in R^{m\times m}$ are symmetric matrices. Here we denote $M\succ 0$ (respectively, $M\succeq 0$) if *M* is symmetric positive definite (respectively, positive semidefinite). The unknown $X\in R^{n\times n}$ is a symmetric solution to SARE (1a)-(1b). Let S^n be the set of all symmetric n×n real matrices. For any $X,Y\in S^n$, we write $X\succeq Y$ if $X-Y\succeq 0$.

In essence, SARE (1a)-(1b) is a rational Riccati-type matrix equation associated with the operator $R:\mathrm{dom}R\rightarrow S^n$

$$R(X)=P(X)-S(X)Q(X)^{-1}S(X)^{\top},$$

where the affine linear operators $P:S^n\rightarrow S^n$, $Q:S^n\rightarrow S^m$, $S:S^n\rightarrow R^{n\times m}$, and domR are defined by

$$\mathcal{P}(X)=A^{\top}X+XA+C^{\top}XC+H,$$

$$\mathcal{Q}(X)=R+D^{\top}XD,$$

$$\mathcal{S}(X)=XB+C^{\top}XD+S,$$

$$\mathrm{dom}\,\mathcal{R}=\{X\in\mathcal{S}^n \mid \mathcal{Q}(X)\succ 0\}.$$

We say that *X* is the maximal solution (or the greatest solution) of SARE (1a)-(1b) if it satisfies (1a)-(1b) and $X\succeq P$ for any $P\in S^n$ satisfying $R(P)\geq 0$ and (1b), *i.e.*, *X* is the maximal solution of $R(X)\geq 0$ with the constraint

(1b). Furthermore, it is easily seen that SARE (1a)-(1b) also contains the continuous-time algebraic Riccati equation (CARE)

$$A^\top X + XA - XBR^{-1}B^\top X + H = 0 \tag{2}$$

with R>0, C=0, D=0 and S=0, and the discrete-time algebraic Riccati equation (DARE)

$$X - C^\top XC + (C^\top XD + S)(R + D^\top XD)^{-1}(D^\top XC + S^\top) - H = 0 \tag{3}$$

with A=−I/2 and B=0, as special cases.

Matrix equations of the type (1a)-(1b) are encountered in the *indefinite* linear quadratic (LQ) control problem [1], and the disturbance attenuation problem, which is in deterministic case the H_∞ control theory, for linear stochastic systems with both state- and input-dependent white noise. For example, see [2, 3, 4]. For simplicity, we only consider one-dimensional Wiener process of white noise in this paper; it is straightforward but tedious to extend all perturbation results presented in this paper for multi-dimensional cases. In the aforementioned applications of linear stochastic systems, a symmetric solution *X*, called a *stabilizing solution*, to SARE (1a)-(1b) ought to be determined for the design of optimal controllers. This stabilizing solution plays a very important role in many applications of linear system control theory. The definition of a stabilizing solution to SARE (1a)-(1b) is given as follows. (See also [[3], Definition 5.2].)

Definition 1.1 Let $X \in S^n$ be a solution to SARE (1a)-(1b), Φ=A+BF and Ψ=C+DF, where $F = -Q(X)^{-1}S(X)^\top$. The matrix *X* is called a stabilizing solution for $\mathscr{R}$ if the spectrum of the associated operator Lc with respect to *X* defined by

$$\mathcal{L}_c(W) = \Phi^\top W + W\Phi + \Psi^\top W\Psi, \quad W \in \mathcal{S}^n, \tag{4}$$

is contained in the open left half plane, *i.e.*, $\sigma(L_c) \subset C_-$.

Note that if C=D=0 in (1a)-(1b), then it is easily seen from Definition 1.1 that the matrix $X \in S^n$ is a stabilizing solution to SARE (1a)-(1b) or, equivalently, CARE (2) if and only if $\sigma(\Phi) \subset C_-$. Therefore, Definition 1.1 is a natural generalization of the definition of a stabilizing solution to CARE (2) in classical linear control theory. Moreover, a necessary and sufficient condition for the existence of the stabilizing solution to a more general SARE is derived in Theorem 7.2 of [3]. See also [[1], Theorem 10]. In this case, it is also shown that if SARE (1a)-(1b) has a stabilizing solution $X \in \text{dom}(R)$, then it is necessarily a maximal solution and thus unique [1, 3].

The standard CARE (2) and DARE (3) are widely studied and play very important roles in both classical LQ and H∞ control problems for deterministic linear systems [5, 6, 7]. In the past four decades, an extensive amount of numerical methods were studied and developed for solving the CARE and DARE (see [8, 9, 10] and the references therein). There are two major methodologies among these numerical methods or algorithms. One is the so-called *Schur method* or *invariant subspace method*, which was first proposed by Laub [11]. According to this methodology, the unique and non-negative definite stabilizing solution of the CARE (or DARE) can be obtained by computing the stable invariant subspace (or deflating subspace) of the associated Hamiltonian matrix (or symplectic matrix pencil). Some variants of the invariant subspace method, which preserve the structure of the Hamiltonian matrix (or symplectic matrix pencil) by special orthogonal transformations in the whole computational process, are considered by Mehrmann and his coauthors [12, 13, 14, 15, 16, 17, 18]. The other methodology comes from the *iterative method*, for example, it is referred to as Newton's method [6], matrix sign function method [19], disk function method [20], and structured doubling algorithms [21, 22] and references therein. So far there has been no sources in applying the invariant subspace methods for solving SARE (1a)-(1b), since the structures of associated Hamiltonian matrix or symplectic matrix pencil are not available. Only the iterative methods, *e.g.*, Newton's method [3] and the interior-point algorithm presented in [1], can be applied to computing the numerical solutions of SARE (1a)-(1b). Recently, normwise residual bounds were proposed for assessing the accuracy of a computed solution to SARE (1a)-(1b) [23].

Due to the effect of roundoff errors or the measurement errors of experimental data, small perturbations are often incorporated in the coefficient matrices of SARE (1a)-(1b), and hence we obtain the perturbed SARE

$$\widetilde{A}^{\top}\widetilde{X}+\widetilde{X}\widetilde{A}+\widetilde{C}^{\top}\widetilde{X}\widetilde{C}-(\widetilde{X}\widetilde{B}+\widetilde{C}^{\top}\widetilde{X}\widetilde{D}+\widetilde{S})(\widetilde{R}+\widetilde{D}^{\top}\widetilde{X}\widetilde{D})^{-1}(\widetilde{B}^{\top}\widetilde{X}+\widetilde{D}^{\top}\widetilde{X}\widetilde{C}+\widetilde{S}^{\top})$$
$$+\widetilde{H}=0, \tag{5a}$$

$$\widetilde{R}+\widetilde{D}^{\top}\widetilde{X}\widetilde{D}\succ 0, \tag{5b}$$

where $\widetilde{A}, \widetilde{B}, \widetilde{C}, \widetilde{D}, \widetilde{H}, \widetilde{R}$ and $\widetilde{S}$ are perturbed coefficient matrices of compatible sizes. The main question is under what conditions perturbed SARE (5a)-(5b) still has a stabilizing solution $X^{\sim}\in S^{n}$. Moreover, how sensitive is the stabilizing solution $X\in \mathrm{dom}(R)$ of original SARE (1a)-(1b) with respect to small changes in the coefficient matrices? This is related to the conditioning of SARE (1a)-(1b). Therefore, we will try to answer these questions for SARE (1a)-(1b) in this paper. For CARE (2) and DARE (3), the normwise non-local and local perturbation bounds have been widely studied in the literature. See, *e.g.*, [24, 25, 26]. Also, computable residual bounds were derived for measuring the accuracy of a computed solution to CARE (2) and DARE (3), respectively [27, 28]. To our best knowledge, these issues have not been taken into account for constrained SARE (1a)-(1b) in the literature.

To facilitate our discussion, we use $\|\cdot\|_F$ to denote the Frobenius norm and $\|\cdot\|$ to denote the operator norm induced by the Frobenius norm. For $A=(A_1,\ldots,A_n)=(a_{ij})\in R^{m\times n}$ and $B\in R^{p\times q}$, the Kronecker product of A and B is defined by $A\otimes B=(a_{ij}B)\in R^{mp\times nq}$, and the operator vec(A) is denoted by $\text{vec}(A)=(A^\top_1,\ldots,A^\top n)^\top$. It is known that

$$\text{vec}(ABC)=\left(C^\top\otimes A\right)\text{vec}(B),\qquad \text{vec}\left(A^\top\right)=P_{n,m}\,\text{vec}(A),$$

where $A\in R^{n\times m}$, $B\in R^{m\times \ell}$, $C\in R^{\ell\times k}$, and $P_{n,m}$ is the Kronecker permutation matrix which maps vec(A) into $\text{vec}(A^\top)$ for a rectangle matrix A, *i.e.*,

$$P_{n,m}=\sum_{i,j=1}^{n,m}E_{i,j,n\times m}\otimes E_{j,i,m\times n},$$

where the $n\times m$ matrix $E_{i,j,n\times m}$ has 1 as its (i,j) entry and 0's elsewhere.

This paper is organized as follows. In Section 2, a perturbation equation is derived from SAREs (1a)-(1b) and (5a)-(5b) without dropping any higher-order terms. By using Brouwer fixed point theorem, we obtain a perturbation bound for the stabilizing solution of SARE (5a)-(5b) in Section 3. In order to guarantee the existence of the stabilizing solution of perturbed SARE (5a)-(5b), some stability analysis of the operator L_c is established in Section 4. A theoretical formula of the normwise condition number of the stabilizing solution to SARE (1a)-(1b) is derived in Section 5. Finally, in Section 6, a numerical example is given to illustrate the sharpness and tightness of our perturbation bounds, and Section 7 concludes the paper.

PERTURBATION EQUATION

Assume that $X\in S^n$ is the unique stabilizing solution to SARE (1a)-(1b) and $\tilde{X}\in S^n$ is a symmetric solution of perturbed SARE (5a)-(5b), that is,

$$\mathcal{R}(X):=A^\top X+XA+C^\top XC-\Xi(X)+H=0, \tag{6}$$

$$\widetilde{\mathcal{R}}(\widetilde{X}) := \widetilde{A}^\top \widetilde{X} + \widetilde{X}\widetilde{A} + \widetilde{C}^\top \widetilde{X}\widetilde{C} - \widetilde{\Xi}(\widetilde{X}) + \widetilde{H} = 0, \tag{7}$$

where the two operator $\Xi:S^n \to S^n$ and $\Xi^{\sim}:S^n \to S^n$ are given by

$$\Xi(X) = \mathcal{S}(X)\mathcal{Q}(X)^{-1}\mathcal{S}(X)^\top,$$

$$\widetilde{\Xi}(\widetilde{X}) = \widetilde{\mathcal{S}}(\widetilde{X})\widetilde{\mathcal{Q}}(\widetilde{X})^{-1}\widetilde{\mathcal{S}}(\widetilde{X})^\top, \tag{8}$$

and two affine linear operators $S^{\sim}:S^n \to S^n$, $Q^{\sim}:S^n \to S^m$ are defined by

$$\widetilde{\mathcal{S}}(\widetilde{X}) = \widetilde{X}\widetilde{B} + \widetilde{C}^\top \widetilde{X}\widetilde{D} + \widetilde{S},$$

$$\widetilde{\mathcal{Q}}(\widetilde{X}) = \widetilde{R} + \widetilde{D}^\top \widetilde{X}\widetilde{D}$$

for all $X^{\sim} \in S^n$. Let

$\Delta X = X^{\sim} - X$.

The purpose of this section is to derive a perturbation equation of ΔX from SAREs (1a)-(1b) and (5a)-(5b). For the sake of perturbation analysis, we adopt the following notations:

$$\Delta A = \widetilde{A} - A, \qquad \Delta B = \widetilde{B} - B, \qquad \Delta C = \widetilde{C} - C, \qquad \Delta D = \widetilde{D} - D,$$

$$\Delta S = \widetilde{S} - S, \qquad \Delta R = \widetilde{R} - R, \qquad \Delta H = \widetilde{H} - H \tag{9}$$

and

$$\delta Q = \Delta R + D^\top X \Delta D + \Delta D^\top X D + \Delta D^\top X \Delta D,$$

$$\delta S = \Delta S + C^\top X \Delta D + \Delta C^\top X D + \Delta C^\top X \Delta D + X \Delta B. \tag{10}$$

Moreover, let

$$F = -\mathcal{Q}(X)^{-1}\mathcal{S}(X)^\top, \qquad \widetilde{F} = -\widetilde{\mathcal{Q}}(X)^{-1}\widetilde{\mathcal{S}}(X)^\top,$$

$$\Psi = C + DF, \qquad \widetilde{\Psi} = \widetilde{C} + \widetilde{D}\widetilde{F},$$

$$\Phi = A + BF, \qquad \widetilde{\Phi} = \widetilde{A} + \widetilde{B}\widetilde{F}, \tag{11}$$

and by the definition of Ψ, we define

$$K := X\Psi. \tag{12}$$

Note that $S^{\sim}(X)=S(X)+\delta S$ and $Q^{\sim}(X)=Q(X)+\delta Q$. Substituting (11) into (8), we observe that

$$\Xi(X) = -\mathcal{S}(X)F,$$

$$\widetilde{\Xi}(\widetilde{X}) = (\widetilde{\mathcal{S}}(X) + \Delta X\widetilde{B} + \widetilde{C}^\top \Delta X\widetilde{D})(\widetilde{\mathcal{Q}}(X) + \widetilde{D}^\top \Delta X\widetilde{D})^{-1}$$

$$\times (\widetilde{\mathcal{S}}(X) + \Delta X\widetilde{B} + \widetilde{C}^\top \Delta X\widetilde{D})^\top. \tag{13}$$

Thus far, we have not specified the relation between R(X) and $\widetilde{\mathcal{R}}(\widetilde{X})$. Such a tedious task can be turned into a breeze by repeatedly applying the matrix identities [29]

$$(I+U)^{-1} = I - U(I+U)^{-1}, \qquad V(I+UV)^{-1} = (I+VU)^{-1}V. \tag{14}$$

To begin with, assume that ΔR and ΔD are sufficiently small so that Q˜(X) is invertible. We see that the product

$$\begin{aligned}
&(\widetilde{S}(X) + \Delta X\widetilde{B} + \widetilde{C}^\top \Delta X\widetilde{D})(\widetilde{Q}(X) + \widetilde{D}^\top \Delta X\widetilde{D})^{-1} \\
&\quad = (\widetilde{S}(X) + \Delta X\widetilde{B} + \widetilde{C}^\top \Delta X\widetilde{D}) \\
&\qquad \times [I - \widetilde{Q}(X)^{-1}\widetilde{D}^\top \Delta X\widetilde{D}(I + \widetilde{Q}(X)^{-1}\widetilde{D}^\top \Delta X\widetilde{D})^{-1}]\widetilde{Q}(X)^{-1} \\
&\quad = -\widetilde{F}^\top + \Delta X\widetilde{B}(I + \widetilde{Q}(X)^{-1}\widetilde{D}^\top \Delta X\widetilde{D})^{-1}\widetilde{Q}(X)^{-1} \\
&\qquad + \widetilde{\Psi}^\top \Delta X\widetilde{D}(I + \widetilde{Q}(X)^{-1}\widetilde{D}^\top \Delta X\widetilde{D})^{-1}\widetilde{Q}(X)^{-1}.
\end{aligned}$$

It follows that

$$\begin{aligned}
\widetilde{\Xi}(\widetilde{X}) &= -\widetilde{S}(X)\widetilde{F} - \widetilde{F}^\top\widetilde{B}^\top \Delta X - \widetilde{F}^\top\widetilde{D}^\top \Delta X\widetilde{C} - \widetilde{\Psi}^\top \Delta X\widetilde{D}\widetilde{F} - \Delta X\widetilde{B}\widetilde{F} \\
&\quad + (\widetilde{\Psi}^\top \Delta X\widetilde{D} + \Delta X\widetilde{B})(I + \widetilde{Q}(X)^{-1}\widetilde{D}^\top \Delta X\widetilde{D})^{-1}\widetilde{Q}(X)^{-1} \\
&\quad \times (\widetilde{D}^\top \Delta X\widetilde{\Psi} + \widetilde{B}^\top \Delta X)
\end{aligned}$$

since $\widetilde{F}^\top\widetilde{S}(X)^\top = \widetilde{S}(X)\widetilde{F}$. Next, from (11) we can see that

$$\begin{aligned}
&\widetilde{\Phi}^\top \Delta X + \Delta X\widetilde{\Phi} = \widetilde{A}^\top \Delta X + \Delta X\widetilde{A} + \widetilde{F}^\top\widetilde{B}^\top \Delta X + \Delta X\widetilde{B}\widetilde{F}, \\
&\widetilde{\Psi}^\top \Delta X\widetilde{\Psi} = \widetilde{C}^\top \Delta X\widetilde{C} + \widetilde{F}^\top\widetilde{D}^\top \Delta X\widetilde{C} + \widetilde{\Psi}^\top \Delta X\widetilde{D}\widetilde{F}.
\end{aligned} \tag{15}$$

Applying (15), we obtain the linear equation

$$\widetilde{\mathcal{R}}(\widetilde{X}) - \mathcal{R}(X) = \widetilde{\Phi}^\top \Delta X + \Delta X\widetilde{\Phi} + \widetilde{\Psi}^\top \Delta X\widetilde{\Psi} - E - h_2(\Delta X) = 0, \tag{16}$$

where

$$\begin{aligned}
E &:= -(\Delta A^\top X + X\Delta A + \widetilde{C}^\top X\widetilde{C} - C^\top XC + \widetilde{S}(X)\widetilde{F} - S(X)F + \Delta H), \\
h_2(\Delta X) &:= \widetilde{\Psi}^\top \Delta X\widetilde{D}(I + \widetilde{Q}(X)^{-1}\widetilde{D}^\top \Delta X\widetilde{D})^{-1}\widetilde{Q}(X)^{-1}\widetilde{D}^\top \Delta X\widetilde{\Psi} \\
&\quad + \Delta X\widetilde{B}(I + \widetilde{Q}(X)^{-1}\widetilde{D}^\top \Delta X\widetilde{D})^{-1}\widetilde{Q}(X)^{-1}\widetilde{B}^\top \Delta X \\
&\quad + \Delta X\widetilde{B}(I + \widetilde{Q}(X)^{-1}\widetilde{D}^\top \Delta X\widetilde{D})^{-1}\widetilde{Q}(X)^{-1}\widetilde{D}^\top \Delta X\widetilde{\Psi} \\
&\quad + \widetilde{\Psi}^\top \Delta X\widetilde{D}(I + \widetilde{Q}(X)^{-1}\widetilde{D}^\top \Delta X\widetilde{D})^{-1}\widetilde{Q}(X)^{-1}\widetilde{B}^\top \Delta X.
\end{aligned}$$

It follows from (16) that

$$\widetilde{\Phi}^\top \Delta X + \Delta X\widetilde{\Phi} + \widetilde{\Psi}^\top \Delta X\widetilde{\Psi} = E + h_2(\Delta X). \tag{17}$$

Equipped with this fact, we now are going to derive a perturbation equation in terms of ΔX by using ΔA, ΔB, ΔC, ΔD, ΔS, ΔR, δS, and δQ. It should be noted that

$$\begin{aligned}
\widetilde{\Psi} &= (\Delta C + C) - (\Delta D + D)(Q(X) + \delta Q)^{-1}(S(X) + \delta S)^\top \\
&= \Psi + \Delta\Psi,
\end{aligned}$$

with

$$\Delta\Psi := \Delta C - \Delta D Q(X)^{-1}\mathcal{S}(X)^{\top} - \Delta D Q(X)^{-1}\delta S^{\top} - D Q(X)^{-1}\delta S^{\top}$$
$$+ (\Delta D + D) Q(X)^{-1}\delta Q Q(X)^{-1}\left(I + Q(X)^{-1}\delta Q\right)^{-1}\left(\mathcal{S}(X)^{\top} + \delta S^{\top}\right) \tag{18}$$

and

$$\widetilde{\Phi} = (\Delta A + A) - (\Delta B + B)\left(Q(X) + \delta Q\right)^{-1}\left(\mathcal{S}(X) + \delta S\right)^{\top}$$
$$= \Phi + \Delta\Phi,$$

with

$$\Delta\Phi := \Delta A - \Delta B Q(X)^{-1}\mathcal{S}(X)^{\top} - \Delta B Q(X)^{-1}\delta S^{\top} - B Q(X)^{-1}\delta S^{\top}$$
$$+ (\Delta B + B) Q(X)^{-1}\delta Q Q(X)^{-1}\left(I + Q(X)^{-1}\delta Q\right)^{-1}\left(\mathcal{S}(X)^{\top} + \delta S^{\top}\right). \tag{19}$$

It then is natural to express the left-hand side of (17) by $\Delta\Phi$ and $\Delta\Psi$ such that

$$\widetilde{\Phi}^{\top}\Delta X + \Delta X\widetilde{\Phi} + \widetilde{\Psi}^{\top}\Delta X\widetilde{\Psi} = \Phi^{\top}\Delta X + \Delta X\Phi + \Psi^{\top}\Delta X\Psi - h_1(\Delta X),$$

with

$$h_1(\Delta X) := -\left(\Delta\Phi^{\top}\Delta X + \Delta X\Delta\Phi + \Psi^{\top}\Delta X\Delta\Psi + \Delta\Psi^{\top}\Delta X\Psi + \Delta\Psi^{\top}\Delta X\Delta\Psi\right).$$

Observe further that

$$\widetilde{C}^{\top}X\widetilde{C} - C^{\top}XC = C^{\top}X\Delta C + \Delta C^{\top}XC + \Delta C^{\top}X\Delta C,$$

$$\widetilde{\mathcal{S}}(X)\widetilde{F} - \mathcal{S}(X)F = -\left(\mathcal{S}(X) + \delta S\right)\left(Q(X) + \delta Q\right)^{-1}\left(\mathcal{S}(X) + \delta S\right)^{\top}$$
$$+ \mathcal{S}(X)Q(X)^{-1}\mathcal{S}(X)^{\top}$$
$$= F^{\top}\delta S^{\top} + \delta S F - \delta S\left(I + Q(X)^{-1}\delta Q\right)^{-1}Q(X)^{-1}\delta S^{\top}$$
$$- F^{\top}\delta Q\left(I + Q(X)^{-1}\delta Q\right)^{-1}Q(X)^{-1}\delta S^{\top}$$
$$- \delta S Q(X)^{-1}\delta Q\left(I + Q(X)^{-1}\delta Q\right)^{-1}F$$
$$+ F^{\top}\delta Q F - F^{\top}\delta Q\left(I + Q(X)^{-1}\delta Q\right)^{-1}Q(X)^{-1}\delta Q F.$$

Upon substituting (10) into δSF and $F^{\top}\delta QF$, we have

$$\delta S F = \Delta S F + C^{\top}X\Delta D F + \Delta C^{\top}X D F + \Delta C^{\top}X\Delta D F + X\Delta B F,$$

$$F^{\top}\delta Q F = F^{\top}\Delta R F + F^{\top}D^{\top}X\Delta D F + F^{\top}\Delta D^{\top}X D F + F^{\top}\Delta D^{\top}X\Delta D F,$$

so that the structure of E in (17) can be partitioned into linear equations

$$E_1 := -\left(K^{\top}\Delta D F + F^{\top}\Delta D^{\top}K + K^{\top}\Delta C + \Delta C^{\top}K + F^{\top}\Delta R F + F^{\top}\Delta S^{\top} + \Delta S F + \Delta H\right),$$

$$E_2 := -\left[\Delta A^{\top}X + X\Delta A + \Delta C^{\top}X\Delta C + F^{\top}\Delta B^{\top}X + X\Delta B F\right.$$
$$+ F^{\top}\Delta D^{\top}X\Delta C + \Delta C^{\top}X\Delta D F + F^{\top}\Delta D^{\top}X\Delta D F$$
$$\left.- \left(F^{\top}\delta Q + \delta S\right)\left(I + Q(X)^{-1}\delta R\right)^{-1}Q(X)^{-1}\left(\delta Q F + \delta S^{\top}\right)\right],$$

that is, $E=E_1+E_2$.

Lemma 2.1 *Let X be the stabilizing solution of SARE* (1a)-(1b) *and* X˜ *be a symmetric solution of perturbed SARE* (5a)-(5b). *If* ΔX=X˜−X, *then* ΔX *satisfies the equation*

$$\Phi^{\top}\Delta X + \Delta X\Phi + \Psi^{\top}\Delta X\Psi = E_1 + E_2 + h_1(\Delta X) + h_2(\Delta X), \quad (20)$$

where

$$E_1 = -\big(K^{\top}\Delta DF + F^{\top}\Delta D^{\top}K + K^{\top}\Delta C + \Delta C^{\top}K + F^{\top}\Delta RF$$
$$+ F^{\top}\Delta S^{\top} + \Delta SF + \Delta H\big), \quad (21a)$$

$$E_2 = -\big[\Delta A^{\top}X + X\Delta A + \Delta C^{\top}X\Delta C + F^{\top}\Delta B^{\top}X + X\Delta BF$$
$$+ F^{\top}\Delta D^{\top}X\Delta C + \Delta C^{\top}X\Delta DF + F^{\top}\Delta D^{\top}X\Delta DF$$
$$- \big(F^{\top}\delta Q + \delta S\big)\big(I + Q(X)^{-1}\delta R\big)^{-1}Q(X)^{-1}\big(\delta QF + \delta S^{\top}\big)\big],$$

$$h_1(\Delta X) = -\big(\Delta\Phi^{\top}\Delta X + \Delta X\Delta\Phi + \Psi^{\top}\Delta X\Delta\Psi + \Delta\Psi^{\top}\Delta X\Psi + \Delta\Psi^{\top}\Delta X\Delta\Psi\big), \quad (21b)$$

$$h_2(\Delta X) = \widetilde{\Psi}^{\top}\Delta X\widetilde{D}\Omega\widetilde{D}^{\top}\Delta X\widetilde{\Psi} + \Delta X\widetilde{B}\Omega\widetilde{B}^{\top}\Delta X$$
$$+ \Delta X\widetilde{B}\Omega\widetilde{D}^{\top}\Delta X\widetilde{\Psi} + \widetilde{\Psi}^{\top}\Delta X\widetilde{D}\Omega\widetilde{B}^{\top}\Delta X, \quad (21c)$$

where $\Omega = (I + \widetilde{Q}(X)^{-1}\widetilde{D}^{\top}\Delta X\widetilde{D})^{-1}\widetilde{Q}(X)^{-1}$, *the matrices* ΔA, ΔB, *and so on are given by* (9)-(12).

Note that E1 and E2 are not dependent on ΔX, $h_1(\Delta X)$ is a linear function of ΔX, and $h_2(\Delta X)$ is a function of ΔX with degree at most 2. Assume that the linear operator L_c of (4) is invertible. It is easy to see that the perturbed equation (20) is true if and only if

$$\Delta X = \mathcal{L}_c^{-1}E_1 + \mathcal{L}_c^{-1}E_2 + \mathcal{L}_c^{-1}h_1(\Delta X) + \mathcal{L}_c^{-1}h_2(\Delta X). \quad (22)$$

Thus far, we have not specified the condition for the existence of the solution ΔX in (22). In the subsequent discussion, we shall limit our attention to identifying the condition of the existence of a fixed point of (23), that is, to determine an upper bound on the size of ΔX.

PERTURBATION BOUNDS

Let $f: S^n \to S^n$ be a continuous mapping defined by

$$f(Y) = \mathcal{L}_c^{-1}E_1 + \mathcal{L}_c^{-1}E_2 + \mathcal{L}_c^{-1}h_1(Y) + \mathcal{L}_c^{-1}h_2(Y) \quad \text{for } Y \in S^n. \quad (23)$$

We see that any fixed point of the mapping f is a solution to the perturbed equation (22). Our approach in this section is to present an upper bound for the existence of some fixed points ΔX. It starts with the discussion that the mapping f given by (23) satisfies

$$\|f(\Delta X)\|_F \le \|\mathcal{L}_c^{-1}E_1\|_F + \|\mathcal{L}_c^{-1}E_2\|_F + \|\mathcal{L}_c^{-1}h_1(\Delta X)\|_F + \|\mathcal{L}_c^{-1}h_2(\Delta X)\|_F.$$

Define linear operators $M:R^{n\times n}\to S^n$, $N:R^{n\times m}\to S^n$, $T:S^m\to S^n$ and $H:R^{n\times m}\to S^n$ by

$$\mathcal{M}\Delta C = \mathcal{L}_c^{-1}(K^\top \Delta C + \Delta C^\top K), \tag{24a}$$

$$\mathcal{N}\Delta D = \mathcal{L}_c^{-1}(K^\top \Delta DF + F^\top \Delta D^\top K), \tag{24b}$$

$$\mathcal{T}\Delta R = \mathcal{L}_c^{-1}(F^\top \Delta RF), \tag{24c}$$

$$\mathcal{H}\Delta S = \mathcal{L}_c^{-1}(F^\top \Delta S^\top + \Delta SF), \tag{24d}$$

and the scalars ω, μ, v, τ, η by

$$\omega = \left\| \mathcal{L}_c^{-1} \right\|, \qquad \mu = \|\mathcal{M}\|, \qquad \nu = \|\mathcal{N}\|, \qquad \tau = \|\mathcal{T}\|, \qquad \eta = \|\mathcal{H}\|. \tag{25}$$

From (21a) we then have

$$\left\| \mathcal{L}_c^{-1}E_1 \right\|_F \leq \mu\|\Delta C\|_F + \nu\|\Delta D\|_F + \tau\|\Delta S\|_F + \eta\|\Delta R\|_F + \omega\|\Delta H\| \equiv \varepsilon_1. \tag{26}$$

We now move into more specific details pertaining to the discussion of the fixed point of the continuous mapping *f*. Before doing so, we need to describe an important property of the norm of the product of two matrices and repeatedly employ it in the following discussion. For the proof, the reader is referred to [[30], Theorem 3.9].

Lemma 3.1 *Let A and B be two matrices in* $R^{n\times n}$. *Then* $\|AB\|_F \leq \|A\|_2\|B\|_F$ *and* $\|AB\|_F \leq \|A\|_F\|B\|_2$.

It immediately follows that the matrices δQ and δS, defined by (10), satisfy

$$\|\delta Q\|_F \leq \|\Delta R\|_F + 2\|XD\|_2\|\Delta D\|_F + \|X\|_2\|\Delta D\|_F^2 \equiv \delta_r,$$

$$\begin{aligned}\|\delta S\|_F \leq{}& \|\Delta S\|_F + \|XC\|_2\|\Delta D\|_F + \|XD\|_2\|\Delta C\|_F \\ &+ \|X\|_2\|\Delta D\|_F\|\Delta C\|_F + \|X\|_2\|\Delta B\|_F \equiv \delta_s.\end{aligned} \tag{27}$$

Assume that the scalar δ_r satisfies

$$1 - \left\| Q(X)^{-1} \right\|_2 \delta_r > 0. \tag{28}$$

Then $\|\mathcal{L}_c^{-1}E_2\|_F$ is bounded by

$$\begin{aligned}\left\| \mathcal{L}_c^{-1}E_2 \right\|_F \leq{}& 2\omega\|X\|_2\left(\|\Delta A\|_F + \|F\|_2\|\Delta B\|_F\right) + \omega\|X\|_2\left(\|\Delta C\|_F + \|F\|_2\|\Delta D\|_F\right)^2 \\ &+ \frac{\omega\|Q(X)^{-1}\|_2(\|F\|_2\delta_r + \delta_s)^2}{1 - \|Q(X)^{-1}\|_2\delta_r} \equiv \varepsilon_2.\end{aligned} \tag{29}$$

From (21b) we see that

$$\left\| h_1(\Delta X) \right\|_F \leq \left(2\|\Delta\Phi\|_F + 2\|\Psi\|_2\|\Delta\Psi\|_F + \|\Delta\Psi\|_F^2\right)\|\Delta X\|_F,$$

and also from (18) and (19) we have

$$\|\Delta\Phi\|_F \le \|\Delta A\|_F + \|F\|_2\|\Delta B\|_F + \left(\left\|BQ(X)^{-1}\right\|_2 + \left\|Q(X)^{-1}\right\|_2\|\Delta B\|_F\right)\delta_s$$
$$+ \frac{\|Q(X)^{-1}\|_2(\|B\|_2 + \|\Delta B\|_F)(\|F\|_2 + \|Q(X)^{-1}\|_2\delta_s)\delta_r}{1 - \|Q(X)^{-1}\|_2\delta_r} \equiv \delta_\Phi, \quad (30)$$

$$\|\Delta\Psi\|_F \le \|\Delta C\|_F + \|F\|_2\|\Delta D\|_F + \left(\left\|DQ(X)^{-1}\right\|_2 + \left\|Q(X)^{-1}\right\|_2\|\Delta D\|_F\right)\delta_s$$
$$+ \frac{\|Q(X)^{-1}\|_2(\|D\|_2 + \|\Delta D\|_F)(\|F\|_2 + \|Q(X)^{-1}\|_2\delta_s)\delta_r}{1 - \|Q(X)^{-1}\|_2\delta_r} \equiv \delta_\Psi. \quad (31)$$

It follows that

$$\left\|\mathcal{L}_c^{-1}h_1(\Delta X)\right\|_F \le \omega\delta\|\Delta X\|_F, \quad (32)$$

where the positive scalar δ is defined by

$$\delta = 2\delta_\Phi + 2\psi\delta_\Psi + \delta_\Psi^2 \quad \text{with } \|\Psi\|_2 = \psi. \quad (33)$$

Also, from (28) and Lemma 3.1 we know that $\widetilde{Q}(X) = Q(X) + \delta Q = Q(X)^{-1}(I + Q(X)^{-1}\delta Q)$ and $\|Q(X)^{-1}\delta Q\|_F \le \|Q(X)^{-1}\|_2\delta_r < 1$. This implies that Q˜(X) is nonsingular,

$$\|\tilde{B}\|_2^2\left\|\widetilde{Q}(X)^{-1}\right\|_2 = \|B + \Delta B\|_2^2\left\|\left(Q(X) + \delta Q\right)^{-1}\right\|_2$$
$$= \|B + \Delta B\|_2^2\left\|\left(I + Q(X)^{-1}\delta Q\right)^{-1}Q(X)^{-1}\right\|_2$$
$$\le \frac{\|Q(X)^{-1}\|_2(\|B\|_2 + \|\Delta B\|_F)^2}{1 - \|Q(X)^{-1}\|_2\delta_r} \equiv \gamma_B. \quad (34)$$

Similarly, we have

$$\|\tilde{D}\|_2^2\left\|\widetilde{Q}(X)^{-1}\right\|_2 \le \frac{\|Q(X)^{-1}\|_2(\|D\|_2 + \|\Delta D\|_F)^2}{1 - \|Q(X)^{-1}\|_2\delta_r} \equiv \gamma_D. \quad (35)$$

Assume that

$$1-\gamma_D\|\Delta X\|_F>0. \quad (36)$$

It then follows from Lemma 3.1 and (21c) that

$$\left\|h_2(\Delta X)\right\|_F \le \frac{(\|\widetilde{\Psi}\|_2\|\widetilde{D}\|_2 + \|\widetilde{B}\|_2)^2\|\widetilde{Q}(X)^{-1}\|_2\|\Delta X\|_F^2}{1 - \|\widetilde{D}\|_2^2\|\widetilde{Q}(X)^{-1}\|_2\|\Delta X\|_F}, \quad (37)$$

and from (9), (11) and (31) that

$$\|\widetilde{B}\|_2 \le \|B\|_2 + \|\Delta B\|_2 \equiv \alpha_B,$$

$$\|\widetilde{D}\|_2 \le \|D\|_2 + \|\Delta D\|_2 \equiv \alpha_D,$$

$$\|\widetilde{\Psi}\|_2 \le \|\Psi\|_2 + \|\Delta\Psi\|_F \le \|\Psi\|_2 + \delta_\Psi \equiv \tilde{\psi}. \quad (38)$$

Upon substituting (34), (35) and (38) into (37), we see that

$$\left\|\mathcal{L}_c^{-1}h_2(\Delta X)\right\|_F \le \frac{\omega(\tilde{\psi}^2\gamma_D + 2\tilde{\psi}\alpha_B\alpha_D + \gamma_B)\|\Delta X\|_F^2}{1 - \gamma_D\|\Delta X\|_F}.$$

Finally, by (26), (29) and (32), we arrive at the statement

$$\|f(\Delta X)\|_F \leq \varepsilon + \omega\delta\|\Delta X\|_F + \frac{\omega\alpha\|\Delta X\|_F^2}{1-\gamma_D\|\Delta X\|_F}, \tag{39}$$

where

$$\alpha \equiv \tilde{\psi}^2\gamma_D + 2\tilde{\psi}\alpha_B\alpha_D + \gamma_B, \qquad \varepsilon \equiv \varepsilon_1 + \varepsilon_2. \tag{40}$$

Consider the quadratic equation

$$(\gamma_D - \omega\delta\gamma_D + \omega\alpha)\xi^2 - (1-\omega\delta+\varepsilon\gamma_D)\xi + \varepsilon = 0. \tag{41}$$

It is true that if

$$\delta<1/\omega, \tag{42a}$$

$$\varepsilon \leq \frac{(1-\omega\delta)^2}{\gamma_D - \omega\delta\gamma_D + 2\omega\alpha + \sqrt{(\gamma_D-\omega\delta\gamma_D+2\omega\alpha)^2 - \gamma_D^2(1-\omega\delta)^2}}, \tag{42b}$$

then the positive scalar ξ_* denoted by

$$\xi_* = \frac{2\varepsilon}{(1-\omega\delta+\varepsilon\gamma_D) + \sqrt{(1-\omega\delta+\varepsilon\gamma_D)^2 - 4(\gamma_D-\omega\delta\gamma_D+\omega\alpha)\varepsilon}} \tag{43}$$

is a solution to (41). Let $S^n_{\xi*}$ be a compact subset of S^n given by $\mathcal{S}^n_{\xi_*} = \{\Delta X \in \mathcal{S}^n : \|\Delta X\|_F \leq \xi_*\}$.

It can be seen that in (39)

$$\|f(\Delta X)\|_F \leq \xi_* \quad \text{if } \Delta X \in \mathcal{S}^n_{\xi_*}.$$

It then follows from the Brouwer fixed-point theorem (see [31]) that the continuous mapping f has a fixed point $\Delta X_* \in S^n_{\xi*}$, that is, condition (22) automatically holds.

Observe also that if $\Delta X \in S^n_{\xi*}$, then

$$\begin{aligned}
1-\gamma_D\|\Delta X\|_F &\geq (1-\gamma_D\xi_*) \quad \text{(by (43))}\\
&\geq 1 - \frac{2\varepsilon\gamma_D}{1-\omega\delta+\varepsilon\gamma_D} = \frac{1-\omega\delta-\varepsilon\gamma_D}{1-\omega\delta+\varepsilon\gamma_D} \quad \text{(by (42b))}\\
&\geq \frac{1-\omega\delta-\frac{(1-\omega\delta)^2\gamma_D}{\gamma_D-\omega\delta\gamma_D+2\omega\alpha}}{1-\omega\delta+\varepsilon\gamma_D} \quad \text{(by (42a))}\\
&= \frac{2(1-\omega\delta)\omega\alpha}{(1-\omega\delta+\varepsilon\gamma_D)(\gamma_D-\omega\delta\gamma_D+2\omega\alpha)} \geq 0.
\end{aligned}$$

This implies that assumption (36) is true, if assumption (42a)-(42b) is true.

STABILITY ANALYSIS

We have shown that the mapping *f* given by (23) has a Hermitian fixed point $\Delta X*$. This further implies that perturbed SARE (1a)-(1b) has a Hermitian solution $X^{\sim}=X+\Delta X*$. In this section, we want to discuss the stability of the solution $X^{\sim}$, *i.e.*, show that the solution $X^{\sim}$ is the unique maximal solution to SARE (1a)-(1b). Let Υ and Π be two operators defined by

$$\Upsilon(W) = \Phi^{\top} W + W\Phi, \qquad \Pi(W) = \Psi^{\top} W\Psi, \quad W \in \mathcal{S}^n,$$

with the notations Φ and Ψ given in Definition 1.1. It follows that the operator L_c defined by (4) can also be written as

$$\mathcal{L}_c(W) = \Upsilon(W) + \Pi(W), \quad W \in \mathcal{S}^n. \tag{44}$$

We then have the following important result addressing the condition for a linear operator to be stable. To see a few necessary and sufficient conditions on the stability, we refer to the results and proofs given in [3].

Theorem 4.1 *The linear operator* $L_c=Y+\Pi$ *given by* (44) *is stable, i.e.,* $\sigma(L_c)\subset C_-$, *if and only if* $\sigma(\Phi)\subset C_-$ *and* $\det(Y+\tau\Pi)\neq 0$ *for all* $\tau\in[0,1]$.

When small perturbations $Z_1,Z_2\in R^{n\times n}$ are taken into consideration, the perturbed operator of L_c can be expressed by

$$\widetilde{\mathcal{L}}_c(W) = \widetilde{\Upsilon}(W) + \widetilde{\Pi}(W), \tag{45}$$

where $\widetilde{\Upsilon}(W) = (\Phi + Z_1)^{\top} W + W(\Phi + Z_1)$ and $\widetilde{\Pi}(W) = (\Psi + Z_2)^{\top} W(\Psi + Z_2)$ for all $W\in S^n$. Define the quantity

$$\ell(\theta) = \left\|(\Upsilon + \theta\Pi)^{-1}\right\|$$

for $\theta\in[0,1]$ and

$$\beta(\mathcal{L}_c) = \min_{(Z_1,Z_2)\in\mathcal{Z}} \max\left\{\|Z_1\|, \|Z_2\|\right\},$$

where the set $Z=\{(Z_1,Z_2)\in R^{n\times n}\times R^{n\times n} \mid \det(\widetilde{\Upsilon} + \theta\widetilde{\Pi})=0 \text{ for some } \theta\in[0,1]\}$. It should be noted that if $\sigma(\Phi)\subset C_-$, $\Psi=0$ and $Z_2=0$, then

$$\beta(L_c)\leq\beta(\Phi), \tag{46}$$

where the value $\beta(\Phi)$ is defined by [27]

$$\beta(\Phi) = \min\left\{\|Z_1\| \;\middle|\; \max_{1\leq j\leq n} \operatorname{Re}\lambda_j(\Phi + Z_1) = 0, Z_1 \in \mathbb{R}^{n\times n}\right\}. \tag{47}$$

Here, $\lambda_j(\Phi+Z_1)$ $(j=1,\ldots,n)$ denote the eigenvalues of $\Phi+Z_1$.

The connection between $\beta(L_c)$ and the maximum of the scalar function $\ell(\theta)$ on $[0,1]$ can be established in the following form.

Theorem 4.2 [23]

Suppose that the linear operator L_c *given by* (44) *is stable, and let*

$$\ell_c = \max_{\theta\in[0,1]} \ell(\theta), \qquad \psi = \|\Psi\|. \tag{48}$$

Then

$$\beta(\mathcal{L}_c) \geq \frac{\ell_c^{-1}}{(\psi+1)+\sqrt{(\psi+1)^2+\ell_c^{-1}}}.$$

We now apply Theorem 4.2 to (46) and obtain that

$$\beta(\Phi) \geq \beta(\mathcal{L}_c) \geq \frac{\ell_c^{-1}}{(\psi+1)+\sqrt{(\psi+1)^2+\ell_c^{-1}}}.$$

Hence, if a perturbation matrix $Z_1 \in R^{n\times n}$ satisfies

$$\|Z_1\| < \frac{\ell_c^{-1}}{(\psi+1)+\sqrt{(\psi+1)^2+\ell_c^{-1}}},$$

then (47) implies that the matrix $\Phi+Z_1$ must be c-stable.

We now turn to a key stability test of the operator L_c, the striking tool of our stability analysis.

Theorem 4.3 [23]

Suppose that the linear operator L_c *is stable, and let the scalars* ℓ_c *and* ψ *be defined as in*(48). *If the perturbation matrices* $Z_1, Z_2 \in R^{n\times n}$ *satisfy*

$$\max\{\|Z_1\|, \|Z_2\|\} < \frac{\ell_c^{-1}}{(\psi+1)+\sqrt{(\psi+1)^2+\ell_c^{-1}}}, \tag{49}$$

then $\Phi+Z_1$ *is c-stable and the perturbed linear operator* $(\widetilde{\mathcal{L}}_c)$ *defined by* (45) *is also stable, i.e.,* $\sigma(\widetilde{\mathcal{L}}_c) \subset C_-$.

Upon substituting $\tilde{X}$ for X in $\tilde{S}(X)$ and $\tilde{Q}(X)$ of (11), we shall have

$$\widetilde{Q}(\widetilde{X}) = \widetilde{R} + \widetilde{D}^\top \widetilde{X}\widetilde{D} = Q(X) + \delta Q + \widetilde{D}^\top \Delta X\widetilde{D} \equiv Q(X) + \Delta R, \tag{50}$$

$$\widetilde{S}(\widetilde{X}) = \widetilde{S} + \widetilde{C}^\top \widetilde{X}\widetilde{D} + \widetilde{X}\widetilde{B} = S(X) + \delta S + \Delta X\widetilde{B} + \widetilde{C}^\top \Delta X\widetilde{D}$$

$$\equiv S(X) + \Delta S. \tag{51}$$

Also, corresponding to $\tilde{X}$, the perturbed $\Psi_{\tilde{X}}$ and $\Phi_{\tilde{X}}$ of Ψ and Φ, respectively, can be expressed in terms of the formulae

$$\Phi_{\tilde{X}} = (\Delta A + A) - (\Delta B + B)(\Gamma + \Delta S)^{-1}(\Lambda + \Delta R)^{\top} := \Phi + \Delta\Phi,$$

$$\Psi_{\tilde{X}} = (\Delta C + C) - (\Delta D + D)(\Gamma + \Delta R)^{-1}(\Lambda + \Delta S)^{\top} := \Psi + \Delta\Psi, \qquad (52)$$

with

$$\Delta\Phi := \Delta A - \Delta B Q(X)^{-1}S(X)^{\top} - \Delta B Q(X)^{-1}\Delta S^{\top} - BQ(X)^{-1}\Delta S^{\top}$$
$$+ (\Delta B + B)Q(X)^{-1}\Delta R Q(X)^{-1}\left(I + Q(X)^{-1}\Delta R\right)^{-1}\left(S(X)^{\top} + \Delta S^{\top}\right),$$
$$\Delta\Psi := \Delta C - \Delta D Q(X)^{-1}S(X)^{\top} - \Delta D Q(X)^{-1}\Delta S^{\top} - DQ(X)^{-1}\Delta S^{\top}$$
$$+ (\Delta D + D)Q(X)^{-1}\Delta R Q(X)^{-1}\left(I + Q(X)^{-1}\Delta R\right)^{-1}\left(S(X)^{\top} + \Delta S^{\top}\right).$$

Let $\alpha_C := \|C\|_2 + \|\Delta C\|_2$. Since $\|\Delta X\|_F \leq \xi_*$, it follows from (38), (50) and (51) that

$$\|\Delta R\|_F \leq \delta_r + \alpha_D^2\xi_* := c_r,$$

$$\|\Delta S\|_F \leq \delta_s + (\alpha_B + \alpha_C\alpha_D)\xi_* := c_s.$$

Thus $\|\Delta\Phi\|_F$ and $\|\Delta\Psi\|_F$ are bounded by the inequalities

$$\|\Delta\Phi\|_F \leq \|\Delta A\|_F + \|F\|_2\|\Delta B\|_F + \frac{\alpha_B\|Q(X)^{-1}\|_2(c_s + c_r\|F\|_2)}{1 - c_r\|Q(X)^{-1}\|_2},$$

$$\|\Delta\Psi\|_F \leq \|\Delta C\|_F + \|F\|_2\|\Delta D\|_F + \frac{\alpha_D\|Q(X)^{-1}\|_2(c_s + c_r\|F\|_2)}{1 - c_r\|Q(X)^{-1}\|_2}.$$

Here, the above upper bounds are obtained by simplifying those given by (30) and (31). Let

$$f = \|F\|_2, \qquad \gamma = \left\|Q(X)^{-1}\right\|_2,$$
$$\alpha_C = \|C\|_2 + \|\Delta C\|_F, \qquad \zeta_1 = \max\left\{\|\Delta A\|_F, \|\Delta C\|_F\right\},$$
$$\zeta_2 = \max\left\{\|\Delta B\|_F, \|\Delta D\|_F\right\}, \qquad \zeta_3 = \max\{\alpha_B, \alpha_D\}, \qquad (53)$$

where α_B and α_D are defined by (38) and Θ is defined to be the right-hand side of (49), that is,

$$\Theta = \frac{\ell_c^{-1}}{(\psi + 1) + \sqrt{(\psi + 1)^2 + \ell_c^{-1}}}. \qquad (54)$$

We then have

$$\max\left\{\|\Delta\Phi\|_F, \|\Delta\Psi\|_F\right\} \leq \zeta_1 + f\zeta_2 + \frac{\gamma\zeta_3(\delta_s + \delta_r f) + \gamma\zeta_3(\alpha_B + \alpha_C\alpha_D + \alpha_D^2 f)\xi_*}{1 - \delta_r f - \alpha_D^2 f\xi_*}.$$

It follows that if the condition

$$\zeta_1 + f\zeta_2 + \frac{\gamma\zeta_3(\delta_s + \delta_r f) + \gamma\zeta_3(\alpha_B + \alpha_C\alpha_D + \alpha_D^2 f)\xi_*}{1 - \delta_r f - \alpha_D^2 f\xi_*} < \Theta$$

or, equivalently,

$$\xi_* < \frac{(\Theta - \zeta_1 - f\zeta_2)(1-\delta_r f) - \gamma\zeta_3(\delta_s + \delta_r f)}{(\Theta - \zeta_1 - f\zeta_2)\alpha_D^2 f + \gamma\zeta_3(\alpha_B + \alpha_C\alpha_D + \alpha_D^2 f)}$$

holds, then corresponding to Theorem 4.3, the perturbed linear operator $\tilde{L}_c$ with respect to $\tilde{X}$ is stable. In other words, the matrix $\tilde{X} \in S^n$ must be the unique stabilizing (and maximal) solution to perturbed SARE (5a)-(5b).

We now have all the materials needed for the existence of a stabilizing solution of (5a)-(5b).

Theorem 4.4 (Perturbation bound)

Let X be the stabilizing solution of (1a)-(1b). *Let* ω, δ_r, δ_s, δ, γ_D, α_B, α_D, α, ε, f, γ, α_C, ζ_1, ζ_2, ζ_3, Θ *be the scalars defined by* (25), (27), (33), (35), (38), (40), (53) *and* (54), *respectively. Define*

$$\xi_* = \frac{2\varepsilon}{(1-\omega\delta+\varepsilon\gamma_D) + \sqrt{(1-\omega\delta+\varepsilon\gamma_D)^2 - 4(\gamma_D - \omega\delta\gamma_D + \omega\alpha)\varepsilon}}.$$

If the perturbed quantities of the coefficients of (5a)-(5b) *are sufficiently small, for example,* $\varepsilon \ll 1$, *such that*

$$1 - \|Q(X)^{-1}\|_2 \delta_r > 0,$$

$$1 - \omega\delta > 0,$$

$$\frac{(1-\omega\delta)^2}{\gamma_D - \omega\delta\gamma_D + 2\omega\alpha + \sqrt{(\gamma_D - \omega\delta\gamma_D + 2\omega\alpha)^2 - \gamma_D^2(1-\omega\delta)^2}} - \varepsilon \geq 0,$$

$$\frac{(\Theta - \zeta_1 - f\zeta_2)(1-\delta_r f) - \gamma\zeta_3(\delta_s + \delta_r f)}{(\Theta - \zeta_1 - f\zeta_2)\alpha_D^2 f + \gamma\zeta_3(\alpha_B + \alpha_C\alpha_D + \alpha_D^2 f)} - \xi_* > 0,$$

then perturbed SARE (5a)-(5b) *has the unique stabilizing solution* $\tilde{X}$, *and*

$$\frac{\|\widetilde{X} - X\|_F}{\|X\|_F} \leq \frac{\xi_*}{\|X\|_F}. \tag{55}$$

CONDITION NUMBER OF THE SARE

In the study of a computational problem, a fundamental issue is to determine the condition number of a problem to be the ratio of the relative change in the solution to the relative change in the argument. Applying the theory of condition number given by Rice [32], we define the condition number c(X) of the stabilizing solution *X* of SARE (1a)-(1b) by

$$c(X) = \lim_{\delta \to 0^+} \sup_{\Omega_\delta} \frac{\|\Delta X\|_F}{\kappa\delta}, \tag{56}$$

where the set of perturbed matrices Ω_δ is defined by

$$\Omega_\delta = \Omega_\delta(\kappa_A, \kappa_B, \kappa_C, \kappa_D, \kappa_S, \kappa_R, \kappa_H)$$
$$= \{(\Delta A, \Delta B, \Delta C, \Delta D, \Delta S) \in \mathbb{R}^{n\times n} \times \mathbb{R}^{n\times m} \times \mathbb{R}^{n\times n} \times \mathbb{R}^{n\times m} \times \mathbb{R}^{n\times m},$$
$$(\Delta R, \Delta H) \in \mathcal{S}^n \times \mathcal{S}^m \mid 0 < \delta_p \le \delta\}, \quad (57)$$

with

$$\delta_p = \left\| \left(\frac{\Delta A}{\kappa_A}, \frac{\Delta B}{\kappa_B}, \frac{\Delta C}{\kappa_C}, \frac{\Delta D}{\kappa_D}, \frac{\Delta S}{\kappa_S}, \frac{\Delta R}{\kappa_R}, \frac{\Delta H}{\kappa_H} \right) \right\|_F,$$

and κ_A, κ_B, κ_C, κ_D, κ_S, κ_R, κ_H, κ are positive parameters. Then (56) gives the absolute condition number cabs(X) if

$(\kappa_A, \kappa_B, \kappa_C, \kappa_D, \kappa_S, \kappa_R, \kappa_H, \kappa)=(1,1,1,1,1,1,1,1)$

and gives the relative condition number $c_{rel}(X)$ if

$$(\kappa_A, \kappa_B, \kappa_C, \kappa_D, \kappa_S, \kappa_R, \kappa_H, \kappa) = (\|A\|_F, \|B\|_F, \|C\|_F, \|D\|_F, \|S\|_F, \|R\|_F, \|H\|_F, \|X\|_F).$$

It follows from (22) and (24a)-(24d) that

$$\Delta X = \mathcal{P}\Delta A + \mathcal{Q}\Delta B + \mathcal{M}\Delta C + \mathcal{N}\Delta D + \mathcal{H}\Delta S + \mathcal{T}\Delta R + \mathcal{L}_c^{-1}\Delta H + O(\delta_p^2),$$

where the linear operators P:$\mathbb{R}^{n\times n}\to S^n$ and Q:$\mathbb{R}^{n\times m}\to S^n$ are defined by

$$\mathcal{P}\Delta A = \mathcal{L}_c^{-1}(X\Delta A + \Delta A^\top X), \quad (58a)$$

$$\mathcal{Q}\Delta B = \mathcal{L}_c^{-1}(X\Delta BF + F^\top \Delta B^\top X). \quad (58b)$$

In order to derive the explicit expression for the condition number c(X) of the stabilizing solution X of (1a)-(1b), we require a theorem concerning the form of the optimal solution. This theorem can be regarded as a theoretical extension of the results discussed in [25, 33]. Most strategies have been established earlier by using much heavier machinery. Since this theorem is most relevant to our stability analysis, we briefly outline a direct proof with ideas from [34] to make this presentation more self-contained.

Theorem 5.1 *Let* L:$\mathbb{R}^{n\times n}\times\mathbb{R}^{m\times m}\to\mathbb{R}^{k\times k}$ *be a linear operator and*

$$\mathcal{L}(Z_1, Z_2)^\top = \mathcal{L}(Z_1, Z_2^\top) \quad (59)$$

for all $Z_1 \in \mathbb{R}^{n\times n}$ *and* $Z_2 \in \mathbb{R}^{m\times m}$. *Then the optimal solution* $(Z_1\star, Z_2\star)$ *to the problem*

$$\max_{\|(Z_1,Z_2)\|_F=1} \|\mathcal{L}(Z_1, Z_2)\|_F \quad (60)$$

exists for some $Z_1\star \in R^{n\times n}$ *and* $Z_2\star = \pm Z^{\top}_{2\star} \in R^{m\times m}$. *Furthermore, if the linear operator* $L(0,Z_2)$ *is a positive operator with respect to any* $Z_2 \in C^{m\times m}$, *that is, for all* $Z_2 \in C^{m\times m}$, *we have*

$$\mathcal{L}(0,Z_2) \succeq 0 \quad \textit{if}\, Z_2 \succeq 0.$$

Then there exists an optimal solution $(Z_{1\star}, Z_{2\star}) \in R^{n\times n} \times R^{m\times m}$ *to problem* (60) *such that* $Z_{2\star}$ *is symmetric.*

Proof Since $\mathscr{L}$ is a linear operator on a finite dimensional space, it is clear that the optimal solution of (60) exists. Assume that $(Z_{1\star}, Z_{2\star})$ solves this optimization problem. Let $\sigma_{max} = \|L(Z_{1\star}, Z_{2\star})\|_F$ and $L \in R^{k2\times(n2+m2)}$ be the matrix representation of the operator $\mathscr{L}$ such that

$$\mathrm{vec}\big(\mathcal{L}(Z_1,Z_2)\big) = L\begin{bmatrix}\mathrm{vec}(Z_1)\\ \mathrm{vec}(Z_2)\end{bmatrix}. \tag{61}$$

By (59) and (61), we have

$$\begin{bmatrix}\mathrm{vec}(Z_{1\star})\\ \mathrm{vec}(Z_{2\star})\end{bmatrix}^{\top} L^{\top}L \begin{bmatrix}\mathrm{vec}(Z_{1\star})\\ \mathrm{vec}(Z_{2\star})\end{bmatrix} = \begin{bmatrix}\mathrm{vec}(Z_{1\star})\\ \mathrm{vec}(Z_{2\star}^{\top})\end{bmatrix}^{\top} L^{\top}L \begin{bmatrix}\mathrm{vec}(Z_{1\star})\\ \mathrm{vec}(Z_{2\star}^{\top})\end{bmatrix} = \sigma_{\max}^2. \tag{62}$$

Note that

$$\ell := \|Z_{1\star}\|_F^2 + \left\|\frac{1}{2}(Z_{2\star} + Z_{2\star}^{\top})\right\|_F^2 = 0, \quad \text{only if}\, Z_{1\star} = 0, Z_{2\star} = -Z_{2\star}^{\top}.$$

It follows that if $Z_{2\star}^{\top} \neq -Z_{2\star}$, by (62), we see that $\frac{1}{\sqrt{\ell}}(Z_{1\star}, \frac{1}{2}(Z_{2\star} + Z_{2\star}^{\top}))$ is another optimal solution for (60). This proves the first part of the theorem.

For the second part, if there exists a symmetric optimal solution, then it completes the proof. Otherwise, from the first part, we know that there exists an optimal solution $(Z_{1\star}, Z_{2\star})$ with $Z_{1\star}=0$, $Z_{2\star} = -Z^{\top}_{2\star} \in R^{m\times m}$ and $\|(Z_{1\star}, Z_{2\star})\|_F = 1$ to (60). Let $i = \sqrt{-1}$. We have the following matrix decomposition:

$$Z_{2\star} = Q^{\top}\,\mathrm{diag}\left(\begin{bmatrix}0 & -\omega_1\\ \omega_1 & 0\end{bmatrix}, \dots, \begin{bmatrix}0 & -\omega_k\\ \omega_k & 0\end{bmatrix}, 0_r\right) Q,$$

where $0r$ is a zero matrix with size $r\times r$, Q is an $m\times m$ orthogonal matrix, and $\omega_j>0$ for $1\leq j\leq k$. Let

$$\widetilde{Z}_{2\star} := Q^{\top} \operatorname{diag}\left(\begin{bmatrix} \omega_1 & 0 \\ 0 & \omega_1 \end{bmatrix}, \ldots, \begin{bmatrix} \omega_k & 0 \\ 0 & \omega_k \end{bmatrix}, 0_r\right) Q$$

be a real symmetric matrix. Since $-\widetilde{Z}_{2\star} \preceq iZ_{2\star} \preceq \widetilde{Z}_{2\star}$, it is true that $\mathcal{L}(0,\widetilde{Z}_{2\star}) \succeq \mathcal{L}(0,iZ_{2\star}) \succeq \mathcal{L}(0,-\widetilde{Z}_{2\star}) = -\mathcal{L}(0,\widetilde{Z}_{2\star})$.

Using the fact that $\|iZ_{2\star}\|_F = \|\widetilde{Z}_{2\star}\|_F$, we see that $\|(0,\widetilde{Z}_{2\star})\|_F = \|(0,iZ_{2\star})\|_F = \|(0,Z_{2\star})\|_F = 1$ and

$$\left\|\mathcal{L}(0,\widetilde{Z}_{2\star})\right\|_F \geq \left\|\mathcal{L}(0,iZ_{2\star})\right\|_F = \left\|\mathcal{L}(0,Z_{2\star})\right\|_F.$$

If $W_1 \succeq W_2 \succeq -W_1$, then $\|W_1\|_F \geq \|W_2\|_F$, which implies that $(0, \widetilde{Z}_{2\star})$ is a symmetric optimal solution to (60) (see [[25], Lemma A.1]). This completes the proof.

With the existence theory established above, it is interesting to note that the condition number c(X) defined by (56) can be written as

$$c(X) = \frac{1}{\kappa} \lim_{\delta\to 0^+} \sup_{\Omega_\delta} \frac{\|\mathcal{P}\Delta A + \mathcal{Q}\Delta B + \mathcal{M}\Delta C + \mathcal{N}\Delta D + \mathcal{H}\Delta S + \mathcal{T}\Delta R + \mathcal{L}_c^{-1}\Delta H\|_F}{\delta}$$
$$= \frac{1}{\kappa} \max_{\delta_p > 0} \frac{\|\mathcal{P}\Delta A + \mathcal{Q}\Delta B + \mathcal{M}\Delta C + \mathcal{N}\Delta D + \mathcal{H}\Delta S + \mathcal{T}\Delta R + \mathcal{L}_c^{-1}\Delta H\|_F}{\delta_p}. \qquad (63)$$

Note that the second equality in (63) is only an application of linearity of the norm. (For the proof, see Lemma A.1.) Observe further that the inverse operator $\mathcal{L}_c^{-1}$ of (4) satisfies

$$\left[\mathcal{L}_c^{-1}(W)\right]^{\top} = \mathcal{L}_c^{-1}\left(W^{\top}\right)$$

since $[L_c(W)]^{\top} = L_c(W^{\top})$ for all $W \in C^{n\times n}$. It follows that

$$[\mathcal{T}\Delta R]^{\top} = \mathcal{T}\Delta R^{\top}, \qquad [\mathcal{P}\Delta A]^{\top} = \mathcal{P}\Delta A,$$

$$[\mathcal{M}\Delta C]^{\top} = \mathcal{M}\Delta C, \qquad [\mathcal{N}\Delta D]^{\top} = \mathcal{N}\Delta D,$$

Also, it is known that the inverse operator $\mathcal{L}_c^{-1}$ is positive [3, Corollary 3.8]. It follows that Tis also a positive operator. Now, applying Theorem 5.1 to the operator $P\Delta A+Q\Delta B+M\Delta C+N\Delta D+H\Delta S+T\Delta R+\mathcal{L}_c^{-1}\Delta H$ in (63), we obtain the equality

$$c(X) = \frac{1}{\kappa}\max_{\widetilde{\Omega}} \frac{\|\kappa_A \mathcal{P}\Delta A + \kappa_B \mathcal{Q}\Delta B + \kappa_C \mathcal{M}\Delta C + \kappa_D \mathcal{N}\Delta D + \kappa_S \mathcal{H}\Delta S + \kappa_R \mathcal{T}\Delta R + \kappa_H \mathcal{L}_c^{-1}\Delta H\|_F}{\|(\Delta A, \Delta B, \Delta C, \Delta D, \Delta S, \Delta R, \Delta H)\|_F},$$

where the extended set Ω˜ is defined by

$$\widetilde{\Omega} = \{(\Delta A, \Delta B, \Delta C, \Delta D, \Delta S, \Delta R, \Delta H) \in \mathbb{R}^{n\times n} \times \mathbb{R}^{n\times m} \times \mathbb{R}^{n\times n} \times \mathbb{R}^{n\times m} \times \mathbb{R}^{n\times n} \times \mathbb{R}^{n\times n} \times \mathbb{R}^{m\times m} \mid \|(\Delta A, \Delta B, \Delta C, \Delta D, \Delta S, \Delta R, \Delta H)\|_F > 0\}.$$

On the other hand, observe that the matrix representation of the operation L_c in (4) can be written in terms of $L_c = I\otimes\Phi + \Phi^\top\otimes I + \Psi^\top\otimes\Psi$. Corresponding to (24a)-(24d) and (58a)-(58b), we let

$$M_c = L_c^{-1}\left(I\otimes K^\top + (K^\top\otimes I)P_{n,n}\right),$$

$$N_c = L_c^{-1}\left(F^\top\otimes K^\top + (K^\top\otimes F^\top)P_{m,n}\right),$$

$$T_c = L_c^{-1}\left(F^\top\otimes F^\top\right),$$

$$H_c = L_c^{-1}\left(F^\top\otimes I + (I\otimes F^\top)P_{m,n}\right),$$

$$P_c = L_c^{-1}\left(I\otimes X + (X\otimes I)P_{n,n}\right),$$

$$Q_c = L_c^{-1}\left(F^\top\otimes X + (X\otimes F^\top)P_{m,n}\right)$$

and

$$\mathcal{U} = \left(\kappa_A P_c, \kappa_B Q_c, \kappa_C M_c, \kappa_D N_c, \kappa_S H_c, \kappa_R T_c, \kappa_H L_c^{-1}\right).$$

It follows that

$$c(X) = \frac{1}{\kappa}\max_{V\in\widetilde{\Omega}} \frac{\|\mathcal{U}\,\mathrm{vec}(V)\|_2}{\|\mathrm{vec}(V)\|_2} = \frac{\|\mathcal{U}\|_2}{\kappa}.$$

Based on the above discussion, we have the following result.

Theorem 5.2 *The condition number* c(X) *given by* (56) *has the explicit expression* ‖U‖2κ. *In particular, we have the relative condition number*

$$c_{\mathrm{rel}}(X) = \frac{\|(\|A\|_F P_c, \|B\|_F Q_c, \|C\|_F M_c, \|D\|_F N_c, \|S\|_F H_c, \|R\|_F T_c, \|H\|_F L_c^{-1})\|_2}{\|X\|_F}. \qquad (64)$$

NUMERICAL EXPERIMENT

In this section we want to demonstrate the sharpness of perturbation bound (55) and its relationship with the relative condition number (64). Based on Newton's iteration [3], a numerical example, done with 2×2 coefficient matrices, is illustrated. The numerical algorithm is described in Algorithm

1. The corresponding stopping criterion is determined when the value of the *Normalized Residual* (NRes)

$$\text{NRes} = \frac{\|\widetilde{\mathcal{P}}(\widetilde{X}) - \widetilde{\mathcal{S}}(\widetilde{X})\widetilde{\mathcal{Q}}(\widetilde{X})^{-1}\widetilde{\mathcal{S}}(\widetilde{X})^\top\|}{\|\widetilde{\mathcal{P}}(\widetilde{X})\| + \|\widetilde{\mathcal{S}}(\widetilde{X})\|\|\widetilde{\mathcal{Q}}(\widetilde{X})^{-1}\|\|\widetilde{\mathcal{S}}(\widetilde{X})^\top\|}$$

is less than or equal to a prescribed tolerance.

Algorithm 1: SARE $[\widetilde{X}] = \text{SARE}(\widetilde{A}, \widetilde{B}, \widetilde{C}, \widetilde{D}, \widetilde{R}, \widetilde{S}, \widetilde{H})$

Input: Matrices $\widetilde{A}, \widetilde{C} \in \mathbb{R}^{n\times n}$, $\widetilde{B}, \widetilde{D}, \widetilde{S} \in \mathbb{R}^{n\times m}$, $\widetilde{H} \in \mathcal{S}^n$, $\widetilde{R} \in \mathcal{S}^m$
Output: Matrix $\widetilde{X} \in \mathcal{S}^n$
begin
  Choose $\widetilde{X}_0 \in \mathbb{R}^{n\times n}$;
  for $i \leftarrow 1, \ldots$ **do**
    $\widetilde{F}_k = -(\widetilde{R} + \widetilde{D}^\top \widetilde{X}_k \widetilde{D})^{-1}(\widetilde{B}^\top \widetilde{X}_k \widetilde{C} + \widetilde{S}^\top)$; $\widetilde{\Phi}_k = \widetilde{A} + \widetilde{B}\widetilde{F}_k$; $\widetilde{\Psi}_k = \widetilde{C} + \widetilde{D}\widetilde{F}_k$;
    $\widetilde{\mathcal{R}}(\widetilde{X}_k)$ as defined by (7);
    Updating $\widetilde{X}_{k+1}$ by solving
    $\widetilde{\Phi}_k^\top \widetilde{X}_{k+1} + \widetilde{X}_{k+1}\widetilde{\Phi}_k + \widetilde{\Psi}_k^\top \widetilde{X}_{k+1}\widetilde{\Psi}_k = \widetilde{\Phi}_k^\top \widetilde{X}_k + \widetilde{X}_k\widetilde{\Phi}_k + \widetilde{\Psi}_k^\top \widetilde{X}_k \widetilde{\Psi}_k - \widetilde{\mathcal{R}}(\widetilde{X}_k)$;
  end
end

Example 1 Given a parameter r=10^{-m}, for some m>0, let the matrices *A*, *B*, *C*, *D* be defined by

$$A = \begin{bmatrix} -4 & 0 \\ 0 & -1 \end{bmatrix}, \quad B = \begin{bmatrix} 1 & 0 \\ 0 & -\sqrt{r} \end{bmatrix}, \quad C = I_2, \quad D = \begin{bmatrix} 1 & 0 \\ 0 & 0 \end{bmatrix},$$

and the matrices *S*, *R*, *H* be defined by

$$S = \begin{bmatrix} 1 & 0 \\ 0 & 0 \end{bmatrix}, \quad R = \begin{bmatrix} 3 & 0 \\ 0 & r \end{bmatrix}, \quad H = \begin{bmatrix} -13/2 & 0 \\ 0 & 1 \end{bmatrix}.$$

It is easily seen that the unique stabilizing and maximal solution is

$$X = \begin{bmatrix} -1 & 0 \\ 0 & (\sqrt{5}-1)/2 \end{bmatrix}.$$

Let the perturbed coefficient matrices ΔA, ΔB, ΔC, ΔD, ΔS, ΔR and ΔH be generated using the MATLAB command randn with the weighted coefficient 10^{-j}. That is, the matrices ΔA, ΔB, ΔC, ΔD, ΔS, ΔR and ΔH are generated in forms of randn(2)×10^{-j}, respectively. Since ΔR and ΔH are required to be symmetric, we need to fine-tune the perturbed matrices ΔR and ΔH by redefining ΔR and ΔH as $\Delta R+\Delta R^\top$ and $\Delta H+\Delta H^\top$, respectively. Now, let $(\widetilde{A}, \widetilde{B}, \widetilde{C}, \widetilde{D}, \widetilde{S}, \widetilde{R}, \widetilde{H})$=(A+ΔA,B+ΔB,C+ΔC,D+ΔD,S+ΔS,R+ΔR,H+ΔH), which are coefficient matrices of SARE (5a)-(5b).

Firstly, we would like to evaluate the accuracy of the perturbation bound with the fixed parameter $r=10^{-2}$, *i.e.*, m=2, and different weighted coefficients, 10^{-j}, for j=5,…,9. It can be seen from Table 1 that the values of the relative errors are closely bounded by our perturbation bounds of (55). In other words, (55) does provide a sharp upper bound of the relative errors of the stabilizing solution *X*.

Table 1: Relative errors and perturbation bounds

j	**Relative error**	$\frac{\xi_*}{\|X\|_F}$
5	6.28×10^{-5}	5.74×10^{-4}
6	6.34×10^{-6}	5.22×10^{-5}
7	1.14×10^{-6}	8.42×10^{-6}
8	1.61×10^{-7}	7.42×10^{-7}
9	1.05×10^{-8}	5.58×10^{-8}

Secondly, we want to investigate how ill-conditioned matrices affect the quantities of perturbation bounds. In this sense, the weighted coefficients are fixed to be 10^{-15}, *i.e.*, j=15. The relationships among relative errors, perturbation bounds, and relative condition numbers are shown in Table 2. Due to the singularity of the matrix *R* caused by parameter *r*, the accuracy of the perturbation bounds is highly affected by the singularity. When the value of *m* increases, the perturbation bound is still tight to the relative error. Also, it can be seen that the number of accurate digits of the perturbation bounds is reduced proportionally to the increase of the quantities of the relative condition numbers. In other words, if the accurate digits of the perturbation bound are added to the digits in the relative condition numbers, this number is almost equal to 16. (While using IEEE double-precision, the machine precision is around 2.2×10^{-16}.) This implies that the derived perturbation bound of (55) is fairly sharp.

Table 2: Relative errors, perturbation bounds and relative condition numbers

m	**Relative error**	$\frac{\xi_*}{\|X\|_F}$	$c_{rel}(X)$
1	5.94×10^{-15}	1.64×10^{-14}	5.84×10^{1}
2	5.28×10^{-14}	9.47×10^{-14}	5.56×10^{2}
3	4.75×10^{-13}	7.85×10^{-13}	5.56×10^{3}

4	4.58×10^{-12}	7.40×10^{-12}	5.56×10^{4}
5	4.85×10^{-11}	7.26×10^{-11}	5.56×10^{5}
6	4.56×10^{-10}	7.22×10^{-10}	5.57×10^{6}
7	4.69×10^{-9}	8.61×10^{-9}	5.57×10^{7}

CONCLUSION

While doing numerical computation, it is important in practice to have an accurate method for estimating the relative error and the condition number of the given problems. In this paper, we focus on providing a tight perturbation bound of the stabilizing solution to SARE (1a)-(1b) under small changes in the coefficient matrices. Also, some sufficient conditions are presented for the existence of the stabilizing solution to the perturbed SARE. The corresponding condition number of the stabilizing solution is provided in this work. We highlight and compare the practical performance of the derived perturbation bound and condition number through a numerical example. Numerical results show that our perturbation bound is very sensitive to the condition number of the stabilizing solution. As a consequence, they provide good measurement tools for the sensitivity analysis of SARE (1a)-(1b).

APPENDIX

We provide here a proof of the condition given by (63).

Lemma A.1 *Let* P, Q, $\mathcal{M}$, N, $\mathcal{H}$, T, L_c^{-1} *be the operators defined by* (58a)-(58b), (24a)-(24d) *and* (4), *and let* Ω_δ, δ_p *be defined by* (57). *Then the following equality holds*:

$$\lim_{\delta\to 0^+} \sup_{\Omega_\delta} \frac{\|\mathcal{P}\Delta A + \mathcal{Q}\Delta B + \mathcal{M}\Delta C + \mathcal{N}\Delta D + \mathcal{H}\Delta S + \mathcal{T}\Delta R + \mathcal{L}_c^{-1}\Delta H\|_F}{\delta}$$

$$= \max_{\delta_p>0} \frac{\|\mathcal{P}\Delta A + \mathcal{Q}\Delta B + \mathcal{M}\Delta C + \mathcal{N}\Delta D + \mathcal{H}\Delta S + \mathcal{T}\Delta R + \mathcal{L}_c^{-1}\Delta H\|_F}{\delta_p}. \quad (65)$$

Proof For any δ>0, 0<δ_p≤δ, we see that

$$\frac{\|\mathcal{P}\Delta A + \mathcal{Q}\Delta B + \mathcal{M}\Delta C + \mathcal{N}\Delta D + \mathcal{H}\Delta S + \mathcal{T}\Delta R + \mathcal{L}_c^{-1}\Delta H\|_F}{\delta}$$

$$= \frac{\delta_p}{\delta} \frac{\|\mathcal{P}\Delta A + \mathcal{Q}\Delta B + \mathcal{M}\Delta C + \mathcal{N}\Delta D + \mathcal{H}\Delta S + \mathcal{T}\Delta R + \mathcal{L}_c^{-1}\Delta H\|_F}{\delta_p}$$

$$\le \max_{\delta_{\tilde{p}}>0} \frac{\|\mathcal{P}\Delta \tilde{A} + \mathcal{Q}\Delta \tilde{B} + \mathcal{M}\Delta \tilde{C} + \mathcal{N}\Delta \tilde{D} + \mathcal{H}\Delta \tilde{S} + \mathcal{T}\Delta \tilde{R} + \mathcal{L}_c^{-1}\Delta \tilde{H}\|_F}{\delta_{\tilde{p}}},$$

where $\delta_{\tilde{p}} = \|(\frac{\Delta\tilde{A}}{\kappa_A}, \frac{\Delta\tilde{B}}{\kappa_B}, \frac{\Delta\tilde{C}}{\kappa_C}, \frac{\Delta\tilde{D}}{\kappa_D}, \frac{\Delta\tilde{S}}{\kappa_S}, \frac{\Delta\tilde{R}}{\kappa_R}, \frac{\Delta\tilde{H}}{\kappa_H})\|_F$. It follows that

$$\limsup_{\delta\to 0^+} \sup_{\Omega_\delta} \frac{\|\mathcal{P}\Delta A + \mathcal{Q}\Delta B + \mathcal{M}\Delta C + \mathcal{N}\Delta D + \mathcal{H}\Delta S + \mathcal{T}\Delta R + \mathcal{L}_c^{-1}\Delta H\|_F}{\delta}$$

$$\le \max_{\delta_p>0} \frac{\|\mathcal{P}\Delta A + \mathcal{Q}\Delta B + \mathcal{M}\Delta C + \mathcal{N}\Delta D + \mathcal{H}\Delta S + \mathcal{T}\Delta R + \mathcal{L}_c^{-1}\Delta H\|_F}{\delta_p}. \tag{66}$$

On the other hand, for any fixed δ>0, choose any perturbation matrices $(\Delta A_1, \Delta B_1, \Delta C_1, \Delta D_1, \Delta S_1, \Delta R_1, \Delta H_1) \in \Omega_\delta$

and therefore

$$\left\|\left(\frac{\Delta A_1}{\kappa_A}, \frac{\Delta B_1}{\kappa_B}, \frac{\Delta C_1}{\kappa_C}, \frac{\Delta D_1}{\kappa_D}, \frac{\Delta S_1}{\kappa_S}, \frac{\Delta R_1}{\kappa_R}, \frac{\Delta H_1}{\kappa_H}\right)\right\|_F = \delta_{p_1} \le \delta.$$

It is true that $(\frac{\delta}{\delta_{p_1}}\Delta A_1, \frac{\delta}{\delta_{p_1}}\Delta B_1, \frac{\delta}{\delta_{p_1}}\Delta C_1, \frac{\delta}{\delta_{p_1}}\Delta D_1, \frac{\delta}{\delta_{p_1}}\Delta S_1, \frac{\delta}{\delta_{p_1}}\Delta R_1, \frac{\delta}{\delta_{p_1}}\Delta H_1) \in \Omega_\delta$ and this gives the fact that

$$\limsup_{\delta\to 0^+} \sup_{\Omega_\delta} \frac{\|\mathcal{P}\Delta A + \mathcal{Q}\Delta B + \mathcal{M}\Delta C + \mathcal{N}\Delta D + \mathcal{H}\Delta S + \mathcal{T}\Delta R + \mathcal{L}_c^{-1}\Delta H\|_F}{\delta}$$

$$\ge \frac{\|\mathcal{P}\frac{\delta}{\delta_{p_1}}\Delta A_1 + \mathcal{Q}\frac{\delta}{\delta_{p_1}}\Delta B_1 + \mathcal{M}\frac{\delta}{\delta_{p_1}}\Delta C_1 + \mathcal{N}\frac{\delta}{\delta_{p_1}}\Delta D_1 + \mathcal{H}\frac{\delta}{\delta_{p_1}}\Delta S_1 + \mathcal{T}\frac{\delta}{\delta_{p_1}}\Delta R_1 + \mathcal{L}_c^{-1}\frac{\delta}{\delta_{p_1}}\Delta H_1\|_F}{\delta}$$

$$= \frac{\|\mathcal{P}\Delta A_1 + \mathcal{Q}\Delta B_1 + \mathcal{M}\Delta C_1 + \mathcal{N}\Delta D_1 + \mathcal{H}\Delta S_1 + \mathcal{T}\Delta R_1 + \mathcal{L}_c^{-1}\Delta H_1)\|_F}{\delta_{p_1}}.$$

Hence

$$\limsup_{\delta\to 0^+} \sup_{\Omega_\delta} \frac{\|\mathcal{P}\Delta A + \mathcal{Q}\Delta B + \mathcal{M}\Delta C + \mathcal{N}\Delta D + \mathcal{H}\Delta S + \mathcal{T}\Delta R + \mathcal{L}_c^{-1}\Delta H\|_F}{\delta}$$

$$\ge \max_{\delta_p>0} \frac{\|\mathcal{P}\Delta A + \mathcal{Q}\Delta B + \mathcal{M}\Delta C + \mathcal{N}\Delta D + \mathcal{H}\Delta S + \mathcal{T}\Delta R + \mathcal{L}_c^{-1}\Delta H\|_F}{\delta_p}. \tag{67}$$

Comparison of (66) and (67) gives (65).

ACKNOWLEDGEMENTS

The authors wish to thank the editor and two anonymous referees for many interesting and valuable suggestions on the manuscript. This research work is partially supported by the National Science Council and the National Center for Theoretical Sciences in Taiwan. The first author was supported by the National Science Council of Taiwan under Grant NSC 102-2115-M-

150-002. The second author was supported by the National Science Council of Taiwan under Grant NSC 102-2115-M-003-009. The third author was supported by the National Science Council of Taiwan under Grant NSC 101-2115-M-194-007-MY3.

REFERENCES

1. Rami MA, Zhou XY: Linear matrix inequalities, Riccati equations, and indefinite stochastic linear quadratic controls. *IEEE Trans. Autom. Control* 2000, 45(6):1131–1143. 10.1109/9.863597
2. El Bouhtouri A, Hinrichsen D, Pritchard AJ: On the disturbance attenuation problem for a wide class of time invariant linear stochastic systems. *Stoch. Stoch. Rep.* 1999, 65(3–4):255–297.
3. Damm T, Hinrichsen D: Newton's method for a rational matrix equation occurring in stochastic control. 332/334. *Proceedings of the Eighth Conference of the International Linear Algebra Society* 2001, 81–109.
4. Hinrichsen D, Pritchard AJ:Stochastic H∞. *SIAM J. Control Optim.* 1998, 36: 1504–1538. 10.1137/S0363012996301336
5. Lancaster P, Rodman L Oxford Science Publications. In *Algebraic Riccati Equations*. Clarendon, New York; 1995.
6. Mehrmann VL Lecture Notes in Control and Information Sciences 163. In *The Autonomous Linear Quadratic Control Problem: Theory and Numerical Solution*. Springer, Berlin; 1991.
7. Zhou K, Doyle JC, Glover K: *Robust and Optimal Control*. Prentice Hall, Upper Saddle River; 1996.
8. *Benner, P, Laub, AJ, Mehrmann, V: A collection of benchmark examples for the numerical solution of algebraic Riccati equations I: continuous-time case. Technical Report SPC 95_22, Fakultät für Mathematik, TU Chemnitz-Zwickau, 09107 Chemnitz, FRG. http://www.tu-chemnitz.de/sfb393/spc95pr.html (1995).*
9. *Benner, P, Laub, AJ, Mehrmann, V: A collection of benchmark examples for the numerical solution of algebraic Riccati equations II: discrete-time case. Technical Report SPC 95_23, Fakultät für Mathematik, TU Chemnitz-Zwickau, 09107 Chemnitz, FRG. http://www.tu-chemnitz.de/sfb393/spc95pr.html (1995).*
10. Sima V Monographs and Textbooks in Pure and Applied Mathematics 200. In *Algorithms for Linear-Quadratic Optimization*. Dekker, New York; 1996.
11. Laub AJ: A Schur method for solving algebraic Riccati equations. *IEEE Trans. Autom. Control* 1979, 24(6):913–921. 10.1109/TAC.1979.1102178
12. Ammar G, Benner P, Mehrmann V: A multishift algorithm for the numerical solution of algebraic Riccati equations. *Electron. Trans.*

Numer. Anal. 1993, 1: 33–48.

13. Ammar G, Mehrmann V: On Hamiltonian and symplectic Hessenberg forms. *Linear Algebra Appl.* 1991, 149: 55–72.
14. Benner P, Mehrmann V, Xu H: A new method for computing the stable invariant subspace of a real Hamiltonian matrix. *J. Comput. Appl. Math.* 1997, 86: 17–43. 10.1016/S0377-0427(97)00146-5
15. Benner P, Mehrmann V, Xu H: A numerically stable, structure preserving method for computing the eigenvalues of real Hamiltonian or symplectic pencils. *Numer. Math.* 1998, 78(3):329–358. 10.1007/s002110050315
16. Bunse-Gerstner A, Byers R, Mehrmann V: A chart of numerical methods for structured eigenvalue problems. *SIAM J. Matrix Anal. Appl.* 1992, 13: 419–453. 10.1137/0613028
17. Bunse-Gerstner A, Mehrmann V: A symplectic QR like algorithm for the solution of the real algebraic Riccati equation. *IEEE Trans. Autom. Control* 1986, 31(12):1104–1113. 10.1109/TAC.1986.1104186
18. Mehrmann V: A step toward a unified treatment of continuous and discrete time control problems. *Linear Algebra Appl.* 1996, 241–243: 749–779.
19. Byers R: Solving the algebraic Riccati equation with the matrix sign function. *Linear Algebra Appl.* 1987, 85: 267–279.
20. Benner, P: Contributions to the numerical solutions of algebraic Riccati equations and related eigenvalue problems. PhD thesis, Fakultät für Mathematik, TU Chemnitz-Zwickau, Chemnitz, Germany (1997)
21. Chu EK-W, Fan H-Y, Lin W-W: A structure-preserving doubling algorithm for continuous-time algebraic Riccati equations. *Linear Algebra Appl.* 2005, 396: 55–80.
22. Chu EK-W, Fan H-Y, Lin W-W, Wang C-S: Structure-preserving algorithms for periodic discrete-time algebraic Riccati equations. *Int. J. Control* 2004, 77(8):767–788. 10.1080/00207170410001714988
23. Chiang C-Y, Fan H-Y: Residual bounds of the stochastic algebraic Riccati equation. *Appl. Numer. Math.* 2013, 63: 78–87.
24. Konstantinov M, Gu D-W, Mehrmann V, Petkov P Studies in Computational Mathematics 9. In *Perturbation Theory for Matrix Equations*. North-Holland, Amsterdam; 2003.
25. Sun J-G: Perturbation theory for algebraic Riccati equations. *SIAM J. Matrix Anal. Appl.*1998, 19(1):39–65. 10.1137/S0895479895291303

26. Sun J-g: Sensitivity analysis of the discrete-time algebraic Riccati equation. 275/276. *Proceedings of the Sixth Conference of the International Linear Algebra Society* 1998, 595–615. Chemnitz, 1996
27. Sun J-g: Residual bounds of approximate solutions of the algebraic Riccati equation. *Numer. Math.* 1997, 76(2):249–263. 10.1007/s002110050262
28. Sun J-g: Residual bounds of approximate solutions of the discrete-time algebraic Riccati equation. *Numer. Math.* 1998, 78(3):463–478. 10.1007/s002110050321
29. Riedel KS: A Sherman-Morrison-Woodbury identity for rank augmenting matrices with application to centering. *SIAM J. Matrix Anal. Appl.* 1992, 13(2):659–662. 10.1137/0613040
30. Stewart GW, Sun JG Computer Science and Scientific Computing. In *Matrix Perturbation Theory*. Academic Press, Boston; 1990.
31. Ortega JM, Rheinboldt WC Classics in Applied Mathematics 30. In *Iterative Solution of Nonlinear Equations in Several Variables*. SIAM, Philadelphia; 2000. Reprint of the 1970 original
32. Rice JR: A theory of condition. *SIAM J. Numer. Anal.* 1966, 3: 287–310. 10.1137/0703023
33. Sun J-g: Condition numbers of algebraic Riccati equations in the Frobenius norm. *Linear Algebra Appl.* 2002, 350: 237–261. 10.1016/S0024-3795(02)00294-X
34. Xu S: *Matrix Computation in Control Theory*. Higher Education Press, Beijing; 2010. (In Chinese)

CHAPTER 5

A TRIDIAGONAL MATRIX CONSTRUCTION BY THE QUOTIENT DIFFERENCE RECURSION FORMULA IN THE CASE OF MULTIPLE EIGENVALUES

Kanae Akaiwa[1], Masashi Iwasaki[2], Koichi Kondo[3] and Yoshimasa Nakamura[1]

[1]Graduate School of Informatics, Kyoto University, Yoshida-Hommachi, Sakyo-ku, Kyoto 606-8501, Japan

[2]Department of Informatics and Environmental Sciences, Kyoto Prefectural University, 1-5 Nakaragi-cho, Shimogamo, Sakyo-ku, Kyoto 606-8522, Japan

[3]Graduate School of Science and Engineering, Doshisha University, 1-3 Tatara Miyakodani, Kyoto 610-0394, Japan

Citation (APA): Akaiwa, K., Iwasaki, M., Kondo, K., & Nakamura, Y. (2014). A tridiagonal matrix construction by the quotient difference recursion formula in the case of multiple eigenvalues. Pacific Journal of Mathematics for Industry, 6(1), 10. (9 pages).

ABSTRACT

In this paper, we grasp an inverse eigenvalue problem which constructs a tridiagonal matrix with specified multiple eigenvalues, from the viewpoint of the quotient difference (qd) recursion formula. We also prove that the characteristic and the minimal polynomials of a constructed tridiagonal matrix are equal to each other. As an application of the qd formula, we present a procedure for getting a tridiagonal matrix with specified multiple eigenvalues. Examples are given through providing with four tridiagonal matrices with specified multiple eigenvalues.

Keywords: Quotient difference formula, Tridiagonal matrix, Multiple eigenvalues, Characteristic polynomial, Minimal polynomial

INTRODUCTION

One of the important problems in linear algebra is to construct matrices with specified eigenvalues. This is an inverse eigenvalue problem which is classified in Structured Inverse Eigenvalue Problem (SIEP) called in [1]. The main purpose of this paper is to design a procedure for solving an SIEP in the case where the constructed matrix has tridiagonal form with multiple eigenvalues, through reconsidering the quotient difference (qd) formula. It is known that the qd formula has the applications to computing a continued fraction expansion of power series [5], zeros of polynomial [3], eigenvalues of a so-called Jacobi matrix [9] and so on. Though the book [9] refers to an aspect similar to in the following sections, it gives only an anticipated comment without proof in the case of multiple eigenvalues. There is no observation about numerical examples for verifying it. The key point for the purpose is to investigate the Hankel determinants appearing in the determinant solution to the qd formula with the help of the Jordan canonical form. In this paper, we give our focus on the unsettled case in order to design a procedure for constructing a tridiagonal matrix with specified multiple eigenvalues, based on the qd formula. The reason why the sequence of discussions was stopped is expected that multiple-precision arithmetic and symbolic computing around the published year of Rutishauser's works for the qd formula were not sufficiently developed. The qd formula, strictly speaking the differential form of it, for computing tridiagonal eigenvalues acts with high relative accuracy in single-precision or double-precision arithmetic [7], while, actually, that serving for constructing a tridiagonal matrix gives rise to no small errors. Thus, the qd formula serving for constructing a tridiagonal matrix is not so worth in single-precision or double-precision arithmetic. In recent computers, it is not difficult to employ not only single or double precision arithmetic but also arbitrary-precision arithmetic or symbolic computing. In fact, an expression involving only symbolic quantities achieves exact arithmetic on the scientific computing software such as Wolfram Mathematica, Maple and so on. Numerical errors frequently occur in finite-precision arithmetic, so that a constructed tridiagonal matrix probably does not have multiple eigenvalues without symbolic computing. The resulting procedure in this paper is assumed to be carried out on symbolic computing.

This paper is organized as follows. In Section 2, we first give a short explanation of some already obtained properties concerning the qd formula. In Section 3, we observe a tridiagonal matrix whose characteristic polynomial is associated with the minimal polynomial of a general matrix through reconsidering the qd formula. The tridiagonal matrix essentially differs from the Jacobi matrix in that it is not always symmetrized. We also discuss the characteristic and the minimal polynomials of a tridiagonal matrix in Section 4. In Section 5, we design a procedure for constructing a tridiagonal matrix with specified multiple eigenvalues, and then demonstrate four tridiagonal matrices as examples of the resulting procedure. Finally, in Section 6, we give conclusion.

SOME PROPERTIES FOR THE QD RECURSION FORMULA

In this section, we briefly review two theorems in [4] concerning the qd formula from the viewpoint of a generating function, the Hankel determinant and a tridiagonal matrix.

Let us introduce the Hankel determinants $H_1^{(n)}, H_2^{(n)}, \ldots$ given in terms of a complex sequence $\{f_n\}_0^\infty$ as

$$H_s^{(n)} = \begin{vmatrix} f_n & f_{n+1} & \cdots & f_{n+s-1} \\ f_{n+1} & f_{n+2} & \cdots & f_{n+s} \\ \vdots & \vdots & \ddots & \vdots \\ f_{n+s-1} & f_{n+s} & \cdots & f_{n+2s-2} \end{vmatrix},$$

$$s = 1, 2, \ldots, \quad n = 0, 1, \ldots, \tag{1}$$

where $H_{-1}^{(n)}$ =0 and $H_0^{(n)}$ =1 for n=0,1,…. Moreover, let $F(z)$ be a generating function associated with $\{f_n\}_0^{\infty}$ as

$$F(z) = \sum_{n=0}^{\infty} f_n z^n = f_0 + f_1 z + f_2 z^2 + \cdots. \tag{2}$$

Let us consider that $F(z)$ is a rational function with respect to z with a pole of order $l_0 \geq 0$ at infinity and finite poles $z_k \neq 0$ of order l_k for k=1,2,…,L. Then the sum of the orders of the finite poles is $l = l_1 + l_2 + \cdots + l_L$, and $F(z)$ is factorized as

$$F(z) = G_0(z) + \frac{G(z)}{(z - z_1)^{l_1}(z - z_2)^{l_2} \cdots (z - z_L)^{l_L}}, \tag{3}$$

where $G(z)$ is a polynomial of degree at most l, and $G_0(z)$ is a polynomial of degree l_0 if $l_0 > 0$, or $G_0(z) = 0$ if $l_0 = 0$. The following theorem gives the determinant solution to the qd recursion formula

$$\begin{cases} q_s^{(n+1)} + e_{s-1}^{(n+1)} = q_s^{(n)} + e_s^{(n)}, \\ \qquad s = 1, 2, \ldots, \quad n = 0, 1, \ldots, \\ q_s^{(n+1)} e_s^{(n+1)} = q_{s+1}^{(n)} e_s^{(n)}, \quad s = 1, 2, \ldots, \quad n = 0, 1, \ldots. \end{cases} \tag{4}$$

Theorem 1. *([4], pp. 596, 603, 610)* Let $F(z)$ be factorized as in (3). Then it holds that

$$H_s^{(n)} = 0, \quad s = l + 1, l + 2, \ldots, \quad n = l_0 + 1, l_0 + 2, \ldots. \tag{5}$$

Let us assume that

$$H_s^{(n)} \neq 0, \quad s = 1, 2, \ldots, l, \quad n = 0, 1, \ldots. \tag{6}$$

Then the qd formula (4) with the initial settings

$$e_0^{(n)} = 0, \quad q_1^{(n)} = \frac{f_{n+1}}{f_n}, \quad n = 0, 1, \ldots \tag{7}$$

admits the determinant solution

$$q_s^{(n)} = \frac{H_s^{(n+1)} H_{s-1}^{(n)}}{H_s^{(n)} H_{s-1}^{(n+1)}}, \quad s = 1, 2, \ldots, l, \quad n = 0, 1, \ldots, \tag{8}$$

$$e_s^{(n)} = \frac{H_{s+1}^{(n)} H_{s-1}^{(n+1)}}{H_s^{(n)} H_s^{(n+1)}}, \quad s = 0, 1 \ldots, l, \quad n = 0, 1, \ldots. \tag{9}$$

From (9) with (5), it follows that $e^{(n)}_l=0$ for n=0,1,.... Moreover, it turns out that $q^{(n)}_s$ and $e^{(n)}_s$ for $s=l+1,l+2,\ldots$ and n=0,1,... are not given in the same form as (8) and (9).

Let us introduce s-by-s tridiagonal matrices,

$$T_s^{(n)} = \begin{pmatrix} q_1^{(n)} & q_1^{(n)} e_1^{(n)} & & & \\ 1 & q_2^{(n)} + e_1^{(n)} & \ddots & & \\ & \ddots & \ddots & q_{s-1}^{(n)} e_{s-1}^{(n)} \\ & & 1 & q_s^{(n)} + e_{s-1}^{(n)} \end{pmatrix},$$

$$s = 1, 2, \ldots, l, \quad n = 0, 1, \ldots \tag{10}$$

with the qd variables $q^{(n)}_s$ and $e^{(n)}_s$. Let I_s be the s-by-s identity matrix. Then we obtain a theorem for the characteristic polynomial of T(n)l.

Theorem 2. *([4], pp. 626, 635)* Let $F(z)$ be factorized as in (3). Let us assume that $H^{(n)}_s$ satisfies (6). For n=0,1,..., it holds that

$$\det\left(zI_l - T_l^{(n)}\right) = \left(z - z_1^{-1}\right)^{l_1} \left(z - z_2^{-1}\right)^{l_2} \cdots \left(z - z_L^{-1}\right)^{l_L}. \tag{11}$$

TRIDIAGONAL MATRIX ASSOCIATED WITH GENERAL MATRIX

In this section, from the viewpoint of the characteristic and the minimal polynomials, we associate a general M-by-M complex matrix A with a tridiagonal matrix $T^{(n)}_l$.

Let $\lambda_1,\lambda_2,\ldots,\lambda_N$ be the distinct eigenvalues of A, which are numbered as $|\lambda_1|\geq|\lambda_2|\geq\cdots\geq|\lambda_N|$. It is noted that some of $|\lambda_1|,|\lambda_2|,\ldots,|\lambda_N|$ may equal to each other in the case where some of $\lambda_1,\lambda_2,\ldots,\lambda_N$ are negative eigenvalues or complex eigenvalues. Let M_k be the algebraic multiplicity of λ_k, where $M=M_1+M_2+\cdots+M_N$. For the identity matrix $I_M \in R^{M\times M}$, let $\phi_A(z)=\det(zI_M-A)$ be the characteristic polynomial of A, namely,

$$\phi_A(z) = (z-\lambda_1)^{M_1}(z-\lambda_2)^{M_2}\cdots(z-\lambda_N)^{M_N}. \tag{12}$$

Let us prepare the sequence $\{f_n\}_0^\infty$ given by

$$f_n = \boldsymbol{w}^H A^n \boldsymbol{u}, \quad n = 0, 1, \ldots \tag{13}$$

for some nonzero M-dimensional complex vectors $\boldsymbol{u}$ and $\boldsymbol{w}$, where the superscript H denotes the Hermitian transpose. Originally, $f_0, f_1, \ldots$ were called the Schwarz constants, but they are usually today called the moments or the Markov parameters [2]. Since the matrix power series $\sum_{n=0}^\infty (zA)^n$ is a Neumann series (cf. [6]), F(z)= $\sum_{n=0}^\infty \boldsymbol{w}^H (zA)^n \boldsymbol{u}$ converges absolutely in the disk $D{:}|z|<|\lambda_1|^{-1}$. Moreover, we derive $F(z)=\boldsymbol{w}^H(I_M-zA)^{-1}\boldsymbol{u}$ which implies that $F(z)$ is a rational function with the denominator $\det(I_M-zA)=z^M\phi_A(z^{-1})$ as follows.

$$F(z) = \frac{\tilde{G}(z)}{(1-\lambda_1 z)^{M_1}(1-\lambda_2 z)^{M_2}\cdots(1-\lambda_N z)^{M_N}}, \tag{14}$$

where $\tilde{G}(z)$ is some polynomial with respect to z. It is remarkable that the numerator $\tilde{G}(z)$ may have the same factors as the denominator $(1-\lambda_{1z})^{M1}(1-\lambda_{2z})^{M2}\cdots(1-\lambda_{Nz})^{MN}$. In other words, $F(z)$ has the poles $\lambda_1^{-1}, \lambda_2^{-1}, \ldots, \lambda_N^{-1}$ whose orders are equal to or less than $M_1, M_2, \ldots, M_N$, respectively.

Let us introduce the Jordan canonical form of A in order to investigate the poles of $F(z)$ with (13) even in the case where A has multiple eigenvalues. Let $\mathcal{M}_k$ be the geometric multiplicity of λ_k which indicates the dimension of eigenspace $\mathrm{Ker}(A-\lambda_k I_M)$. It is noted that $\mathcal{M}_k$ is equal to or less than the algebraic multiplicity M_k. The matrix A has $\mathcal{M}_k$ eigenvectors corresponding to λ_k, and then the eigenvectors, denoted by $v_{k,1}, v_{k,2}, \ldots, v_{k,\mathcal{M}_k}$, satisfy

$$A\boldsymbol{v}_{k,j} = \lambda_k \boldsymbol{v}_{k,j}, \quad j = 1, 2, \ldots, \mathcal{M}_k. \tag{15}$$

Hereinafter, for $j=1,2,\ldots,\mathcal{M}_k$, let $\boldsymbol{v}_{k,j}(1)=\boldsymbol{v}_{k,j}$. Moreover, for $j=1,2,\ldots,\mathcal{M}_k$, let $\boldsymbol{v}_{k,j}(2), \boldsymbol{v}_{k,j}(3), \ldots, \boldsymbol{v}_{k,j}(m_{k,j})$ denote the generalized eigenvectors associated with the eigenvectors $\boldsymbol{v}_{k,j}(1)$, where $m_{k,j}$ is the maximal integer such that $\boldsymbol{v}_{k,j}(1), \boldsymbol{v}_{k,j}(2), \ldots, \boldsymbol{v}_{k,j}(m_{k,j})$ are linearly independent. Of course, $m_{k,1}+m_{k,2}+\cdots+m_{k,\mathcal{M}_k}$. Then, the generalized eigenvectors $\boldsymbol{v}_{k,j}(2), \boldsymbol{v}_{k,j}(3), \ldots, \boldsymbol{v}_{k,j}(m_{k,j})$ satisfy

$$A\boldsymbol{v}_{k,j}(i) = \lambda_k \boldsymbol{v}_{k,j}(i) + \boldsymbol{v}_{k,j}(i-1),$$
$$i = 2, 3, \ldots, m_{k,j}, \quad j = 1, 2, \ldots, \mathcal{M}_k. \tag{16}$$

From (15) and (16), we derive the Jordan canonical form of A as

$$V^{-1}AV = J \tag{17}$$

with the nonsingular matrix

$$V = (V_1\ V_2\ \cdots\ V_N) \in \mathbb{C}^{M\times M}, \tag{18}$$

and the block diagonal matrix

$$J = \mathrm{diag}(J_1, J_2, \ldots, J_N) \in \mathbb{C}^{M\times M}, \tag{19}$$

where

$$V_k = (V_{k,1}\ V_{k,2}\ \cdots\ V_{k,\mathcal{M}_k}) \in \mathbb{C}^{M\times M_k}, \tag{20}$$

$$V_{k,j} = (\boldsymbol{v}_{k,j}(1)\ \boldsymbol{v}_{k,j}(2)\ \cdots\ \boldsymbol{v}_{k,j}(m_{k,j})) \in \mathbb{C}^{M\times m_{k,j}}, \tag{21}$$

$$J_k = \mathrm{diag}(J_{k,1}, J_{k,2}, \ldots, J_{k,\mathcal{M}_k}) \in \mathbb{C}^{M_k\times M_k}, \tag{22}$$

$$J_{k,j} = \begin{pmatrix} \lambda_k & 1 & & \\ & \lambda_k & \ddots & \\ & & \ddots & 1 \\ & & & \lambda_k \end{pmatrix} \in \mathbb{C}^{m_{k,j}\times m_{k,j}}. \tag{23}$$

Without loss of generality, we may assume that $m_{k,1} \geq m_{k,2} \geq \cdots \geq m_{k}, M_{k}$.

Let $m_k = \max\{m_{k,1}, m_{k,2}, \ldots, m_k, M_k\}$. Since $m_{k,1} \geq m_{k,2} \geq \cdots \geq m_k, M_k$, it is obvious that $m_k = m_{k,1}$. With the help of the Jordan canonical form of A as in (17), we get a proposition for the sequence $\{f_n\}_0^\infty$ in (13).

Proposition1. Let $\boldsymbol{u}$ be the vector given by the linear combination of the eigenvectors and the generalized eigenvectors of A, namely, for some constants $\kappa_{k,j,i}$,

$$\boldsymbol{u} = \sum_{k=1}^{N}\sum_{j=1}^{\mathcal{M}_k}\sum_{i=1}^{m_{k,j}} \kappa_{k,j,i}\boldsymbol{v}_{k,j}(i). \tag{24}$$

Moreover, for a vector $\boldsymbol{w}$, let

$$c_{k,i} = \sum_{j=1}^{\mathcal{M}_k}\sum_{i'=i}^{m_{k,j}} \kappa_{k,j,i'}\boldsymbol{v}_{k,j}^{H}(i' - i + 1)\boldsymbol{w}. \tag{25}$$

Then, the sequence $\{f_n\}_0^\infty$ in (13) can be expressed by

$$f_n = \sum_{k=1}^{N} \sum_{i=1}^{m_k} \binom{n}{i-1} c_{k,i} \lambda_k^{n-i+1}, \tag{26}$$

where the binomial coefficients are 0 if $n<i-1$. Also, for suitable $\boldsymbol{u}$ and $\boldsymbol{w}$, it holds that

$$c_{k,i} \neq 0, \quad i = 1, 2, \ldots, m_k, \quad k = 1, 2, \ldots, N. \tag{27}$$

Proof.

From $V^{-1}AV=J$ in (17), it holds that $A^n=VJ^nV^{-1}$. By combining it with (13) and (24), we derive

$$\begin{aligned} f_n &= \mathbf{w}^H VJ^n V^{-1} \boldsymbol{u} \\ &= \sum_{k=1}^{N} \sum_{j=1}^{\mathcal{M}_k} \sum_{i=1}^{m_{k,j}} \kappa_{k,j,i} \mathbf{w}^H VJ^n V^{-1} \mathbf{v}_{k,j}(i). \end{aligned} \tag{28}$$

Let $\rho_{k,j,i}$ be the column number in which $\mathbf{v}_{k,j}(i)$ arranges. Then it is obvious that $V^{-1}\mathbf{v}_{k,j}(i)=\boldsymbol{e}_{k,j}(i)$ where $\boldsymbol{e}_{k,j}(i)$ denotes a unit vector such that the $\rho_{k,j,i}$th entry is 1 and the others are 0. Thus, it follows that

$$f_n = \sum_{k=1}^{N} \sum_{j=1}^{\mathcal{M}_k} \sum_{i=1}^{m_{k,j}} \kappa_{k,j,i} \mathbf{w}^H VJ^n \boldsymbol{e}_{k,j}(i). \tag{29}$$

Since J is the block diagonal matrix, the matrix J^n and its small blocks $(J_k)^n$ are also so. It also turns out that $(J_{k,j})^n$ is upper triangle. So, it is worth noting that $J^n \boldsymbol{e}_{k,j}(i)$ becomes the $\rho_{k,j,i}$th column vector of J^n and the zero-entries arrange in except for its $\rho_{k,j,1}$th, $\rho_{k,j,2}$th, …, $\rho_{k,j,i}$th rows. The Jordan blocks $J_{k,j}$ can be decomposed as

$$J_{k,j} = \lambda_k I_{m_{k,j}} + E_{m_{k,j}}, \tag{30}$$

$$E_{m_{k,j}} = \begin{pmatrix} 0 & 1 & & \\ & 0 & \ddots & \\ & & \ddots & 1 \\ & & & 0 \end{pmatrix} \in \mathbb{R}^{m_{k,j} \times m_{k,j}}. \tag{31}$$

It is emphasized that $E_{mk,j}$ is a nilpotent matrix whose i'th power becomes the zero-matrix O for $i' \geq m_{k,j}$. Thus, $(J_{k,j})^n$ can be expressed as

$$(J_{k,j})^n = \sum_{i'=1}^{m_{k,j}} \binom{n}{i'-1} \lambda_k^{n-i'+1} (E_{m_{k,j}})^{i'-1}, \tag{32}$$

where $(E_{mk,j})^0 = I_{mk,j}$. Let us introduce an $m_{k,j}$-dimensional unit vector $\boldsymbol{e}(i)$ which is regarded as a part of $\boldsymbol{e}_{k,j}(i)$. Then, by taking account that $(E_{m_{k,j}})^{i'-1}\boldsymbol{e}(i) = \boldsymbol{e}(i - i' + 1)$ in (32), we derive

$$J^n \boldsymbol{e}_{k,j}(i) = \sum_{i'=1}^{i} \binom{n}{i'-1} \lambda_k^{n-i'+1} \boldsymbol{e}_{k,j}(i - i' + 1). \tag{33}$$

Since it holds that $V\boldsymbol{e}_{k,j}(i-i'+1) = \boldsymbol{v}_{k,j}(i-i'+1)$, by combining it with (29) and (33), we therefore have

$$f_n = \sum_{k=1}^{N} \sum_{j=1}^{\mathcal{M}_k} \sum_{i=1}^{m_{k,j}} \sum_{i'=1}^{i} \kappa_{k,j,i} \boldsymbol{w}^H \boldsymbol{v}_{k,j}(i - i' + 1) \times \binom{n}{i'-1} \lambda_k^{n-i'+1}. \tag{34}$$

By writing down two summations, we get

$$f_n = \sum_{k=1}^{N} \sum_{j=1}^{\mathcal{M}_k} \Bigg[\kappa_{k,j,1} \boldsymbol{w}^H \boldsymbol{v}_{k,j}(1) \binom{n}{0} \lambda_k^n + \kappa_{k,j,2} \left(\boldsymbol{w}^H \boldsymbol{v}_{k,j}(2) \binom{n}{0} \lambda_k^n + \boldsymbol{w}^H \boldsymbol{v}_{k,j}(1) \binom{n}{1} \lambda_k^{n-1} \right) + \kappa_{k,j,3} \left(\boldsymbol{w}^H \boldsymbol{v}_{k,j}(3) \binom{n}{0} \lambda_k^n + \boldsymbol{w}^H \boldsymbol{v}_{k,j}(2) \binom{n}{1} \lambda_k^{n-1} + \boldsymbol{w}^H \boldsymbol{v}_{k,j}(1) \binom{n}{2} \lambda_k^{n-2} \right)$$

$$
\begin{aligned}
&+\cdots\\
&+\kappa_{k,j,m_{k,j}}\left(\boldsymbol{w}^H\boldsymbol{v}_{k,j}(m_{k,j})\binom{n}{0}\lambda_k^n\right.\\
&+\boldsymbol{w}^H\boldsymbol{v}_{k,j}(m_{k,j}-1)\binom{n}{1}\lambda_k^{n-1}\\
&\left.\left.+\cdots+\boldsymbol{w}^H\boldsymbol{v}_{k,j}(1)\binom{n}{m_{k,j}-1}\lambda_k^{n-m_{k,j}+1}\right)\right].
\end{aligned}
$$

Moreover, by paying our attention to the binomial coefficients, we can rewrite f_n as

$$
\begin{aligned}
f_n=\sum_{k=1}^{N}&\left[\binom{n}{0}\lambda_k^n\sum_{j=1}^{\mathcal{M}_k}\left(\kappa_{k,j,1}\boldsymbol{w}^H\boldsymbol{v}_{k,j}(1)+\kappa_{k,j,2}\boldsymbol{w}^H\boldsymbol{v}_{k,j}(2)\right.\right.\\
&\left.+\cdots+\kappa_{k,j,m_{k,j}}\boldsymbol{w}^H\boldsymbol{v}_{k,j}(m_{k,j})\right)\\
&+\binom{n}{1}\lambda_k^{n-1}\sum_{j=1}^{\mathcal{M}_k}\left(\kappa_{k,j,2}\boldsymbol{w}^H\boldsymbol{v}_k(j,1)+\kappa_{k,j,3}\boldsymbol{w}^H\boldsymbol{v}_k(j,2)\right.\\
&\left.+\cdots+\kappa_{k,j,m_{k,j}}\boldsymbol{w}^H\boldsymbol{v}_{k,j}(m_{k,j}-1)\right)\\
&+\cdots\\
&\left.+\binom{n}{m_{k,j}-1}\lambda_k^{n-m_{k,j}+1}\sum_{j=1}^{\mathcal{M}_k}\kappa_{k,j,m_{k,j}}\boldsymbol{w}^H\boldsymbol{v}_{k,j}(1)\right]\\
=\sum_{k=1}^{N}&\left[\binom{n}{0}\lambda_k^n\sum_{j=1}^{\mathcal{M}_k}\sum_{i=1}^{m_{k,j}}\kappa_{k,j,i}\boldsymbol{w}^H\boldsymbol{v}_{k,j}(i-1+1)\right.\\
&+\binom{n}{1}\lambda_k^{n-1}\sum_{j=1}^{\mathcal{M}_k}\sum_{i=2}^{m_{k,j}}\kappa_{k,j,i}\boldsymbol{w}^H\boldsymbol{v}_{k,j}(i-2+1)\\
&+\cdots\\
&+\binom{n}{m_{k,j}-1}\lambda_k^{n-m_{k,j}+1}\\
&\left.\times\sum_{j=1}^{\mathcal{M}_k}\sum_{i=m_{k,j}}^{m_{k,j}}\kappa_{k,j,i}\boldsymbol{w}^H\boldsymbol{v}_{k,j}(i-m_{k,j}+1)\right].
\end{aligned}
$$

From $m_k \geq m_{k,j}$ and $w^H v_{k,j}(i-i'+1)=v^H_{k,j}(i-i'+1)w$, it follows that

$$f_n = \sum_{k=1}^{N} \sum_{i'=1}^{m_k} \left(\sum_{j=1}^{\mathcal{M}_k} \sum_{i=i'}^{m_{k,j}} \kappa_{k,j,i} \boldsymbol{v}_{k,j}^H (i - i' + 1) \boldsymbol{w} \right) \times \binom{n}{i' - 1} \lambda_k^{n-i'+1}. \tag{35}$$

The exchange of i for i' in (35) brings us to (25) and (26).

For example, let us consider the case where the constants $\kappa_{k,j,i}$ are all 1. Then $\boldsymbol{u}$ becomes the sum of all the eigenvectors and generalized eigenvectors. Moreover, let $\boldsymbol{w}=V^{-H}\boldsymbol{\alpha}$ in (25) where $\boldsymbol{\alpha}$ is an M-dimensional vector with all the entries 1. Then it holds that $\kappa_{k,j,i} \boldsymbol{v}_{k,j}^H (i' - i + 1) \boldsymbol{w} = \boldsymbol{e}_{k,j}^\top (i' - i + 1) \boldsymbol{\alpha} = 1$. Thus, it is concluded that $c_{k,i} \neq 0$. The above discussion suggests that there exists at least a pair of $\boldsymbol{u}$ and $\boldsymbol{w}$ for satisfying (27).

Proposition 1 leads to a theorem concerning the generating function $F(z)$ with the moments $f_n = \boldsymbol{w}^H A^n \boldsymbol{u}$.

Theorem 3. Let $F(z)$ be the generating function with the moments $f_n = \boldsymbol{w}^H A^n \boldsymbol{u}$. Then, $F(z)$ converges absolutely in the disk $D: |z| < |\lambda_1|^{-1}$, and $F(z)$ is expressed as

$$F(z) = \sum_{k=1}^{N} \sum_{i=1}^{m_k} \frac{c_{k,i} z^{i-1}}{(1 - \lambda_k z)^i}. \tag{36}$$

Especially, if $\lambda_N = 0$, then $F(z)$ is expressed as

$$F(z) = c_{N,1} + c_{N,2} z + \cdots + c_{N,m_N} z^{m_N - 1} + \sum_{k=1}^{N-1} \sum_{i=1}^{m_k} \frac{c_{k,i} z^{i-1}}{(1 - \lambda_k z)^i}. \tag{37}$$

Let us assume that (27) holds for suitable $\boldsymbol{u}$ and $\boldsymbol{w}$. If $\lambda_N \neq 0$, then $F(z)$ has the finite poles $\lambda_1^{-1}, \lambda_2^{-1}, \ldots, \lambda_N^{-1}$ of the orders $m_1, m_2, \ldots, m_N$, respectively, and the sum of the orders is $m = m_1 + m_2 + \cdots + m_N$. If $\lambda_N = 0$, then $F(z)$ has the pole of the order $m_N - 1$ at infinity and the finite poles $\lambda_1^{-1}, \lambda_2^{-1}, \ldots, \lambda_N^{-1}$ of the orders $m_1, m_2, \ldots, m_N$, respectively, and the sum of the orders of all the finite poles is $m - m_N$.

Proof.

By substituting f_n in (26) into $F(z)$ in (2), we get

$$F(z) = \sum_{k=1}^{N}\sum_{i=1}^{m_k} c_{k,i}\left(\sum_{n=0}^{\infty} z^n \binom{n}{i-1}\lambda_k^{n-i+1}\right)$$
$$= \sum_{k=1}^{N}\sum_{i=1}^{m_k} c_{k,i}\left(\sum_{n=i-1}^{\infty} z^n \binom{n}{i-1}\lambda_k^{n-i+1}\right) \tag{38}$$

By letting $n=n'+i-1$, we derive

$$F(z) = \sum_{k=1}^{N}\sum_{i=1}^{m_k} c_{k,i}z^{i-1}\left(\sum_{n'=0}^{\infty}\binom{n'+i-1}{i-1}(\lambda_k z)^{n'}\right). \tag{39}$$

It is noted that, for $|z|<1$,

$$\sum_{n'=0}^{\infty}\binom{n'+i-1}{i-1}z^{n'} = \frac{1}{(1-z)^i}. \tag{40}$$

From (39) and (40), it turns out that $F(z)$ converges absolutely in the disk $D:|z|<|\lambda_1|^{-1}$. Simultaneously, we have (36) for $z\in D$. It is obvious that (36) with $\lambda_N=0$ becomes (37). Moreover, (36) and (37) immediately lead to the latter half concerning the poles of $F(z)$.

Let $\psi_A(z)$ be the polynomial whose degree is the smallest such that $\psi_A(A)=O$. Here $\psi_A(z)$ is called the minimal polynomial of A. Let us recall here that the maximal dimension of the Jordan blocks Jk,1,Jk,2,…,Jk,MkJk,1,Jk,2,… ,Jk,Mk corresponding to λ_k is m_k. So, $\psi_A(z)$ is representable as

$$\psi_A(z) = (z-\lambda_1)^{m_1}(z-\lambda_2)^{m_2}\cdots(z-\lambda_N)^{m_N}. \tag{41}$$

Therefore, we have the main theorem in this section for the relationship between the minimal polynomial of a general matrix A and the characteristic polynomial of a tridiagonal matrix $T^{(n)}_1$.

Theorem 4. Let $F(z)$ be given by the generating function with the moments $f_n=\boldsymbol{w}^H A^n \boldsymbol{u}$. Let us assume that (6) and (27) hold for suitable $\boldsymbol{u}$ and $\boldsymbol{w}$. If $\lambda_1\neq0,\lambda_2\neq0,\ldots,\lambda_N\neq0$, then it holds that

$$\det\left(zI_m - T_m^{(n)}\right) = \psi_A(z), \quad n=0,1,\ldots, \tag{42}$$

otherwise,

$$\det\left(zI_{m-m_N} - T_{m-m_N}^{(n)}\right) = \frac{\psi_A(z)}{z^{m_N}}, \quad n=0,1,\ldots. \tag{43}$$

Proof.

It is remarkable that three integers L,l,l_k and a complex z_k associated with the tridiagonal matrix $T^{(n)}_1$ in Theorem 2 are given in terms of three integers N,m,m_k and a complex λ_k associated with a general matrix A. If $\lambda_N \neq 0$, then it follows from the latter half of Theorem 3that $L=N$, $l=m$, $l_0,l_1=m_1,l_2=m_2,\ldots,l_N=m_N$ and $z_k = \lambda_k^{-1}$. So, from (11) and (41), we derive (42). Similarly, if $\lambda_N=0$, then $L=N-1$, $l=m-m_N$, $l_0=m_N-1,l_1=m_1,l_2=m_2,\ldots,l_{N-1}=m_{N-1}$ and $z_k = \lambda_k^{-1}$. Thus (11) and (41) lead to (43).

Incidentally, the editors in ([9], pp. 444–445) give a simple example with short comments concerning the minimal polynomial, the Jordan canonical form of A and the multiple poles of $F(z)$.

MINIMAL POLYNOMIAL OF TRIDIAGONAL MATRIX

In this section, with the help of the Jordan canonical form, we clarify the relationship of the characteristic polynomial of the tridiagonal matrix $T^{(n)}_1$ to the minimal one.

For simplicity, let us here adopt the following abbreviations for matrices $T^{(n)}_S$,

$$T_s = \begin{pmatrix} u_1 & v_1 & & \\ 1 & u_2 & \ddots & \\ & \ddots & \ddots & v_{s-1} \\ & & 1 & u_s \end{pmatrix}, \qquad s = 0, 1, \ldots, l, \tag{44}$$

where $l=m$ if $\lambda_N \neq 0$ or $l=m-m_N$ if $\lambda_N=0$. Let $p_0(z)=1$ and $p_s(z)=\det(zI_s-T_s)$ for $s=1,2,\ldots,l$. Then $p_l(z)$ is just the characteristic polynomial of T_l, namely,

$$\phi_T(z) = (z-\lambda_1)^{m_1}(z-\lambda_2)^{m_2}\cdots(z-\lambda_L)^{m_L}, \tag{45}$$

where $L=N$ if $\lambda_N \neq 0$ or $L=N-1$ if $\lambda_N=0$. The following proposition gives the Jordan canonical form of the tridiagonal matrix T_l.

Proposition 2. There exists a nonsingular matrix P such that

$$P^{-1}(T_l)^\top P = \hat{J}, \tag{46}$$

$$\hat{J} = diag(J_{1,1}, J_{2,1}, \ldots, J_{L,1}) \in \mathbb{C}^{l\times l}, \tag{47}$$

where $J_{1,1}, J_{2,1}, \ldots, J_{L,1}$ are of the same form as (23).

Proof.

The characteristic polynomials $p_0(z), p_1(z), \ldots, p_l(z)$ satisfy

$$\begin{cases} zp_0(z) = u_1 p_0(z) + p_1(z), \\ zp_s(z) = v_s p_{s-1}(z) + u_{s+1} p_s(z) + p_{s+1}(z), \\ \quad s = 1, 2, \ldots, l-1. \end{cases} \tag{48}$$

This is easily derived from the expansion of $\det(zI_s - T_s)$ by the sth row minors. By taking the 0th, the 1st, …, the (m_k-1)th derivatives with respect to z in (48), we get

$$\begin{cases} zD^i p_0(z) + iD^{i-1} p_0(z) = u_1 D^i p_0(z) + D^i p_1(z), \\ \quad i = 0, 1, \ldots, m_k - 1, \\ zD^i p_s(z) + iD^{i-1} p_s(z) \\ \quad = v_s D^i p_{s-1}(z) + u_{s+1} D^i p_s(z) + D^i p_{s+1}(z), \\ \quad i = 0, 1, \ldots, m_k - 1, \quad s = 1, 2, \ldots, l-1, \end{cases} \tag{49}$$

where $D^i p_s(z)$ denotes the ith derivative of $p_s(z)$ with respect to z. Let $\mathrm{p}_{k,i} = (D^i p_0(\lambda_k), D^i p_1(\lambda_k), \ldots, D^i p_{l-1}(\lambda_k))^\top \in C^l$. Then, by substituting $z=\lambda_k$ in (49) and by taking account that $D^i p_l(\lambda_k) = D^i(z-\lambda_1)^{m1}(z-\lambda_2)^{m2}\cdots(z-\lambda_l)ml|_z=\lambda_k=0$ for $i=0,1,\ldots,m_k-1$, we obtain

$$\lambda_k \boldsymbol{p}_{k,i} + i\boldsymbol{p}_{k,i-1} = (T_l)^\top \boldsymbol{p}_{k,i}, \quad i = 0, 1, \ldots, m_k - 1. \tag{50}$$

Moreover, it follows that

$$\begin{cases} (T_l)^\top P_{k,0} = \lambda_k P_{k,0}, \\ (T_l)^\top P_{k,i} = \lambda_k P_{k,i} + P_{k,i-1}, \\ \quad i = 1, 2, \ldots, m_k - 1, \end{cases} \tag{51}$$

where $P_{k,i} = (1/i!)\boldsymbol{p}_{k,i}$. Thus, by letting $\mathrm{P} = (\mathrm{P}_{1,0}\mathrm{P}_{1,1}\cdots\mathrm{P}_{1,m1-1}|\mathrm{P}_{2,0}\mathrm{P}_{2,1}\cdots\mathrm{P}_{2,m2-1}|\cdots|\mathrm{P}_{L,0}\mathrm{P}_{L,1}\cdots\mathrm{P}_{L,mL-1}) \in C^{l\times l}$, we have $(\mathrm{T}_l)^\top \mathrm{P} = \mathrm{P}\hat{J}$.

Here, it remains to prove that P is nonsingular. Of course, $P_{k,i} \neq O$ since the $(i+1)$th row of $P_{k,i}$ is $D^i p_i(\lambda_k)/i! = 1$. Let $W_{k,i} = \mathrm{Ker}((T_l)^\top - \lambda_k I_l)^i$ for $i=1,2,\ldots m_k - 1$, which indicates the generalized eigenspace of $(T_l)^\top$ corresponding to λ_k. Then it is obvious from (51) that $((T_l)^\top - \lambda_k I_l)P_{k,0} = O$and $P_{k,0} \in W_{k,1}$. Eq. (51) with $i=1$ also leads to that $((T_l)^\top - \lambda_k I_l)^2 P_{k,1} = O$ and $P_{k,1} \in W_{k,2}$. Simultaneously, it is observed that $P_{k,1} \notin W_{k,1}$. Let us assume that $P_{k,1} \in W_{k,1}$, namely, $(T_l)^\top P_{k,1} = \lambda_k P_{k,1}$. Then, from (51), we derive $P_{k,0} = O$, which contradicts with $P_{k,0} \neq O$. Thus, it follows that $P_{k,1} \notin W_{k,1}$. Similarly, by induction for $i=2,3,\ldots,m_k-1$ in $P_{k,i}$, we have

$$P_{k,i} \notin W_{k,1}, W_{k,2}, \ldots, W_{k,i}, \quad P_{k,i} \in W_{k,i+1},$$
$$i = 1, 2, \ldots, m_k - 1. \tag{52}$$

From (52), it turns out that $P_{k,i}$ for $i=0,1,\ldots,m_k-1$ and $k=1,2,\ldots,L$ are linearly independent. Therefore, it is concluded that P is nonsingular and the Jordan canonical form of $(T_l)^\top$ is given by (46).

Proposition 2 suggests that the minimal polynomial of $(T_l)^\top$ becomes

$$\psi_T(z) = (z - \lambda_1)^{m_1}(z - \lambda_2)^{m_2} \cdots (z - \lambda_L)^{m_L}, \tag{53}$$

which is equal to the characteristic polynomial of T_l in (45). If $m_1=m_2=\cdots=m_L=1$, then it is obvious that T_l is diagonalizable. Otherwise, T_l is not diagonalizable. This is because multiplicity of roots in minimal polynomial coincides with maximal size of the Jordan blocks. To sum up, we have a theorem for the properties of the tridiagonal matrix T_l.

Theorem 5. The minimal polynomial of T_l is equal to the characteristic one. Also, T_l is diagonalizable tridiagonal matrix if and only if it has no multiple eigenvalues.

PROCEDURE FOR CONSTRUCTING TRIDIAGONAL MATRIX AND ITS EXAMPLES

In this section, based on the discussions in the previous sections, we first design a procedure for constructing a tridiagonal matrix with specified multiple eigenvalues. We next give four kinds of examples for demonstrating that the resulting procedure can provide with tridiagonal matrices with multiple eigenvalues. Examples have been carried out with our computer with OS: Mac OS X 10.8.5, CPU: Intel Core i7 2 GHz, RAM: 8 GB. We also use the scientific computing software Wolfram Mathematica 9.0. In every example, all the entries of $\boldsymbol{u}$ are simply set to 1 and those of $\boldsymbol{w}$ are not artificial. The readers will realize that the settings of $\boldsymbol{u}$ and $\boldsymbol{w}$ are not so difficult for satisfying (6) and (27).

Let us here consider the relationship of five theorems in the previous sections. Theorem 2shows that the eigenvalues of $T^{(n)}_1$ in the tridiagonal form as (10) are equal to the poles of the generating function $F(z)$ and the multiplicity of the eigenvalues coincide with the those of the poles of $F(z)$. Theorems 3 and 4 claim that the minimal polynomial of a general matrix A, denoted by $\psi_A(z)$ is just the denominator of $F(z)$ involving $f_n=\boldsymbol{w}^H A^n \boldsymbol{u}$, and it coincides with the characteristic polynomial of $T^{(n)}_1$ denoted by

$\phi_T(z)$, except for the factor corresponding to zero-eigenvalues. With the help of Theorem 1, we thus realize that the nonzero eigenvalues of $T^{(0)}_1$ with the entries involving $q_1^{(0)}, q_2^{(0)}, \ldots, q_l^{(0)}$ and $e_1^{(0)}, e_2^{(0)}, \ldots, e_{l-1}^{(0)}$ become roots of the minimal polynomial $\psi_A(z)$ in the case where $q_1^{(0)}, q_2^{(0)}, \ldots, q_l^{(0)}$ and $e_1^{(0)}, e_2^{(0)}, \ldots, e_{l-1}^{(0)}$ are given by the qd formula (4) under the initial settings $e^{(n)}_0=0$ and $q^{(n)}_1=f_{n+1}/f_n$ with $f_n=\boldsymbol{w}^H A^n \boldsymbol{u}$. See also Figure 1 for the diagram for getting $q^{(n)}_s$ and $e^{(n)}_s$ by the qd formula (4). A procedure for constructing $T= T^{(0)}_1$ with the same nonzero eigenvalues as A is therefore as follows.

1. Set $l=m$ if $\lambda_N \neq 0$ or $l=m-m_N$ if $\lambda_N=0$.
2. Choose $\boldsymbol{u}$ and $\boldsymbol{w}$ as in (6) and (27).
3. Compute $f_n=\boldsymbol{w}^H A^n \boldsymbol{u}$ for $n=0,1,\ldots,2l-1$.
4. Set $e^{(n)}_0=0$ for $n=0,1,\ldots,2l-3$.
5. Compute $q^{(n)}_1=f_{n+1}/f_n$ for $n=0,1,\ldots,2l-2$.
6. Repeat (a) and (b) for $s=2,3,\ldots,l$.
   (a) Compute $e_{s-1}^{(n)} = q_{s-1}^{(n+1)} + e_{s-2}^{(n+1)} - q_{s-1}^{(n)}$ for $n=0,1,\ldots,2l-2s+1$.
   (b) Compute $q_s^{(n)} = q_{s-1}^{(n+1)} e_{s-1}^{(n+1)} / e_{s-1}^{(n)}$ for $n=0,1,\ldots,2l-2s$.
7. Construct a tridiagonal matrix by arranging $q_1^{(0)}, q_2^{(0)}, \ldots, q_l^{(0)}$ and $e_1^{(0)}, e_2^{(0)}, \ldots, e_{l-1}^{(0)}$.

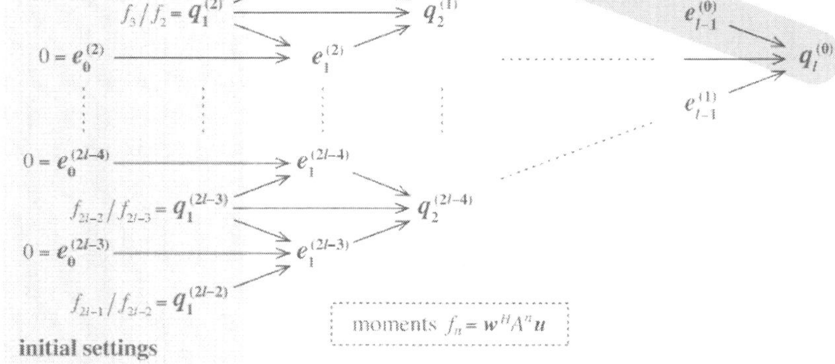

Figure 1: The qd diagram for a tridiagonal matrix construction.

According to Theorem 5, the minimal and the characteristic polynomials of the resulting tridiagonal matrix T are equal to each other. Moreover, T is diagonalizable if and only if it has no multiple eigenvalues.

It is necessary to control the eigenvalues of A for getting T as a tridiagonal matrix with specified eigenvalues. It is easy to specify the eigenvalues of the diagonal matrix and those of the Jordan matrix.

First, in the procedure, let us consider the case where

$A=\mathrm{diag}(2,2,2,1,1,1)\in R^{6\times 6}$

which is a diagonal matrix with two eigenvalues 1 and 2 each of multiplicity 3. Obviously, the characteristic and the minimal polynomials are factorized as $(z-1)^3(z-2)^3$ and $(z-1)(z-2)$, respectively. So, the integers l and m are immediately determined as $l=2$ and $m=6$. Moreover, by letting $\boldsymbol{u}=(1,1,1,1,1,1)^\top$ and $\boldsymbol{w}=(1,1,1,1,1,1)^\top$, we derive a tridiagonal matrix as

$$T = \begin{pmatrix} \frac{3}{2} & \frac{1}{4} \\ 1 & \frac{3}{2} \end{pmatrix} \in \mathbb{R}^{2\times 2}$$

whose characteristic and minimal polynomials are both factorized as $(z-1)(z-2)$. The tridiagonal matrix T is a diagonalizable matrix with the distinct eigenvalues 1 and 2.

Next, let us adopt a bidiagonal matrix, which can be regarded as the Jordan matrix, with eigenvalues 2 of multiplicity 6 as A, namely,

$$A = \begin{pmatrix} 2 & 1 & & & & \\ & 2 & 1 & & & \\ & & 2 & 1 & & \\ & & & 2 & 1 & \\ & & & & 2 & 1 \\ & & & & & 2 \end{pmatrix} \in \mathbb{R}^{6\times 6},$$

in the procedure. Since the characteristic polynomial of A is equal to the minimal one, the integers l and m are determined as $l=m=6$. Then the procedure with $\boldsymbol{u}=(1,1,1,1,1,1)^\top$ and $\boldsymbol{w}=(1,1,0,1,0,1)^\top$ constructs a tridiagonal matrix, which can not be symmetrized,

$$T = \begin{pmatrix} \frac{11}{4} & \frac{3}{16} & & & & \\ 1 & \frac{11}{12} & -\frac{4}{9} & & & \\ & 1 & \frac{10}{3} & 3 & & \\ & & 1 & 0 & -8 & \\ & & & 1 & \frac{29}{8} & -\frac{1}{64} \\ & & & & 1 & \frac{11}{8} \end{pmatrix} \in \mathbb{R}^{6\times 6}.$$

The characteristic and the minimal polynomials of A and T are all the same polynomial with respect to z, which is factored as $(z-2)^6$. So, the tridiagonal matrix T is not diagonalizable.

Let us prepare the Jordan matrix

$$A = \begin{pmatrix} 3 & 1 & & & & & & \\ & 3 & 1 & & & & & \\ & & 3 & & & & & \\ & & & 3 & 1 & & & \\ & & & & 3 & & & \\ & & & & & 3 & & \\ & & & & & & 2 & 1 \\ & & & & & & & 2 \end{pmatrix} \in \mathbb{R}^{8\times 8}.$$

The matrix A has multiple eigenvalues such as $\lambda_1=3$, $\lambda_2=3$, $\lambda_3=3$, $\lambda_4=3$, $\lambda_5=3$, $\lambda_6=3$, $\lambda_7=2$, $\lambda_8=2$. It is noted that $|\lambda_1|=|\lambda_2|=|\lambda_3|=|\lambda_4|=|\lambda_5|=|\lambda_6|>|\lambda_7|=|\lambda_8|>0$. The characteristic and the minimal polynomials of A are factorized as $(z-2)^2(z-3)^6$ and $(z-2)^2(z-3)^3$, respectively. So, let $l=5$ and $m=8$ in the procedure. Then, the settings $\boldsymbol{u}=(1,1,1,1,1,1,1,1)^\top$ and $\boldsymbol{w}=(1,1,1,1,1,1,1,1)^\top$ bring us to a tridiagonal matrix, which can not be symmetrized,

$$T = \begin{pmatrix} \frac{13}{4} & \frac{1}{16} & & & \\ 1 & \frac{17}{4} & -\frac{13}{2} & & \\ & 1 & -\frac{11}{26} & \frac{116}{169} & \\ & & 1 & \frac{1232}{377} & -\frac{13}{841} \\ & & & 1 & \frac{77}{29} \end{pmatrix} \in \mathbb{R}^{5\times 5}$$

whose characteristic and minimal polynomials are both factorized as $(z-2)^2(z-3)^3$, which is just equal to the minimal one of A. The tridiagonal matrix T is not a diagonalizable matrix with eigenvalues 2 and 3 of multiplicity 2 and 3, respectively.

Finally, let us A be set as the Jordan matrix with complex eigenvalues $2+i$ and $2-i$ each of multiplicity 2 and distinct real eigenvalues 1 and 2, namely,

$$A = \begin{pmatrix} 2+i & 1 & & & & \\ & 2+i & & & & \\ & & 2-i & 1 & & \\ & & & 2-i & & \\ & & & & 2 & \\ & & & & & 1 \end{pmatrix} \in \mathbb{C}^{6\times 6},$$

in this procedure. By taking account that the characteristic and the minimal polynomials of $A$are equal to each other, let $l=m=6$ in the procedure. Under the settings $\boldsymbol{u}=(1,1,1,1,1,1)^{\top}$ and $\boldsymbol{w}=(1,1,1,1,1,1)^{\top}$, the resulting matrix T is a real tridiagonal matrix, which can not be symmetrized,

$$T = \begin{pmatrix} \frac{13}{6} & -\frac{19}{36} & & & & \\ 1 & \frac{443}{114} & -\frac{1920}{361} & & & \\ & 1 & -\frac{363}{760} & -\frac{209}{1600} & & \\ & & 1 & \frac{1187}{440} & -\frac{240}{121} & \\ & & & 1 & \frac{37}{66} & \frac{11}{36} \\ & & & & 1 & \frac{13}{6} \end{pmatrix} \in \mathbb{R}^{6\times 6}.$$

The characteristic and the minimal polynomials of A and $T$are all the same polynomial with respect to z, which is factorized as $(z-2+i)^2(z-2-i)^2(z-2)(z-1)$. So, the tridiagonal matrix T is not a diagonalizable matrix with the same complex multiple eigenvalues and real distinct ones as A.

CONCLUSION

In this paper, we clarify that the qd recursion formula is applicable to constructing a tridiagonal matrix with specified multiple eigenvalues. We first investigate the denominator of the generating function associated with the sequence given from two suitable vectors and the powers of a general matrix A, through considering the Jordan canonical form of A. Accordingly, it is observed that the minimal polynomial of A coincides with the characteristic polynomial of a tridiagonal matrix T, denoted by $\phi_T(z)$, or the polynomial $z^{mL}\phi T(z)$ for the multiplicity m_L of the zero-eigenvalues of A. Next, by taking account of the Jordan canonical form of T, we show that the characteristic and the minimal polynomials of T are equal to each other.

We finally present a procedure for constructing a tridiagonal matrix with specified multiple eigenvalues, and then give four examples for the resulting procedure.

ACKNOWLEDGEMENTS

The authors would like to thank the reviewer for his/her careful reading and beneficial suggestions. This work is supported by JSPS KAKENHI Grant Number 23654032.

REFERENCES

1. Chu, MT., Golub, GH.: Inverse Eigenvalue Problem: Theory, Algorithms, and Applications. Oxford University Press, New York (2005).
2. Gutknecht, MH., Parlett, BN.: From qd to LR, or, how were the qd and LR algorithms discovered?IMA J. Numer. Anal. 31, 741–754 (2011).
3. Henrici, P., Watkins, BO.: Finding zeros of a polynomial by the Q-D algorithm. Commut. ACM. 8, 570–574 (1965).
4. Henrici, P.: Applied and, Computational Complex Analysis, Vol. 1. John Wiley, New York (1974).
5. Henrici, P.: Applied and Computational Complex Analysis, Vol. 2. John Wiley, New York (1977).
6. Meyer, CD.: Matrix Analysis and Applied Linear Algebra. SIAM, Philadelphia (2000).
7. Parlett, BN.: The new qd algorithms. Acta Numerica. 4, 459–491 (1995).
8. Rutishauser, H.: Bestimmung der Eigenwerte und Eigenvektoren einer Matrix mit Hilfe des Quotienten-Differenzen-Algorithmus. Z. Angew. Math. Phys. 6, 387–401 (1955).
9. Rutishauser, H.: Lectures on Numerical Mathematics. Birkhäuser, Boston (1990).

SECTION III

CHAPTER 6

STABILITY ANALYSIS OF ADDITIVE RUNGE-KUTTA METHODS FOR DELAY-INTEGRO-DIFFERENTIAL EQUATIONS

Hongyu Qin,[1] Zhiyong Wang,[2] Fumin Zhu,[3] and Jinming Wen[4]

[1]Wenhua College, Wuhan 430074, China

[2]School of Mathematical Sciences, University of Electronic Science and Technology of China, Sichuan 611731, China

[3]College of Economics, Shenzhen University, Shenzhen 518060, China

[4]Department of Electrical and Computer Engineering, University of Toronto, Toronto, M5S3G4, Canada

ABSTRACT

This paper is concerned with stability analysis of additive Runge-Kutta methods for delay-integro-differential equations. We show that if the

Citation (APA): Qin, H., Wang, Z., Zhu, F., & Wen, J. (2018). Stability Analysis of Additive Runge-Kutta Methods for Delay-Integro-Differential Equations. International Journal of Differential Equations, 2018. (5 pages).

additive Runge-Kutta methods are algebraically stable, the perturbations of the numerical solutions are controlled by the initial perturbations from the system and the methods.

INTRODUCTION

Spatial discretization of many nonlinear parabolic problems usually gives a class of ordinary differential equations, which have the stiff part and the nonstiff part; see, e.g., [1–5]. In such cases, the most widely used time-discretizations are the special organized numerical methods, such as the implicit-explicit numerical methods [6, 7], the additive Runge-Kutta methods [8–12], and the linearized methods [13, 14]. When applying the split numerical methods to numerically solve the equations, it is important to investigate the stability of the numerical methods.

In this paper, it is assumed that the spatial discretization of time-dependent partial differential equations yields the following nonlinear delay-integro-differential equations:

$$\begin{aligned} y'(t) &= f^{[1]}(t, y(t)) \\ &\quad + f^{[2]}\left(t, y(t), y(t-\tau), \int_{t-\tau}^{t} g(t, s, y(s))\, ds\right), \\ &\qquad t > 0, \\ y(t) &= \psi(t), \quad -\tau \le t \le 0. \end{aligned} \tag{1}$$

Here τ is a positive delay term, (t) is continuous, $f^{[1]}$: $[t_0, +\infty]\times X \rightarrow X$, and $f^{[2]}$: $[t_0,+\infty]\times X\times X\times X \rightarrow X$, such that problem (1) owns a unique solution, where X is a real or complex Hilbert space. Particularly, when $g \equiv 0$, problem (1) is reduced to the nonlinear delay differential equations. When the delay term $\tau = 0$, problem (1) is reduced to the ordinary differential equations.

The investigation on stability analysis of different numerical methods for problem (1) has fascinated generations of researchers. For example, Torelli [15] considered stability of Euler methods for the nonautonomous nonlinear delay differential equations. Hout [16] studied the stability of Runge-Kutta methods for systems of delay differential equations. Baker and Ford [17] discussed stability of continuous Runge-Kutta methods for integrodifferential systems with unbounded delays. Zhang and Vandewalle [18] discussed the stability of the general linear methods for integrodifferential equations with memory. Li and Zhang obtained the stability and convergence of the discontinuous Galerkin methods for nonlinear delay differential equations [19, 20]. More references for this topic can be found in [21–30]. However, few works have been found on the stability of splitting methods for the proposed methods.

In the present work, we present the additive Runge-Kutta methods with some appropriate quadrature rules to numerically solve the nonlinear delay-integrodifferential equations (1). It is shown that if the additive Runge-Kutta methods are algebraically stable, the obtained numerical solutions are globally and asymptotically stable under the given assumptions, respectively. The rest of the paper is organized as follows. In Section 2, we present the numerical methods for problems (1). In Section 3, we consider stability analysis of the numerical schemes. Finally, we present some extensions in Section 4.

THE NUMERICAL METHODS

In this section, we present the additive Runge-Kutta methods with the appropriate quadrature rules to numerically solve problem (1).

The coefficients of the additive Runge-Kutta methods can be organized in Buther tableau as follows (cf. [31]):

$$\begin{array}{c|c|c} c & A^{[1]} & A^{[2]} \\ \hline & \left(b^{[1]}\right)^T & \left(b^{[2]}\right)^T \end{array}, \tag{2}$$

where $c = [c_1, \cdots, c_s]^T$, $b^{[k]} = [b^{[k]}_1, \cdots, b^{[k]}_s]^T$, and $A^{[k]} = (a^{[k]}_{ij})^s_{i,j=1}$ for $k = 1,2$.

Then, the presented ARKMs for problem (1) can be written by

$$y_{n+1} = y_n + h\sum_{j=1}^{s} b_j^{[1]} f^{[1]}\left(t_n + c_j h, y_j^{(n)}\right)$$

$$+ h\sum_{j=1}^{s} b_j^{[2]} f^{[2]}\left(t_n + c_j h, y_j^{(n)}, \tilde{y}_j^{(n)}\right),$$

$$y_i^{(n)} = y_n + h\sum_{j=1}^{s} a_{ij}^{[1]} f^{[1]}\left(t_n + c_j h, y_j^{(n)}\right)$$

$$+ h\sum_{j=1}^{s} a_{ij}^{[2]} f^{[2]}\left(t_n + c_j h, y_j^{(n)}, y_j^{(n-m)}, \widetilde{y}_j^{(n)}\right),$$

$$i = 1, 2, \cdots, s, \tag{3}$$

where y_n and $y(n)_i$ are approximations to the analytic solution $y(t_n)$ and $y(t_n + c_i h)$, respectively, $y_n = \psi(t_n)$ for $n \leq 0$, $y^{(n)}{}_i = \psi(t_n+c_ih)$ for $t_n+c_ih \leq 0$, and $\widetilde{y}_i^{(n)}$ denotes the approximation to $\int_{t_n+c_ih-\tau}^{t_n+c_ih} g(t_n + c_ih, \xi, y(\xi))d\xi$, which can be computed by some appropriate quadrature rules

$$\widetilde{y}_i^{(n)} = h\sum_{k=0}^{m} p_k g\left(t_n + c_i h, t_{n-k} + c_i h, y_i^{(n-k)}\right),$$

$$i = 1, 2, \cdots, s. \tag{4}$$

For example, we usually adopt the repeated Simpson's rule or Newton-Cotes rule, etc., according to the requirement of the convergence of the method (cf. [18]).

STABILITY ANALYSIS

In this section, we consider the numerical stability of the proposed methods. First, we introduce a perturbed problem, whose solution satisfies

$$z'(t) = f^{[1]}(t, z(t)) + f^{[2]}\left(t, z(t), z(t-\tau), \int_{t-\tau}^{t} g(t, s, z(s))\, ds\right), \quad t > 0,$$

$$y(t) = \phi(t), \quad -\tau \leq t \leq 0. \tag{5}$$

It is assumed that there exist some inner product $< \cdot, \cdot >$ and the induced norm $\|\cdot\|$ such that

$$\mathrm{Re}\left\langle y-z, f^{[1]}(t,y)-f^{[1]}(t,z)\right\rangle \le \alpha \|y-z\|^2,$$

$$\mathrm{Re}\left\langle y-z, f^{[2]}(t,y,u_1,v_1)-f^{[2]}(t,z,u_2,v_2)\right\rangle$$

$$\le \beta_1 \|y-z\|^2+\beta_2\|u_1-u_2\|^2+\gamma\|v_1-v_2\|^2,$$

$$\|g(t,v,s_1)-g(t,v,s_2)\| \le \theta\|s_1-s_2\|, \tag{6}$$

where $\alpha < 0$, $\beta_1 < 0$, $\beta_2 > 0$, $\gamma > 0$, and $\theta > 0$ are constants. It is remarkable that the conditions can be equivalent to the assumptions in [32, 33] (see. [34] *Remark* 2.1).

Definition 1 (cf. [9]). An additive Runge-Kutta method is called algebraically stable if the matrices

$$B_\nu := \mathrm{diag}\left(b_1^{[\nu]},\cdots,b_s^{[\nu]}\right), \quad \nu = 1,2,$$

$$M_{\nu\mu} := B_\nu A^{[\mu]} + A^{[\nu]T}B_\mu - b^{[\nu]}b^{[\mu]T} \tag{7}$$

are nonnegative.

Theorem 2. *Assume an additive Runge-Kutta method is algebraically stable and* $\beta_1 + \beta_2 + 4\gamma\tau^2\eta^2\theta^2 < 0$, *where* $\eta = \max\{p_1, p_2, \cdots, p_k\}$. *Then, it holds that*

$$\|y_n - z_n\| \le \sqrt{\left(1+2\sum_{i=1}^{s}\tau b_i^{[2]}\beta_2+4\gamma\tau^2\eta^2\theta^2\right)} \cdot \max_{-\tau\le t\le 0}\|\psi(t)-\phi(t)\|, \tag{8}$$

where y_n *and* z_n *are numerical approximations to problems (1) and (5), respectively.*

Proof. Let $\{y_n, y_i^{(n)}, \overline{y}_i^{(n)})\}$ and $\{z_n, z_i^{(n)}, \overline{z}_i^{(n)})\}$ be two sequences of approximations to problems (1) and (5), respectively, by ARKMs with the same stepsize h and write

$$
\begin{aligned}
U_i^{(n)} &= y_i^{(n)} - z_i^{(n)}, \\
\widetilde{U}_i^{(n)} &= \widetilde{y}_i^{(n)} - \widetilde{z}_i^{(n)}, \\
U_0^{(n)} &= y_n - z_n, \\
W_i^{[1]} &= h\left[f^{[1]}\left(t_n + c_i^{[1]}h, y_i^{(n)}\right)\right. \\
&\quad \left. - f^{[1]}\left(t_n + c_i^{[1]}h, z_i^{(n)}\right)\right], \\
W_i^{[2]} &= h\left[f^{[2]}\left(t_n + c_i^{[2]}h, y_i^{(n)}, y_i^{(n-m)}, \widetilde{y}_i^{(n)}\right)\right. \\
&\quad \left. - f^{[2]}\left(t_n + c_i^{[2]}h, z_i^{(n)}, z_i^{(n-m)}, \widetilde{z}_i^{(n)}\right)\right].
\end{aligned} \tag{9}
$$

With the notation, the ARKMs for (1) and (5) yield

$$
U_0^{(n+1)} = U_0^n + \sum_{\mu=1}^{2}\sum_{j=1}^{s} b_j^{[\mu]} W_j^{[\mu]},
$$

$$
U_i^{(n)} = U_0^{(n)} + \sum_{\mu=1}^{2}\sum_{j=1}^{s} a_{ij}^{[\mu]} W_j^{[\mu]}, \quad i = 1, 2, \cdots, s. \tag{10}
$$

Thus, we have

$$
\begin{aligned}
\left\|U_0^{(n+1)}\right\|^2 &= \left\langle U_0^{(n)} + \sum_{\mu=1}^{2}\sum_{j=1}^{s} b_j^{[\mu]} W_j^{[\mu]}, U_0^{(n)} \right. \\
&\quad \left. + \sum_{\nu=1}^{2}\sum_{i=1}^{s} b_i^{[\nu]} W_i^{[\nu]} \right\rangle = \left\|U_0^{(n)}\right\|^2 + 2\sum_{\mu=1}^{2}\sum_{i=1}^{s} b_i^{[\mu]} \\
&\quad \cdot \operatorname{Re}\left\langle U_0^{(n)}, W_i^{[\mu]} \right\rangle + \sum_{\mu,\nu=1}^{2}\sum_{i,j=1}^{s} b_i^{[\mu]} b_j^{[\nu]} \left\langle W_i^{[\mu]}, W_j^{[\nu]} \right\rangle
\end{aligned}
$$

$$= \left\|U_0^{(n)}\right\|^2 + 2\sum_{\mu=1}^{2}\sum_{i=1}^{s} b_i^{[\mu]}$$

$$\cdot \operatorname{Re}\left\langle U_i^{(n)} - \sum_{\nu=1}^{2}\sum_{j=1}^{s} a_{ij}^{[\nu]} W_j^{[\nu]}, W_i^{[\mu]} \right\rangle$$

$$+ \sum_{\mu,\nu=1}^{2}\sum_{i,j=1}^{s} b_i^{[\mu]} b_j^{[\nu]} \left\langle W_i^{[\mu]}, W_j^{[\nu]} \right\rangle = \left\|U_0^{(n)}\right\|^2$$

$$+ 2\sum_{\mu=1}^{2}\sum_{i=1}^{s} b_i^{[\mu]} \operatorname{Re}\left\langle U_i^{(n)}, W_i^{[\mu]} \right\rangle$$

$$- \sum_{\mu,\nu=1}^{2}\sum_{i,j=1}^{s} \left(b_i^{[\mu]} a_{ij}^{[\nu]} + a_{ji}^{[\mu]} b_j^{[\nu]} - b_i^{[\mu]} b_j^{[\nu]}\right)$$

$$\cdot \left\langle W_i^{[\mu]}, W_j^{[\nu]} \right\rangle. \tag{11}$$

Since that the matrix M is a nonnegative matrix, we obtain

$$- \sum_{\mu,\nu=1}^{2}\sum_{i,j=1}^{s} \left(b_i^{[\mu]} a_{ij}^{[\nu]} + a_{ji}^{[\mu]} b_j^{[\nu]} - b_i^{[\mu]} b_j^{[\nu]}\right) \left\langle W_i^{[\mu]}, W_j^{[\nu]} \right\rangle$$

$$\leq 0. \tag{12}$$

Furthermore, by conditions (6), we find

$$\operatorname{Re}\left\langle U_i^{(n)}, W_i^{[1]} \right\rangle \leq \alpha h \left\|U_i^{(n)}\right\|^2, \tag{13}$$

and

$$\operatorname{Re}\left\langle U_i^{(n)}, W_i^{[2]} \right\rangle \leq \beta_1 h \left\|U_i^{(n)}\right\|^2 + \beta_2 h \left\|U_i^{(n-m)}\right\|^2$$

$$+ \gamma h \left\|\widetilde{U}_i^{(n)}\right\|^2. \tag{14}$$

Together with (11), (12), (13), and (14), we get

$$\left\|U_0^{(n+1)}\right\|^2 \le \left\|U_0^{(n)}\right\|^2 + 2\sum_{i=1}^{s} hb_i^{[1]}\alpha \left\|U_i^{(n)}\right\|^2$$

$$+ 2\sum_{i=1}^{s} hb_i^{[2]} \left(\beta_1 \left\|U_i^{(n)}\right\|^2 + \beta_2 \left\|U_i^{(n-m)}\right\|^2\right.$$

$$\left. + \gamma \left\|\widetilde{U}_i^{(n)}\right\|^2\right) \le \left\|U_0^{(n)}\right\|^2 + 2\sum_{i=1}^{s} hb_i^{[2]} \left(\beta_1 \left\|U_i^{(n)}\right\|^2\right.$$

$$\left. + \beta_2 \left\|U_i^{(n-m)}\right\|^2 + \gamma \left\|\widetilde{U}_i^{(n)}\right\|^2\right). \tag{15}$$

Note that

$$\left\|\widetilde{U}_i^{(n)}\right\|^2 = \left\| h\sum_{k=0}^{m} p_k \left[g\left(t_n + c_i h, t_{n-k} + c_i h, y_i^{n-k}\right)\right.\right.$$

$$\left.\left. - g\left(t_n + c_i h, t_{n-k} + c_i h, z_i^{n-k}\right)\right]\right\|^2 \le (m+1)$$

$$\cdot \eta^2 \theta^2 h^2 \sum_{k=0}^{m} \left\|U_i^{(n-k)}\right\|^2. \tag{16}$$

Then, we obtain

$$\left\|U_0^{(n+1)}\right\|^2 \le \left\|U_0^{(n)}\right\|^2 + 2\sum_{i=1}^{s} hb_i^{[2]} \left(\beta_1 \left\|U_i^{(n)}\right\|^2\right.$$

$$\left. + \beta_2 \left\|U_i^{(n-m)}\right\|^2 + \gamma (m+1)\eta^2\theta^2 h^2 \sum_{k=0}^{m} \left\|U_i^{(n-k)}\right\|^2 \right)$$

$$\le \left\|U_0^{(0)}\right\|^2 + 2\sum_{j=0}^{n}\sum_{i=1}^{s} hb_i^{[2]} \left(\beta_1 \left\|U_i^{(j)}\right\|^2\right.$$

$$\left. + \beta_2 \left\|U_i^{(j-m)}\right\|^2 + \gamma (m+1)\eta^2\theta^2 h^2 \sum_{k=0}^{m} \left\|U_i^{(j-k)}\right\|^2 \right)$$

$$\leq \left\|U_0^{(0)}\right\|^2 + 2\sum_{j=0}^{n}\sum_{i=1}^{s} hb_i^{[2]}\left(\beta_1\left\|U_i^{(j)}\right\|^2 + \beta_2\left\|U_i^{(j)}\right\|^2\right.$$

$$\left.+\gamma(m+1)^2h^2\eta^2\theta^2\left\|U_i^{(j)}\right\|^2\right)$$

$$+2\sum_{j=-m}^{-1}\sum_{i=1}^{s} hb_i^{[2]}\left(\beta_2\left\|U_i^{(j)}\right\|^2\right.$$

$$\left.+\gamma(m+1)^2h^2\eta^2\theta^2\left\|U_i^{(j)}\right\|^2\right) \leq \left\|U_0^{(0)}\right\|^2$$

$$+2\sum_{j=0}^{n}\sum_{i=1}^{s} hb_i^{[2]}\left(\beta_1+\beta_2+4\gamma\tau^2\eta^2\theta^2\right)\left\|U_i^{(j)}\right\|^2$$

$$+2\sum_{j=-m}^{-1}\sum_{i=1}^{s} hb_i^{[2]}\left(\beta_2+4\gamma\tau^2\eta^2\theta^2\right)\left\|U_i^{(j)}\right\|^2 \leq \left\|U_0^{(0)}\right\|^2$$

$$+2\sum_{j=-m}^{-1}\sum_{i=1}^{s} hb_i^{[2]}\left(\beta_2+4\gamma\tau^2\eta^2\theta^2\right)\left\|U_i^{(j)}\right\|^2 \leq \left\|U_0^{(0)}\right\|^2$$

$$+2\sum_{i=1}^{s} mhb_i^{[2]}\left(\beta_2+4\gamma\tau^2\eta^2\theta^2\right)\max_{-m\leq j\leq -1}\left\|U_i^{(j)}\right\|^2. \tag{17}$$

Hence,

$$\left\|U_0^{(n+1)}\right\|^2 \leq C\max_{-\tau\leq t\leq 0}\left\|\psi(t)-\phi(t)\right\|^2, \tag{18}$$

where $C = [(1+2\sum_{i=1}^{s}\tau b_i^{[2]}\beta_2 + 4\gamma\tau^2\eta^2\theta^2)]$. This completes the proof.

Theorem 3. *Assume an additive Runge-Kutta method is algebraically stable and* $\beta_1 + \beta_2 + 4\gamma\tau^2\eta^2\theta^2 < 0$. *Then, it holds that*

$$\lim_{n\to\infty}\left\|U_0^{(n)}\right\| = 0. \tag{19}$$

Proof. Similar to the proof of Theorem 2, it holds that

$$\left\|U_0^{(n+1)}\right\|^2$$

$$\leq \left\|U_0^{(0)}\right\|^2$$

$$+ 2\sum_{j=0}^{n}\sum_{i=1}^{s} hb_i^{[2]}\left(\beta_1 + \beta_2 + 4\gamma\tau^2\eta^2\theta^2\right)\left\|U_i^{(j)}\right\|^2$$

$$+ 2\sum_{j=-m}^{-1}\sum_{i=1}^{s} hb_i^{[2]}\left(\beta_2 + 4\gamma\tau^2\eta^2\theta^2\right)\left\|U_i^{(j)}\right\|^2. \tag{20}$$

Note that $\beta_1 + \beta_2 + 4\gamma\tau^2\eta^2\theta^2 < 0$ and $^{[2]}{}_i > 0$; we have

$$\lim_{n\to\infty}\sum_{i=1}^{s} b_i^{[2]}\left\|U_i^{(n)}\right\| = 0. \tag{21}$$

On the other hand,

$$\left\|W_i^{[1]}\right\| = \left\|h\left[f^{[1]}\left(t_n + c_i^{[1]}h, y_i^{(n)}\right)\right.\right.$$

$$\left.\left. - f^{[1]}\left(t_n + c_i^{[1]}h, z_i^{(n)}\right)\right]\right\| \leq L_1\left\|U_i^{(n)}\right\| \tag{22}$$

$$\left\|W_i^{[2]}\right\| = \left\|h\left[f^{[2]}\left(t_n + c_i^{[2]}h, y_i^{(n)}, y_i^{(n-m)}, \tilde{y}_i^{(n)}\right)\right.\right.$$

$$\left.\left. - f^{[2]}\left(t_n + c_i^{[2]}h, z_i^{(n)}, z_i^{(n-m)}, \tilde{z}_i^{(n)}\right)\right]\right\| \leq L_2\left(\left\|U_i^{(n)}\right\|\right.$$

$$\left. + \left\|U_i^{(n-m)}\right\| + \left\|\tilde{y}_i^{(n)} - \tilde{z}_i^{(n)}\right\|\right). \tag{23}$$

Now, in view of (10), (21), (22), and (23), we obtain

$$\lim_{n\to\infty}\left\|U_0^{(n)}\right\| = 0. \tag{24}$$

This completes the proof.

Remark 4. In [35], Yuan et al. also discussed nonlinear stability of additive Runge-Kutta methods for multidelay-integro-differential equations. However, the main results are different. The main reason is that the results in [35] imply that the perturbations of the numerical solutions tend to infinity when the time increase, while the stability results in present paper indicate that the perturbations of the numerical solutions are independent of the time. Besides, the asymptotical stability of the methods is also discussed in the present paper.

CONCLUSION

The additive Runge-Kutta methods with some appropriate quadrature rules are applied to solve the delay-integro-differential equations. It is shown that if the additive Runge-Kutta methods are algebraically stable, the obtained numerical solutions can be globally and asymptotically stable, respectively. In the future works, we will apply the methods to solve more real-world problems.

ACKNOWLEDGMENTS

This work is supported in part by the National Natural Science Foundation of China (71601125).

REFERENCES

1. V. Thomée, Galerkin Finite Element Methods for Parabolic Problems, Springer, Berlin, Germany, 1997.
2. J. Wu, Theory and Applications of Partial Functional-Differential Equations, Springer, New York, NY, USA, 1996.
3. J. R. Cannon and Y. Lin, "Non-classical H1 projection and Galerkin methods for non-linear parabolic integro-differential equations," Calcolo, vol. 25, pp. 187–201, 1988.
4. D. Li and J. Wang, "Unconditionally optimal error analysis of crank-nicolson galerkin fems for a strongly nonlinear parabolic system," Journal of Scientific Computing, vol. 72, no. 2, pp. 892–915, 2017.
5. B. Li and W. Sun, "Error analysis of linearized semi-implicit galerkin finite element methods for nonlinear parabolic equations," International Journal of Numerical Analysis & Modeling, vol. 10, no. 3, pp. 622–633, 2013.
6. U. M. Ascher, S. J. Ruuth, and B. T. Wetton, "Implicit-explicit methods for time-dependent partial differential equations," SIAM Journal on Numerical Analysis, vol. 32, no. 3, pp. 797–823, 1995.
7. G. Akrivis and B. Li, "Maximum norm analysis of implicit-explicit backward difference formulas for nonlinear parabolic equations," SIAM Journal on Numerical Analysis, 2017.
8. I. Higueras, "Strong stability for additive Runge-Kutta methods," SIAM Journal on Numerical Analysis, vol. 44, no. 4, pp. 1735–1758, 2006.
9. A. Araujo, "A note on B-stability of splitting methods," Computing and Visualization in Science, vol. 26, no. 2-3, pp. 53–57, 2004.
10. C. A. Kennedy and M. H. Carpenter, "Additive Runge-Kutta schemes for convection-diffusion-reaction equations," Applied Numerical Mathematics, vol. 44, no. 1-2, pp. 139–181, 2003.
11. T. Koto, "Stability of IMEX Runge-Kutta methods for delay differential equations," Journal of Computational and Applied Mathematics, vol. 211, pp. 201–212, 2008.
12. H. Liu and J. Zou, "Some new additive Runge-Kutta methods and their applications," Journal of Computational and Applied Mathematics, vol. 190, no. 1-2, pp. 74–98, 2006.

13. D. Li, C. Zhang, and M. Ran, "A linear finite difference scheme for generalized time fractional Burgers equation," Applied Mathematical Modelling, vol. 40, no. 11-12, pp. 6069–6081, 2016.
14. D. Li, J. Wang, and J. Zhang, "Unconditionally convergent L1-Galerkin FEMs for nonlinear time-fractional Schrödinger equations," SIAM Journal on Scientific Computing, vol. 39, no. 6, pp. A3067–A3088, 2017.
15. L. Torelli, "Stability of numerical methods for delay differential equations," Journal of Computational and Applied Mathematics, vol. 25, no. 1, pp. 15–26, 1989.
16. K. J. in't Hout, "Stability analysis of Runge-Kutta methods for systems of delay differential equations," IMA Journal of Numerical Analysis, vol. 17, no. 1, pp. 17–27, 1997.
17. C. T. Baker and A. Tang, "Stability analysis of continuous implicit Runge-Kutta methods for Volterra integro-differential systems with unbounded delays," Applied Numerical Mathematics, vol. 24, no. 2-3, pp. 153–173, 1997.
18. C. Zhang and S. Vandewalle, "General linear methods for Volterra integro-differential equations with memory," SIAM Journal on Scientific Computing, vol. 27, no. 6, pp. 2010–2031, 2006.
19. D. Li and C. Zhang, "Nonlinear stability of discontinuous Galerkin methods for delay differential equations," Applied Mathematics Letters, vol. 23, no. 4, pp. 457–461, 2010.
20. D. Li and C. Zhang, "L∞ error estimates of discontinuous Galerkin methods for delay differential equations," Applied Numerical Mathematics, vol. 82, pp. 1–10, 2014.
21. V. K. Barwell, "Special stability problems for functional differential equations," BIT, vol. 15, pp. 130–135, 1975.
22. A. Bellen and M. Zennaro, "Strong contractivity properties of numerical methods for ordinary and delay differential equations," Applied Numerical Mathematics, vol. 9, no. 3-5, pp. 321–346, 1992.
23. K. Burrage, "High order algebraically stable Runge-Kutta methods," BIT, vol. 18, no. 4, pp. 373–383, 1978.
24. K. Burrage and J. C. Butcher, "Nonlinear stability of a general class of differential equation methods," BIT, vol. 20, no. 2, pp. 185–203, 1980.

25. G. J. Cooper and A. Sayfy, "Additive Runge-Kutta methods for stiff ordinary differential equations," Mathematics of Computation, vol. 40, no. 161, pp. 207–218, 1983.
26. K. Dekker and J. G. Verwer, Stability of Runge-Kutta Methods for Stiff Nonlinear Differential Equations, North-Holland Publishing, Amsterdam, The Netherlands, 1984.
27. L. Ferracina and M. N. Spijker, "Strong stability of singly-diagonally-implicit Runge-Kutta methods," Applied Numerical Mathematics, vol. 58, no. 11, pp. 1675–1686, 2008.
28. K. J. in't Hout and M. N. Spijker, "The θ-methods in the numerical solution of delay differential equations," in The Numerical Treatment of Differential Equations, K. Strehmel, Ed., vol. 121, pp. 61–67, 1991.
29. M. Zennaro, "Asymptotic stability analysis of Runge-Kutta methods for nonlinear systems of delay differential equations," Numerische Mathematik, vol. 77, no. 4, pp. 549–563, 1997. ·
30. D. Li, C. Zhang, and W. Wang, "Long time behavior of non-Fickian delay reaction-diffusion equations," Nonlinear Analysis: Real World Applications, vol. 13, no. 3, pp. 1401–1415, 2012.
31. B. Garcia-Celayeta, I. Higueras, and T. Roldan, "Contractivity/monotonicity for additive Range-kutta methods: Inner product norms," Applied Numerical Mathematics, vol. 56, no. 6, pp. 862–878, 2006.
32. C. Huang, "Dissipativity of one-leg methods for dynamical systems with delays," Applied Numerical Mathematics, vol. 35, no. 1, pp. 11–22, 2000.
33. C. Zhang and S. Zhou, "Nonlinear stability and D-convergence of Runge-Kutta methods for delay differential equations," Journal of Computational and Applied Mathematics, vol. 85, no. 2, pp. 225–237, 1997.
34. C. Huang, S. Li, H. Fu, and G. Chen, "Nonlinear stability of general linear methods for delay differential equations," BIT Numerical Mathematics, vol. 42, no. 2, pp. 380–392, 2002.
35. H. Yuan, J. Zhao, and Y. Xu, "Nonlinear stability and D-convergence of additive Runge-Kutta methods for multidelay-integro-differential equations," Abstract and Applied Analysis, vol. 2012, Article ID 854517, 22 pages, 2012.

CHAPTER 7

A NUMERICAL METHOD FOR PARTIAL DIFFERENTIAL ALGEBRAIC EQUATIONS BASED ON DIFFERENTIAL TRANSFORM METHOD

Murat Osmanoglu and Mustafa Bayram

Department of Mathematical Engineering, Chemical and Metallurgical Faculty, Yildiz Technical University, Esenler 34210, Istanbul, Turkey

ABSTRACT

We have considered linear partial differential algebraic equations (LPDAEs) of the form $Au_t(t, x) + Bu_{xx}(t, x) + Cu(t, x) = f(t, x)$, which has at least one singular matrix of $A, B \in \mathbb{R}^{n\times n}$. We have first introduced a uniform differential time index and a differential space index. The initial conditions and boundary conditions of the given system cannot be prescribed for all components of the solution vector u here. To overcome this, we introduced

Citation (APA): Osmanoglu, M., & Bayram, M. (2013). A numerical method for partial differential algebraic equations based on differential transform method. In Abstract and Applied Analysis (Vol. 2013). Hindawi.(8 pages).

these indexes. Furthermore, differential transform method has been given to solve LPDAEs. We have applied this method to a test problem, and numerical solution of the problem has been compared with analytical solution.

INTRODUCTION

The partial differential algebraic equation was first studied by Marszalek. He also studied the analysis of the partial differential algebraic equations [1]. Lucht et al. [2–4] studied the numerical solution and indexes of the linear partial differential equations with constant coefficients. A study about characteristics analysis and differential index of the partial differential algebraic equations was given by Martinson and Barton [5, 6]. Debrabant and Strehmel investigated the convergence of Runge-Kutta method for linear partial differential algebraic equations [7].

There are numerous LPDAEs applications in scientific areas given, for instance, in the field of Navier-Stokes equations, in chemical engineering, in magnetohydrodynamics, and in the theory of elastic multibody systems [4, 8–12].

On the other hand, the differential transform method was used by Zhou [13] to solve linear and nonlinear initial value problems in electric circuit analysis. Analysis of nonlinear circuits by using differential Taylor transform was given by Köksal and Herdem [14]. Using one-dimensional differential transform, Abdel-Halim Hassan [15] proposed a method to solve eigenvalue problems. The two-dimensional differential transform methods have been applied to the partial differential equations [16–19]. The differential transform method extended to solve differential-difference equations by Arikoglu and Ozkol [20]. Jang et al. have used differential transform method to solve initial value problems [21]. The numerical solution of the differential-algebraic equation systems has been studied by using differential transform method [22, 23].

In this paper, we have considered linear partial differential equations with constant coefficients of the form

$$Au_t(t,x) + Bu_{xx}(t,x) + Cu(t,x) = f(t,x), \quad (t,x) \in J \times \Omega, \tag{1}$$

where $J = [0,\infty)$, $\Omega = [-l, l]$, $l > 0$, and $A, B, C \in \mathrm{R}^{n\times n}$. In (1) at least one of the matrices $A, B \in \mathrm{R}^{n\times n}$ should be singular. If A=0 or B=0, then (1) becomes ordinary differential equation or differential algebraic equation, so we assume that none of the matrices A or B is the zero matrix.

INDEXES OF PARTIAL DIFFERENTIAL ALGEBRAIC EQUATION

Let us consider (1), with initial values and boundary conditions given as follows:

$$u_j(t, \pm l) = 0 \quad \text{for } t \in J,$$

$$u_i(0, x) = g(x) \quad \text{for } x \in \Omega, \tag{2}$$

where $j \in M_{BC} \subseteq \{1, 2, \ldots, n\}$, M_{BC} is the set of indices of components of u for which boundary conditions can be prescribed arbitrarily, and $i \in M_{IC} \subseteq \{1, 2, \ldots, n\}$, M_{IC} is the set of indices of components of u for which initial conditions can be prescribed arbitrarily. The initial boundary value problem (IBVP) (1) has only one solution where a function u is a solution of the problem, if it is sufficiently smooth, uniquely determined by its initial values (IVs) and boundary values (BVs), and if it solves the LPDAE point wise. Definition of the indexes can be given using the following assumptions.

(i) Each component of the vectors u, ut, and f satisfy the following condition:

$$|y(t, x)| \leq Me^{\alpha t}, \quad \alpha \geq 0, \ t \geq 0, \tag{3}$$

where M and α are independent of t and x.

(ii) $(B, \xi A + C)$, $\mathrm{Re}(\xi) > \alpha$, called as the matrix pencil, is regular.

(iii) $(A, \mu kB + C)$ is regular for all k, where μk is an eigenvalue of the operator $\partial^2 / \partial x^2$ together with prescribed BCs.

(iv) The vector (t, x) and the initial vector $g(x)$ are sufficiently smooth.

If we use Laplace transform, from assumption (ii), (1) can be transformed into

$$Bu''_\xi(x) + (\xi A + C)\, u_\xi(x) = f_\xi(x) + Ag(x), \quad \mathrm{Re}(\xi) > \alpha, \tag{4}$$

if B is a singular matrix, then (4) is a DAE depending on the parameter ξ. To characterize M_{BC}, we introduce $j \in M^{(\xi)}{}_{BC} \subseteq \{1, 2, \ldots, n\}$ as the set of indices of components of $u\xi$ for which boundary conditions can be prescribed arbitrarily.

In order to define a spatial index, we need the Kronecker normal form of the DAE (4). Assumption (iii) guarantees that there are nonsingular matrices P_{L}, $Q_{L,\xi} \in \mathbb{C}^{n\times n}$ such that

$$P_{L,\xi}BQ_{L,\xi} = \begin{pmatrix} I_{m_1} & 0 \\ 0 & N_{L,\xi} \end{pmatrix},$$

$$P_{L,\xi}(\xi A + C)Q_{L,\xi} = \begin{pmatrix} R_{L,\xi} & 0 \\ 0 & I_{m_2} \end{pmatrix}, \quad (5)$$

where $RL,\xi \in C^{m1\times m2}$ and $NL,\xi \in R^{m2\times m2}$ is a nilpotent Jordan chain matrix with $m_1 + m_2 = n$. I_k is the unit matrix of order k. The Riesz index (or nilpotency) of NL,ξ is denoted by $]L,\xi$ (i.e. $N]L,\xi\ L,\xi = 0$, $N]L,\xi{-}1\ L,\xi = 0$/).

Here, we will assume that there is a real number $\alpha* \geq \alpha$ such that the index set $M^{(\xi)}{}_{BC}$ is independent of the Laplace parameter ξ, provided $\text{Re}(\xi) \geq \alpha^*$.

Definition 1. Let $\alpha^* \in R^+$ be a number with $\alpha^* \geq \alpha$, such that for all $\xi \in C$ with $\text{Re}(\xi) \geq \alpha^*$

- the matrix pencil $(B, \xi A + C)$ is regular,
- $M^{(\xi)}{}_{BC}$ is independent of ξ, i.e., $M^{(\xi)}{}_{BC} = M_{BC}$,
- the nilpotency of $N_{L,\xi}$ is $]L \geq 1$.

Then $\nu_{d,} = 2\nu_L - 1$ is called the "differential spatial index" of the LPDAE. If $\nu_L = 0$, then the differential spatial index of LPDAE is defined to be zero.

If we use Fourier transform, (1) can be transformed into

$$A\hat{u}_k'(t) + (\mu_k B + C)\hat{u}_k(t) = \hat{f}_k(t) + B\rho_k(t) \quad (6)$$

with $\rho_k(t) = (\rho_{k1}(t), \ldots, \rho_{kn}(t))^T$ and

$$\rho_{ki}(t) = 0 \quad \text{for } i \in \mathfrak{M}_{BC},$$

$$\rho_{kj}(t) = \frac{1}{l}\left[\phi_k'(x)u_j(t,x) - \phi_k(x)u_{x,j}(t,x)\right]_{x=-l}^{x=l} \quad (7)$$

for $j \notin M_{BC}$, which results from partial integration of the term $\int^l_{-l} u_{xx}(t, x)\phi_k(x)dx$.

If A is a singular matrix, then (6) is a DAE depending on the parameter μ_k which can be solved uniquely with suitable ICs under the assumptions (iv) and (v). Analogous to the case of the Laplace transform, the above assumption (iv) implies that there exist regular matrices $P_{F,}$ $Q_{F,k}$ such that

$$P_{F,k}AQ_{F,k} = \begin{pmatrix} I_{n_1} & 0 \\ 0 & N_{F,k} \end{pmatrix},$$

$$P_{F,k}\left(\mu_k B + C\right)Q_{F,k} = \begin{pmatrix} R_{F,k} & 0 \\ 0 & I_{n_2} \end{pmatrix}. \tag{8}$$

With $R_{F,} \in \mathrm{R}^{n1\times n1}$. $NF, \in \mathrm{R}^{n2\times n2}$ is again a nilpotent Jordan chain matrix with Riesz index $\nu_{F,k}$, where $n_1 + n_2 = n$.

To characterize M_{IC}, we introduce $\mathrm{M}^{(k)}{}_{IC} \subseteq \{1, 2, \ldots, n\}$ as the set of indices of components of $\hat{u}_k$ for which initial conditions can be prescribed arbitrarily. Therefore, we always assume in the context of a Fourier analysis of u that $\mathrm{M}^{(k)}{}_{IC}$ is independent of $k \in \mathrm{N}+$, i.e., $\mathrm{M}^{(k)}{}_{IC} = \mathrm{M}_{IC}$.

Definition 2. Assume for $k = 1, 2, \ldots$ tha

(1) the matrix pencil $(A, \mu_k B + C)$ is regular,
(2) $\mathrm{M}^{(k)}{}_{IC}$ is independent of k, i.e., $\mathrm{M}^{(k)}{}_{IC} = \mathrm{M}_{IC}$,
(3) the nilpotency of $N_{F,k}$ is $\nu_{F,k} = \nu_{\mathrm{F}}$

Then the PDAE (1) is said to have uniform differential time index $\nu\, d$, $= \nu_F$.

The differential spatial and time indexes are used to decide which initial and boundary values can be taken to solve the problem.

TWO-DIMENSIONAL DIFFERENTIAL TRANSFORM METHOD

The two-dimensional differential transform of function (x, y) is defined as

$$W(k,h) = \frac{1}{k!h!}\left[\frac{\partial^{k+h} w(x,y)}{\partial x^k \partial y^h}\right]_{\substack{x=0\\y=0}}, \tag{9}$$

where it is noted that upper case symbol $W(k, h)$ is used to denote the two-dimensional differential transform of a function represented by a corresponding lower case symbol $w(x, y)$. The differential inverse transform of (k, h) is defined as

$$w(x,y) = \sum_{k=0}^{\infty}\sum_{h=0}^{\infty} W(k,h)\, x^k y^h. \tag{10}$$

From (9) and (10), we obtain

$$w(x,y)=\sum_{k=0}^{\infty}\sum_{h=0}^{\infty}\frac{x^k y^h}{k!h!}\left[\frac{\partial^{k+h}w(x,y)}{\partial x^k \partial y^h}\right]_{\substack{x=0\\y=0}}. \tag{11}$$

The concept of two-dimensional differential transform is derived from two-dimensional Taylor series expansion, but the method doesn't evaluate the derivatives symbolically.

Theorem 3. Differential transform of the function $(x, y) = u(x, y) \pm V(x, y)$ is

$$W(k, h) = U(k, h) \pm V(k, h), \tag{12}$$

see [17].

Theorem 4. Differential transform of the function $(x, y) = \lambda u(x, y)$ is

$$W(k, h) = \lambda U(k, h), \tag{13}$$

see [17].

Theorem 5. Differential transform of the function $(x, y) = \partial u(x, y)/\partial x$ is

$$W(k, h) = (k+1)\, U(k+1, h), \tag{14}$$

see [17].

Theorem 6. Differential transform of the function $(x, y) = \partial u(x, y)/\partial y$ is

$$W(k, h) = (h+1)\, U(k, h+1), \tag{15}$$

see [17].

Theorem 7. Differential transform of the function $(x, y) = \partial^{r+s}u(x, y)/\partial x^r \partial y^s$ is

$$W(k, h) = (k+1)(k+2)\cdots(k+r)(h+1) \times (h+2)\cdots(h+s)\, U(k+r, h+s), \tag{16}$$

see [17].

Theorem 8. Differential transform of the function $(x, y) = u(x, y) \cdot V(x, y)$ is

$$W(k,h)=\sum_{r=0}^{k}\sum_{s=0}^{h}U(r,h-s)\,V(k-r,s), \tag{17}$$

see [17].

Theorem 9. Differential transform of the function $(x, y) = x^m y^n$ is

$$W(k, h) = \delta(k-m, h-n) = \delta(k-m)\,\delta(h-n), \tag{18}$$

see [17], where

$$\delta(k-m) = \begin{cases} 1, & k = m \\ 0, & k \neq m, \end{cases}$$

$$\delta(h-n) = \begin{cases} 1, & h = n \\ 0, & h \neq n. \end{cases} \tag{19}$$

Theorem 10. Differential transform of the function $(x, y) = g(x + a, y)$ is

$$W(k,h) = \sum_{p=k}^{N} \binom{p}{k} a^{p-k} G(p,h). \tag{20}$$

Proof. From Definition 1, we can write

$$w(x,y) = \sum_{h=0}^{\infty} \sum_{k=0}^{\infty} G(k,h)(x+a)^k y^h$$

$$= G(0,0) + G(0,1)\,y + aG(1,0) + G(1,0)\,x$$

$$+ G(0,2)\,y^2 + G(1,1)\,xy + aG(1,1)\,y$$

$$+ a^2G(2,0) + 2aG(2,0)\,x + G(2,0)\,x^2$$

$$+ G(0,3)\,y^3 + G(1,2)\,xy^2 + aG(1,2)\,y^2$$

$$+ G(2,1)\,x^2y + 2aG(2,1)\,xy + a^2G(2,1)\,y$$

$$+ a^3G(3,0) + 3a^2G(3,0)\,x + 3aG(3,0)\,x^2$$

$$+ G(3,0)\,x^3 + \cdots,$$

$$w(x,y) = \left[G(0,0) + aG(1,0)\right.$$

$$\left. + a^2G(2,0) + a^3G(3,0) + \cdots\right]$$

$$+ x\left[G(1,0) + 2aG(2,0) + 3a^2G(3,0) + \cdots\right]$$

$$+ y\left[G(0,1) + aG(1,1) + a^2G(2,1) + \cdots\right]$$

$$+ x^2\left[G(2,0) + 3aG(3,0) + \cdots\right]$$

$$+ xy\left[G(1,1) + 2aG(2,1) + \cdots\right]$$

$$+ y^2\left[G(0,2) + aG(1,2) + \cdots\right] + \cdots$$

$$= \sum_{p=0}^{\infty} a^p G(p,0) + x \sum_{p=1}^{\infty} p a^{p-1} G(p,0)$$

$$+ y \sum_{p=0}^{\infty} a^p G(p,1) + x^2 \sum_{p=2}^{\infty} \frac{p!}{(p-2)!2!}$$

$$\times a^{p-2} G(p,0) + xy \sum_{p=1}^{\infty} p a^{p-1} G(p,1) + \cdots, \tag{21}$$

where

$$w(x,y) = \sum_{h=0}^{\infty} \sum_{k=0}^{\infty} \sum_{p=k}^{\infty} \binom{p}{k} a^{p-k} G(p,h) x^k y^h \tag{22}$$

hence,

$$W(k,h) = \sum_{p=k}^{N} \binom{p}{k} a^{p-k} G(p,h). \tag{23}$$

Theorem 11. Differential transform of the function $(x, y) = g(x + a, y + b)$ is

$$W(k,h) = \sum_{p=k}^{N} \sum_{q=h}^{N} \binom{q}{h} \binom{p}{k} a^{p-k} b^{q-h} G(p,q). \tag{24}$$

Proof. From Definition 2, we can write

$$w(x,y) = \sum_{h=0}^{\infty} \sum_{k=0}^{\infty} G(k,h)(x+a)^k (y+b)^h$$

$$= G(0,0) + G(1,0)x + aG(1,0)$$

$$+ G(0,1)y + bG(0,1) + G(2,0)x^2$$

$$+ 2aG(2,0)x + a^2 G(2,0) + G(1,1)xy$$

$$+ bG(1,1)x + aG(1,1)y + abG(1,1)$$

$$+ G(0,2)y^2 + 2bG(0,2)y + b^2 G(0,2)$$

$$+ G(3,0)x^3 + 3aG(3,0)x^2 + 3a^2 G(3,0)x$$

$$+ a^3 G(3,0) + G(2,1)x^2 y + \cdots$$

$$
\begin{aligned}
w(x,y) = &\left[G(0,0) + aG(1,0) + bG(0,1) + a^2G(2,0) + abG(1,1) + \cdots\right] \\
&+ \left[G(1,0) + 2aG(2,0) + bG(1,1) + 3a^2G(3,0) + 2abG(2,1) + \cdots\right] x \\
&+ \left[G(0,1) + aG(1,1) + 2bG(0,2) + a^2G(2,1) + \cdots\right] y + \cdots \\
= &\sum_{p=0}^{N}\sum_{q=0}^{N} a^p b^q G(p,q) + x\sum_{p=1}^{N}\sum_{q=0}^{N} p a^{p-1} b^q G(p,q) \\
&+ y\sum_{p=0}^{N}\sum_{q=1}^{N} q a^p b^{q-1} G(p,q) \\
&+ x^2\sum_{p=2}^{N}\sum_{q=0}^{N} \frac{p!}{(p-2)!2!} a^{p-2} b^q G(p,q) \\
&+ xy\sum_{p=1}^{N}\sum_{q=1}^{N} pq a^{p-1} b^{q-1} G(p,q) + \cdots .
\end{aligned} \tag{25}
$$

Hence, we can write

$$
w(x,y) = \sum_{h=0}^{\infty}\sum_{k=0}^{\infty}\sum_{p=k}^{N}\sum_{q=h}^{N} \binom{q}{h}\binom{p}{k} \times a^{p-k} b^{q-h} G(p,q)\, x^k y^h . \tag{26}
$$

Using Definition 2, we obtain

$$
W(k,h) = \sum_{p=k}^{N}\sum_{q=h}^{N} \binom{q}{h}\binom{p}{k} a^{p-k} b^{q-h} G(p,q) . \tag{27}
$$

Theorem 12. Differential transform of the function $(x, y) = \partial^{r+s} g(x + a, y + b)/\partial x^r\, \partial y^s$ is

$$W(k,h) = \frac{(k+r)!}{k!}\frac{(h+s)!}{h!} \times \sum_{p=k+r}^{N}\sum_{q=h+s}^{N}\binom{q}{h+s}\binom{p}{k+r} \times a^{p-k-r}b^{q-h-s}G(p,q). \tag{28}$$

Proof. Let (k, h) be differential transform of the function $g(x+a, y+b)$. From Theorem 7, we can write that differential transform of the function (x, y) is

$$W(k,h) = \frac{(k+r)!}{k!}\frac{(h+s)!}{h!}C(k+r,h+s), \tag{29}$$

from Theorem 4, we can write

$$C(k+r,h+s) = \sum_{p=k+r}^{N}\sum_{q=h+s}^{N}\binom{q}{h+s}\binom{p}{k+r} \times a^{p-k-r}b^{q-h-s}G(p,q). \tag{30}$$

If we substitute (30) into (29), we find

$$W(k,h) = \frac{(k+r)!}{k!}\frac{(h+s)!}{h!} \times \sum_{p=k+r}^{N}\sum_{q=h+s}^{N}\binom{q}{h+s}\binom{p}{k+r} \times a^{p-k-r}b^{q-h-s}G(p,q). \tag{31}$$

APPLICATION

We have considered the following PDAE as a test problem:

$$\begin{pmatrix} 1 & 1 \\ 0 & 0 \end{pmatrix} u_t + \begin{pmatrix} -1 & 0 \\ -1 & 0 \end{pmatrix} u_{xx} + \begin{pmatrix} 1 & 1 \\ -1 & 0 \end{pmatrix} u = f,$$

$$t \in [0, \infty), \quad x \in [-1, 1], \tag{32}$$

with initial values and boundary values

$$u_1(0,x) = x^3 - x, \qquad u_2(0,x) = x^4 - 1,$$

$$u_1(t,1) = u_1(t,-1) = 0, \qquad u_2(t,1) = u_2(t,-1) = 0. \tag{33}$$

The right hand side function f is

$$f = \left(\left(x^4 - 1\right)(\cos t - \sin t) - 6xe^{-t}, -e^{-t}\left(x^3 + 5x\right)\right)^T, \tag{34}$$

and the exact solutions are

$$u_1(t,x) = \left(x^3 - x\right)e^{-t}, \qquad u_2(t,x) = \left(x^4 - 1\right)\cos t. \tag{35}$$

If nonsingular matrices $P_{F,}$, $Q_{F,k}$, $P_{L,\xi}$, and $Q_{L,\xi}$ are chosen such as

$$P_{F,k} = \begin{pmatrix} 1 & \frac{-\mu_k}{\mu_k - 1} \\ 0 & \frac{1}{\mu_k - 1} \end{pmatrix}, \qquad Q_{F,k} = \begin{pmatrix} 0 & -1 \\ 1 & -1 \end{pmatrix},$$

$$P_{L,\xi} = \begin{pmatrix} 0 & \frac{1}{\xi + 1} \\ \frac{1}{\xi + 1} & -\frac{1}{\xi + 1} \end{pmatrix}, \qquad Q_{L,\xi} = \begin{pmatrix} -\xi - 1 & 0 \\ \xi & 1 \end{pmatrix}, \tag{36}$$

matrices $P_{F,k}AQ_{F,k}$ and $P_{L,\xi}BQ_{L,\xi}$ are found as

$$P_{F,k}AQ_{F,k} = \begin{pmatrix} 1 & 0 \\ 0 & 0 \end{pmatrix}, \qquad P_{L,\xi}BQ_{L,\xi} = \begin{pmatrix} 1 & 0 \\ 0 & 0 \end{pmatrix}. \tag{37}$$

From (38), we have $N_{L,} = 0$ and $N_{F,k} = 0$. Then the PDAE (32) has differential spatial index 1 and differential time index 1. So, it is enough to take $M^{(\xi)}_{BC} = \{1\}$ and $M^{(k)}_{IC} = \{2\}$ to solve the problem.

Taking differential transformation of (32), we obtain

$$(k+1)U_1(k+1,h)+(k+1)U_2(k+1,h)$$
$$-(h+1)(h+2)U_1(k,h+2)+U_1(k,h)$$
$$+U_2(k,h)=F_1(k,h), \tag{38}$$
$$-(h+1)(h+2)U_1(k,h+2)-U_1(k,h)=F_2(k,h). \tag{39}$$

The Taylor series of functions f_1 and f_2 about $x = 0$, $t = 0$ are

$$\begin{aligned} f_1(t,x) = & -1+t-6x+\frac{1}{2}t^2+6xt \\ & -\frac{1}{6}t^3-3xt^2-\frac{1}{24}t^4+x^4+xt^3 \\ & -x^4t+\frac{1}{120}t^5-\frac{1}{4}xt^4+\frac{1}{720}t^6 \\ & -\frac{1}{2}x^4t^2+\frac{1}{20}xt^5+\frac{1}{6}x^4t^3 \\ & -\frac{1}{5040}t^7-\frac{1}{120}xt^6+\cdots, \end{aligned} \tag{40}$$

$$\begin{aligned} f_2(t,x) = & -5x+5xt-\frac{5}{2}xt^2-x^3 \\ & +\frac{5}{6}xt^3+x^3t-\frac{1}{2}x^3t^2-\frac{5}{24}xt^4 \\ & +\frac{1}{6}x^3t^3+\frac{1}{24}xt^5-\frac{1}{144}xt^6 \\ & -\frac{1}{24}x^3t^4+\cdots. \end{aligned} \tag{41}$$

The values $F_1(k, h)$ and $F_2(k, h)$ in (39) and (40) are coefficients of polynomials (41) and (42). If we use Theorem 3 for boundary values, we obtain

$$\sum_{i=0}^{7}U_1(j,i)=0, \quad j=0,1,\ldots,7, \tag{42}$$

$$\sum_{i=0}^{7}(-1)^i U_1(j,i) = 0, \quad j = 0, 1, \ldots, 7. \tag{43}$$

In order to write k=0 and h = 0, 1, 2, 3, 4, 5 in (40), we have

$$2U_1(0,2) + U_1(0,0) = 0, \qquad 20U_1(0,5) + U_1(0,3) = 1,$$

$$6U_1(0,3) + U_1(0,1) = 5, \qquad 30U_1(0,6) + U_1(0,4) = 0,$$

$$12U_1(0,4) + U_1(0,2) = 0, \qquad 42U_1(0,7) + U_1(0,5) = 0. \tag{44}$$

If we take j=0 in (43) and (44), we obtain

$$U_1(0,0) + U_1(0,1) + U_1(0,2) + U_1(0,3)$$

$$+ U_1(0,4) + U_1(0,5) + U_1(0,6) + U_1(0,7) = 0,$$

$$U_1(0,0) - U_1(0,1) + U_1(0,2) - U_1(0,3)$$

$$+ U_1(0,4) - U_1(0,5) + U_1(0,6) - U_1(0,7) = 0. \tag{45}$$

From (45) and (46), we find

$$U_1(0,0) = 0, \qquad U_1(0,1) = -1, \qquad U_1(0,2) = 0,$$

$$U_1(0,3) = 1, \qquad U_1(0,4) = 0, \qquad U_1(0,5) = 0,$$

$$U_1(0,6) = 0, \qquad U_1(0,7) = 0. \tag{46}$$

In this manner, from (40), (44), and (45), the coefficients of the u_1 are obtained as follows:

$$U_1(1,0) = 0, \qquad U_1(1,1) = 1, \qquad U_1(1,2) = 0,$$

$$U_1(1,3) = -1, \qquad U_1(1,4) = 0, \qquad U_1(1,5) = 0,$$

$$U_1(1,6) = 0, \qquad U_1(2,0) = 0, \qquad U_1(2,1) = -\frac{1}{2},$$

$$U_1(2,2) = 0, \qquad U_1(2,3) = \frac{1}{2}, \qquad U_1(2,4) = 0,$$

$$U_1(2,5) = 0, \qquad U_1(3,0) = 0, \qquad U_1(3,1) = \frac{1}{6},$$

$$U_1(3,2) = 0, \qquad U_1(3,3) = -\frac{1}{6}, \qquad U_1(3,4) = 0,$$

$$U_1(4,0) = 0, \qquad U_1(4,1) = -\frac{1}{24}, \qquad U_1(4,2) = 0,$$

$$U_1(4,3) = \frac{1}{24}, \qquad U_1(5,0) = 0, \qquad U_1(5,1) = \frac{1}{120},$$

$$U_1(5,2) = 0, \quad U_1(6,0) = 0, \quad U_1(6,1) = -\frac{1}{720},$$
$$U_1(7,0) = 0. \tag{47}$$

Using the initial values for the second component, we obtain the following coefficients:

$$U_2(0,0) = -1, \quad U_2(0,1) = 0, \quad U_2(0,2) = 0,$$
$$U_2(0,3) = 0, \quad U_2(0,4) = 1, \quad U_2(0,5) = 0,$$
$$U_2(0,6) = 0, \quad U_2(0,7) = 0. \tag{48}$$

The coefficients of the u_2 can be found using (47), (48), (49), and taking $k = 0, 1, 2, \ldots$ and $h = 0, 1, 2, \ldots$ in (39) as follows:

$$U_2(1,2) = 0, \quad U_2(1,3) = 0, \quad U_2(1,4) = 0,$$
$$U_2(1,5) = 0, \quad U_2(1,6) = 0, \quad U_2(2,1) = 0,$$
$$U_2(2,2) = 0, \quad U_2(2,3) = 0, \quad U_2(2,4) = -\frac{1}{2},$$
$$U_2(2,5) = 0, \quad U_2(3,0) = 0, \quad U_2(3,1) = 0,$$
$$U_2(3,2) = 0, \quad U_2(3,3) = 0, \quad U_2(3,4) = 0,$$
$$U_2(4,1) = 0, \quad U_2(4,2) = 0, \quad U_2(4,3) = 0,$$
$$U_2(4,0) = -\frac{1}{24}, \quad U_2(5,0) = 0, \quad U_2(5,1) = 0,$$
$$U_2(5,2) = 0, \quad U_2(6,0) = \frac{1}{720}, \quad U_2(6,1) = 0,$$
$$U_2(7,0) = 0. \tag{49}$$

If we write the above values in (39) and (40), then we have

$$\begin{aligned} u_1^*(t,x) = & -x + xt - \frac{1}{2}xt^2 + x^3 + \frac{1}{6}xt^3 \\ & - x^3t + \frac{1}{2}x^3t^2 - \frac{1}{24}xt^4 - \frac{1}{6}x^3t^3 \\ & + \frac{1}{120}xt^5 - \frac{1}{720}xt^6 + \frac{1}{24}x^3t^4 + \cdots, \end{aligned} \tag{50}$$

$$\begin{aligned} u_2^*(t,x) = & -1 + \frac{1}{2}t^2 - \frac{1}{24}t^4 \\ & + x^4 + \frac{1}{720}t^6 - \frac{1}{2}x^4t^2 + \cdots. \end{aligned} \tag{51}$$

Numerical and exact solution of the given problem has been compared in Tables 1 and 2, and simulations of solutions have been depicted in Figures 1, 2, 3, and 4, respectively.

Table 1: The numerical and exact solution of the test problem(32), where $u_1(t, x)$ is the exact solution and $u^*_1(t, x)$ is the numerical solution, for $x = 0.1$

t	$u_1(t, x)$	$u^*_1(t, x)$	$\lvert u_1(t, x) - u^*_1(t, x)\rvert$
0.1	−0.0895789043	−0.0895789043	0
0.2	−0.0810543445	−0.0810543422	0.0000000023
0.3	−0.0733410038	−0.0733409887	0.0000000151
0.4	−0.0663616845	−0.0663616355	0.0000000490
0.5	−0.0600465353	−0.0600464409	0.0000000944
0.6	−0.0543323519	−0.0543322800	0.0000000719
0.7	−0.0491619450	−0.0491621943	0.0000002493
0.8	−0.0444835674	−0.0444849422	0.0000013748
0.9	−0.0402503963	−0.0402546487	0.0000042524
1.0	−0.0364200646	−0.0364305555	0.0000104909

Table 2: The numerical and exact solution of the test problem(32), where $u_2(t, x)$ is exact solution and $u^*_2(t, x)$ is numerical solution, for $x = 0.1$.

t	$u_2(t, x)$	$u^*_2(t, x)$	$\lvert u_2(t, x) - u^*_2(t, x)\rvert$
0.1	−0.9949046649	−0.9949046653	0.0000000004
0.2	−0.9799685711	−0.9799685778	0.0000000067
0.3	−0.9552409555	−0.9552409875	0.0000000320
0.4	−0.9209688879	−0.9209689778	0.0000000899
0.5	−0.8774948036	−0.8774949653	0.0000001637
0.6	−0.8252530813	−0.8252532000	0.0000001187
0.7	−0.7647657031	−0.7647652653	0.0000004378
0.8	−0.6966370386	−0.6966345778	0.0000024608
0.9	−0.6215478073	−0.6215398875	0.0000079198
1.0	−0.5402482757	−0.5402277778	0.0000204979

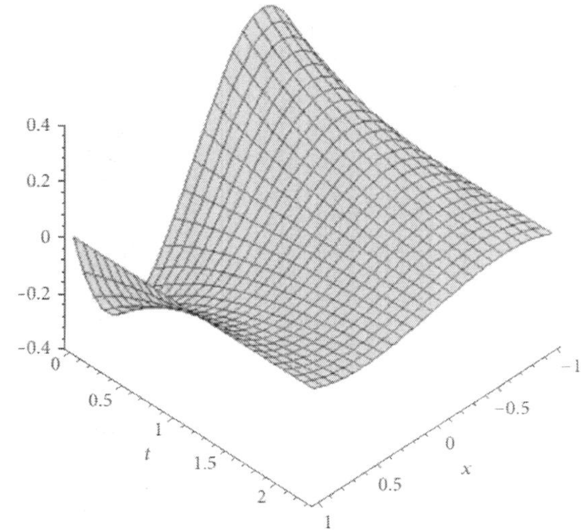

Figure 1: The graphic of the function $u_1(t, x)$ in the test problem (32).

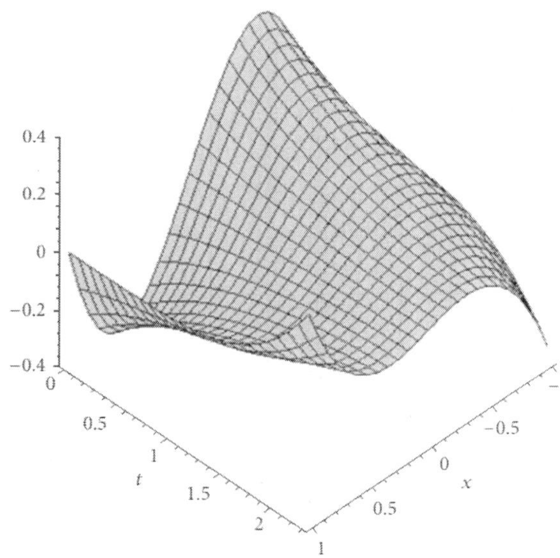

Figure 2: The graphic of the function $u^*_1\ (t, x)$ in the test problem (32).

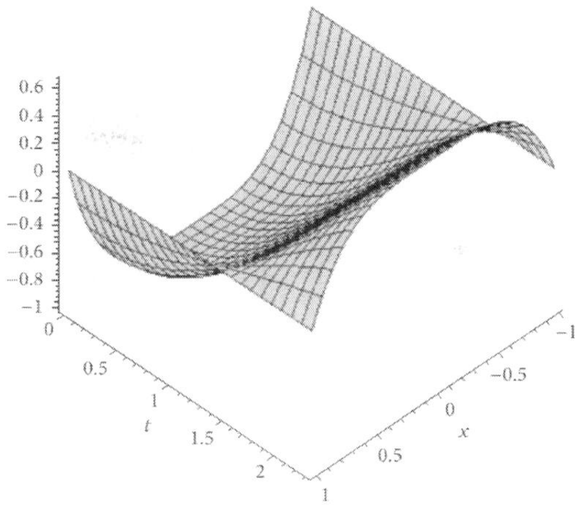

Figure 3: The graphic of the function $u_2(t, x)$ in the test problem (32).

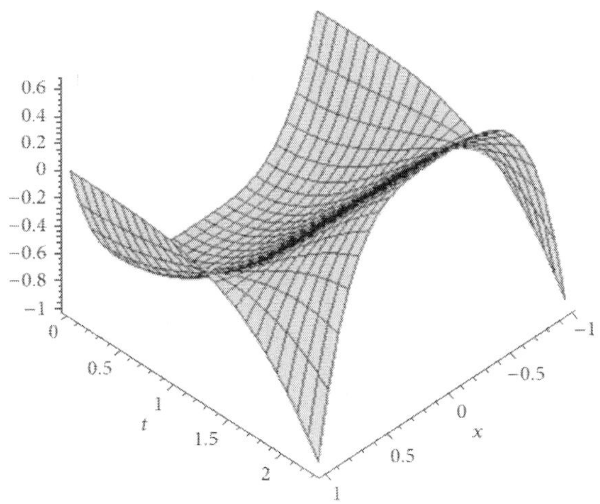

Figure 4: The graphic of the function $u^*_2(t, x)$ in the test problem (32).

CONCLUSION

The computations associated with the example discussed above were performed by using Computer Algebra Techniques [24]. We show the results in Tables 1 and 2 for the solution of (32) by numerical method. The numerical values on Tables 1 and 2 obtained above are in full agreement with the exact solutions of (32). This study has shown that the differential transform method often shows superior performance over series approximants, providing a promising tool for using in applied fields.

REFERENCES

1. W. Marszalek, Analysis of partial differential algebraic equations [Ph.D. thesis], North Carolina State University, Raleigh, NC, USA, 1997.
2. W. Lucht, K. Strehmel, and C. Eichler-Liebenow, "Linear partial differential algebraic equations, Part I: indexes, consistent boundary/initial conditions," Report 17, Fachbereich Mathematik und Informatik, Martin-Luther-Universitat Halle, 1997.
3. W. Lucht, K. Strehmel, and C. Eichler-Liebenow, "Linear partial differential algebraic equations, Part II: numerical solution," Report 18, Fachbereich Mathematik und Informatik, Martin-Luther-Universitat Halle, 1997.
4. W. Lucht, K. Strehmel, and C. Eichler-Liebenow, "Indexes and special discretization methods for linear partial differential algebraic equations," BIT Numerical Mathematics, vol. 39, no. 3, pp. 484–512, 1999.
5. W. S. Martinson and P. I. Barton, "A differentiation index for partial differential-algebraic equations," SIAM Journal on Scientific Computing, vol. 21, no. 6, pp. 2295–2315, 2000.
6. W. S. Martinson and P. I. Barton, "Index and characteristic analysis of linear PDAE systems," SIAM Journal on Scientific Computing, vol. 24, no. 3, pp. 905–923, 2002.
7. K. Debrabant and K. Strehmel, "Convergence of Runge-Kutta methods applied to linear partial differential-algebraic equations," Applied Numerical Mathematics, vol. 53, no. 2-4, pp. 213–229, 2005.
8. N. Guzel and M. Bayram, "On the numerical solution of stiff systems," Applied Mathematics and Computation, vol. 170, no. 1, pp. 230–236, 2005.
9. E. Çelik and M. Bayram, "The numerical solution of physical problems modeled as a system of differential-algebraic equations (DAEs)," Journal of the Franklin Institute, vol. 342, no. 1, pp. 1–6, 2005.
10. M. Kurulay and M. Bayram, "Approximate analytical solution for the fractional modified KdV by differential transform method," Communications in Nonlinear Science and Numerical Simulation, vol. 15, no. 7, pp. 1777–1782, 2010.
11. M. Bayram, "Automatic analysis of the control of metabolic networks," Computers in Biology and Medicine, vol. 26, no. 5, pp. 401–408, 1996.

12. N. Guzel and M. Bayram, "Numerical solution of differential-algebraic equations with index-2," Applied Mathematics and Computation, vol. 174, no. 2, pp. 1279–1289, 2006.
13. J. K. . Zhou, Differential Transformation and Its Application for Electrical CircuIts, Huazhong University Press, Wuhan, China, 1986.
14. M. Köksal and S. Herdem, "Analysis of nonlinear circuits by using differential Taylor transform," Computers and Electrical Engineering, vol. 28, no. 6, pp. 513–525, 2002.
15. I. H. Abdel-Halim Hassan, "On solving some eigenvalue problems by using a differential transformation," Applied Mathematics and Computation, vol. 127, no. 1, pp. 1–22, 2002.
16. F. Ayaz, "On the two-dimensional differential transform method," Applied Mathematics and Computation, vol. 143, no. 2-3, pp. 361–374, 2003.
17. C. K. Chen and S. H. Ho, "Solving partial differential equations by two-dimensional differential transform method," Applied Mathematics and Computation, vol. 106, no. 2-3, pp. 171–179, 1999.
18. M.-J. Jang, C.-L. Chen, and Y.-C. Liu, "Two-dimensional differential transform for partial differential equations," Applied Mathematics and Computation, vol. 121, no. 2-3, pp. 261–270, 2001.
19. X. Yang, Y. Liu, and S. Bai, "A numerical solution of second-order linear partial differential equations by differential transform," Applied Mathematics and Computation, vol. 173, no. 2, pp. 792–802, 2006.
20. A. Arikoglu and I. Ozkol, "Solution of differential-difference equations by using differential transform method," Applied Mathematics and Computation, vol. 181, no. 1, pp. 153–162, 2006.
21. M.-J. Jang, C.-L. Chen, and Y.-C. Liy, "On solving the initial-value problems using the differential transformation method," Applied Mathematics and Computation, vol. 115, no. 2-3, pp. 145–160, 2000.
22. F. Ayaz, "Applications of differential transform method to differential-algebraic equations," Applied Mathematics and Computation, vol. 152, no. 3, pp. 649–657, 2004.
23. H. Liu and Y. Song, "Differential transform method applied to high index differential-algebraic equations," Applied Mathematics and Computation, vol. 184, no. 2, pp. 748–753, 2007.
24. G. Frank, MAPLE V, CRC Press, Boca Raton, Fla, USA, 1996.

SECTION IV

CHAPTER 8

DESIGN AND IMPLEMENTATION OF NUMERICAL LINEAR ALGEBRA ALGORITHMS ON FIXED POINT DSPS

Zoran Nikolic[1], Ha Thai Nguyen[2] and Gene Frantz[3]

[1]DSP Emerging End Equipment, Texas Instruments Inc., 12203 SW Freeway, MS722, Stafford, TX 77477, USA

[2]Coordinated Science Laboratory, Department of Electrical and Computer Engineering, University of Illinois at Urbana-Champaign, 1308 West Main Street, Urbana, IL 61801, USA

[3]Application Specific Products, Texas Instruments Inc., 12203 SW Freeway, MS701, Stafford, TX 77477, USA

ABSTRACT

Numerical linear algebra algorithms use the inherent elegance of matrix formulations and are usually implemented using C/C++ floating point representation. The system implementation is faced with practical constraints

Citation (APA): Nikolić, Z., Nguyen, H. T., & Frantz, G. (2007). Design and implementation of numerical linear algebra algorithms on fixed point DSPs. EURASIP Journal on Advances in Signal Processing, 2007(1), 087046. (22 pages).

because these algorithms usually need to run in real time on fixed point digital signal processors (DSPs) to reduce total hardware costs. Converting the simulation model to fixed point arithmetic and then porting it to a target DSP device is a difficult and time-consuming process. In this paper, we analyze the conversion process. We transformed selected linear algebra algorithms from floating point to fixed point arithmetic, and compared real-time requirements and performance between the fixed point DSP and floating point DSP algorithm implementations. We also introduce an advanced code optimization and an implementation by DSP-specific, fixed point C code generation. By using the techniques described in the paper, speed can be increased by a factor of up to 10 compared to floating point emulation on fixed point hardware.

Keywords: Digital Signal Processor, Floating Point, Code Optimization, Hardware Cost, Practical Constraint

INTRODUCTION

Numerical analysis motivated the development of the earliest computers. During the last few decades linear algebra has played an important role in advances being made in the area of digital signal processing, systems, and control [1]. Numerical algebra tools—such as eigenvalue and singular value decomposition, least squares, updating and downdating—are an essential part of signal processing [2], data fitting, Kalman filters [3], and vision and motion analysis. Computational and implementational aspects of numerical linear algebraic algorithms have strongly influenced the ways in which communications, computer vision, and signal processing problems are being solved. These algorithms depend on high data throughput and high speed computations for real-time performance.

DSPs are divided into two broad categories: fixed point and floating point [4]. Numerical algebra algorithms often rely on floating point arithmetic and long word lengths for high precision, whereas digital hardware implementations of these algorithms need fixed point representation to reduce total hardware costs. In general, the cutting-edge, fixed point families tend to be fast, low power and low cost, while floating point processors offer high precision and wide dynamic range. Fixed point DSP devices are preferred over floating point devices in systems that are constrained by chip size, throughput, price-per-device, and power consumption [5]. Fixed point realizations vastly outperform floating point realizations with regard to these criteria. Figure 1 shows a chart on how DSP performance has increased over the last decade. The performance in this chart is characterized by number of multiply and accumulate (MAC) operations that can execute in parallel. The latest fixed point DSP processors run at clock rates that are approximately three times higher and perform four times more 16 × 16 MAC operations in parallel than floating point DSPs.

Therefore, there is considerable interest in making floating point implementations of numerical linear algebra algorithms amenable to fixed point implementation. In this paper, we investigate whether the fixed point DSPs are capable of handling linear numerical algebra algorithms efficiently and accurately enough to be effective in real time, and we look at how they compare to floating point DSPs.

Today's fixed point processors are entering a performance realm where they can satisfy some floating point needs without requiring a floating point processor. Choosing among floating point and extended-precision fixed point allows designers to balance dynamic range and precision on an as needed basis, thus giving them a new level of control over DSP system implementations. The overlap between fixed point and floating point DSPs is shown in Figure 2(a).

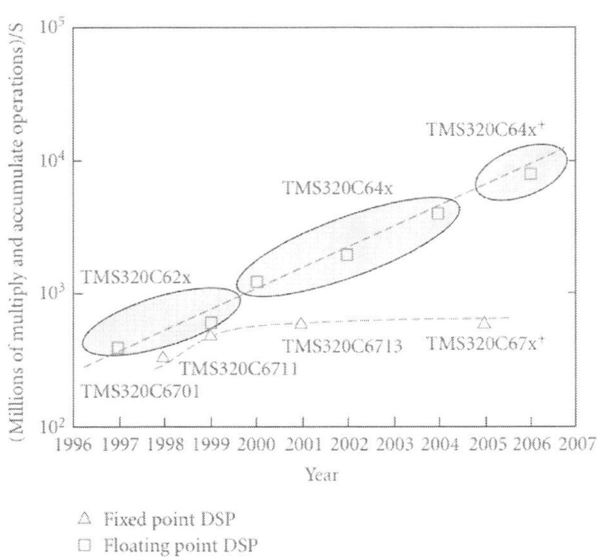

Figure 1: DSP performance trend.

The modeling efficiency level on the floating point is high and the floating point models offer a maximum degree of reusability. Converting the simulation model to fixed point arithmetic and then porting it to a target device is a time consuming and difficult process. DSP devices have very different instruction sets, so an implementation on one device cannot be ported easily to another device if it fails to achieve sufficient quality. Therefore, development cost tends to be lower for floating point systems (Figure 2(b)).

Designers with applications that require only minimal amounts of floating point functionality are caught in an "overlap zone," and they are often forced to move to highercost floating point devices. Today however, fixed point processors are running at high enough clock speeds for designer to combine floating point emulation and fixed point arithmetic in order to meet real-time deadlines. This allows a tradeoff between computational efficiency of floating point and low cost and low power of fixed point. In this paper, we are trying to extend the "overlap zone" and we investigate fixed point implementation of a truly float-intensive application, such as numerical linear algebra.

A typical design flow of a floating point system targeted for implementation on a floating point DSP is shown in Figure 3.

The design flow begins with algorithm implementation in floating point on a PC or workstation. The floating point system description is analyzed by means of simulation without taking the quantization effects into account. The modeling efficiency on the floating point level is high and the floating point models offer a maximum degree of reusability [6, 7]. C/C++ is still the most popular method for describing numerical linear algebra algorithms. The algorithm development in floating point C/C++ can be easily mapped to a floating point target DSP during implementation.

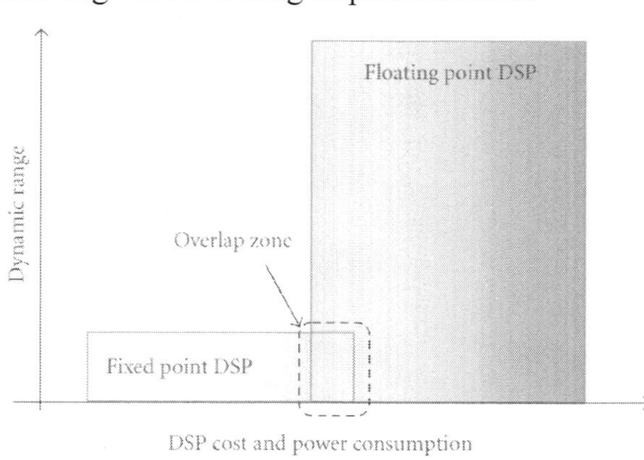

(a)

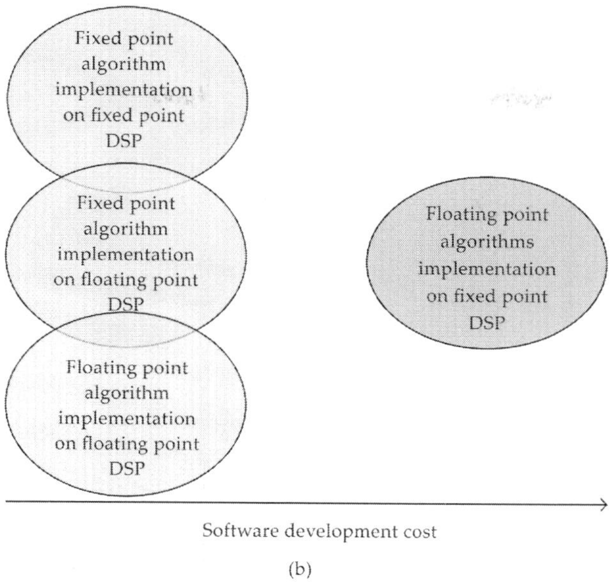

Figure 2: Fixed point and floating point DSP pros and cons.

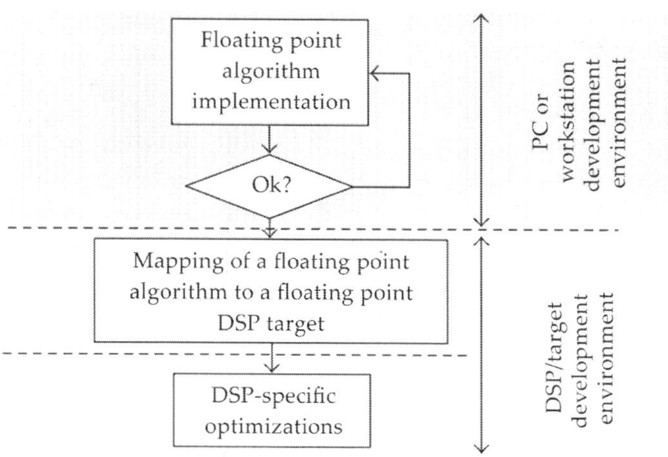

Figure 3: Floating point design process.

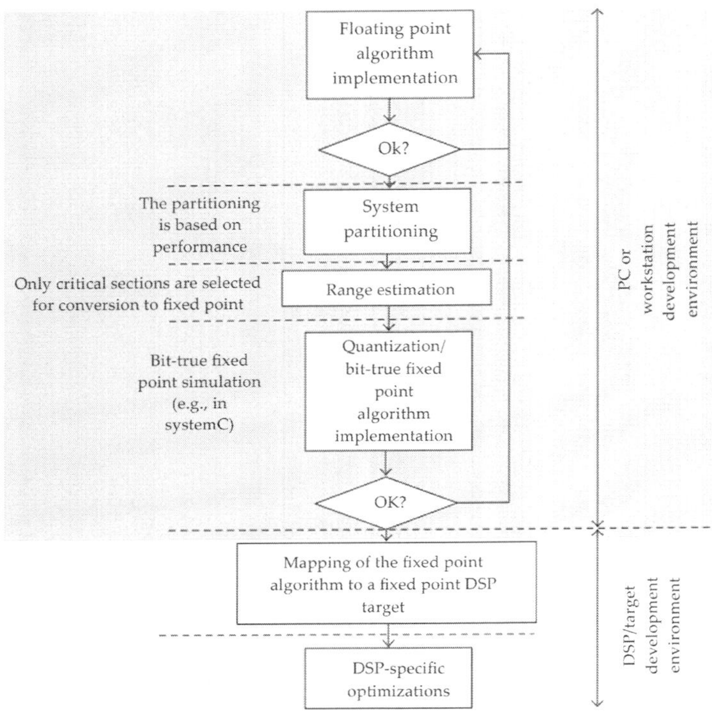

Figure 4: Fixed point design process.

There are several program languages and block diagram based CAD tools that support fixed point data types [6, 8], but C language is still more flexible for the development of digital signal processing programs containing machine vision and control intensive algorithms. Therefore, design flow—in a case when the floating point implementation needs to be mapped to fixed point—is more complicated for two reasons:

- it is difficult to find fixed point system representation that optimally maps to system model developed in floating point;
- C/C++ does not support fixed point formats. Modeling of a bit-true fixed point system in C/C++ is difficult and slow.

A previous approach to alleviate these problems when targeting fixed point DSPs was to use floating point emulation in a high level C/C++ language. In this case, design flow is very similar to the flow presented in Figure 3, with the difference that the target is a fixed point DSP. However, this method sacrifices severely the execution speed because a floating point operation is compiled into several fixed point instructions. To solve these problems, a flow that converts a floating point C/C++ algorithm into a fixed point version is developed.

A typical fixed point design flow is depicted in Figure 4.

To speed up the porting process, only the most time consuming floating point functions can be converted to fixed point arithmetic. The system is divided into subsections and each subsection is benchmarked for performance. Based on the benchmark results functions critical to system performance are identified. To improve overall system performance, only the critical floating point functions can be converted to fixed point representation.

In a next step towards fixed point system implementation, a fixed exponent is assigned to every operand. Determining the optimum fixed point representation can be time-consuming if assignments are performed by trial and error. Often more than 50% of the implementation time is spent on the algorithmic transformation to the fixed point level for complex designs once the floating point model has been specified [9]. The major reasons for this bottleneck are the following:

- the quantization is generally highly dependent on the stimuli applied;
- analytical methods for evaluating the fixed point performance based on signal theory are only applicable for systems with a low complexity [10]. Selecting optimum fixed point representation is a nonlinear process, and exploration of the fixed point design space cannot be done without extensive system simulation;
- due to sensitivity to quantization noise or high signal dynamics, some algorithms are difficult to implement in fixed point. In these cases, algorithmic alternatives need to be employed.

The bit-true fixed point system model is run on a PC or a work station. For efficient modeling of fixed point bit true system representation, language extensions implementing generic fixed point data types are necessary. Fixed point language extensions implemented as libraries in C++ offer a high modeling efficiency [10, 11]. The libraries supply generic fixed point data types and various casting modes for overflow and quantization handling and some of them also offer data monitoring capabilities during simulation time. The simulation speed of these libraries on the other hand is rather poor.

After validation on a PC or workstation, the quantized bit-true system is intended for implementation in software on a programmable fixed point DSP. The implementation needs to be optimized with respect to memory utilization, throughput, and power consumption. Here the bit-true system-level model developed during quantization serves as a "golden" reference for the target implementation which yields bit-by-bit the same results.

Memory, throughput, and word length requirements may not be important issues for off-line implementation of the algorithms, but they can become critical issues for realtime implementations in embedded processors—especially as the system dimension becomes larger [3, 12]. The load that numerical linear algebra algorithms place on real-time DSP implementation is considerable. The system implementation is faced with the practical constraints. Meaningful measures of this load are storage and computation time. The first item impacts the memory requirements of the DSP, whereas the second item helps to determine the rate at which measurements can be accepted. To reach a high level of efficiency, the designer has to keep the special requirements of the DSP target in mind. The performance can be improved by matching the generated code to the target architecture.

The platforms we chose for this evaluation were Very Long Instruction Word (VLIW) DSPs from Texas Instruments. For evaluation of the fixed point design flow we used the C64x+ fixed point CPU core. To evaluate floating point DSP performance we used C67x and C67x+ floating point CPU cores. Our goals were to identify potential numerical algebra algorithms, to convert them to fixed point, and to evaluate their numerical stability on the fixed point of the C64x+. We wanted to create efficient C implementations in order to test whether the C64x+ is fast and accurate enough for this task, and finally to investigate how fixed point realization stacks up against the algorithm implementation on a floating point DSP.

In this paper, we present methods that address the challenges and requirements of fixed point design process. The flow proposed is targeted at converting C/C++ code with floating point operations into C code with integer operations that can then be fed through the native C compiler for various DSPs. The proposed flow relies on the following main concepts:

- range estimation utility used to determine fixed point format. The range estimation software tool presented in this paper, semiautomatically transforms numerical linear algebra algorithms from C/C++ floating point to a bit-true fixed point representation that achieves maximum accuracy. Difference between this tool and existing tools [5, 9, 13–15] is discussed in Section 3;
- software tool support for generic fixed point, data types. This allows modeling of the fixed point behavior of the system. The bit-true fixed point model is simulated and finely tuned on PC or

a work station. When desired precision is achieved, the bit-true fixed point is ported to a DSP;

- seamless design flow from bit-true fixed point simulation on PC down to system implementation, generating optimized input for DSP compilers. The maximum performance is achieved by matching the generated code to the target architecture.

The remainder of this paper is organized as follows: the next subsection gives a brief overview of fixed point arithmetic; Section 2 gives a background on the numerical linear algebra algorithms selection; Section 3 presents dynamic range estimation process; Section 4 presents the quantization and bit-true fixed point simulation tools. Section 5 gives a brief overview of DSP architecture and presents tools for DSP specific optimization and implementation. Results are discussed in Section 6.

Fixed Point Arithmetic

In case of the 32-bit data, the binary point is assumed to be located to the right of bit 0 for an integer format, whereas for a fractional format it is next to the bit 31, the sign bit. It is difficult to represent all the data satisfactorily just by using integer of fractional numbers. The generalized fixed point format allows arbitrary binary point location. The binary point is also called Q point.

We use the standard Q notation Qn where n is the number of fractional bits. The total size of the number is assumed to be the nearest power of 2 greater than or equal to n, or clear from the context unless it is explicitly spelled out. Hence "Q15" refers to a 16-bit signed short with a thought comma point to the right of the leftmost bit. Likewise, an "unsigned Q32" refers to a 32-bit unsigned integer with a thought comma point directly to the left of the leftmost bit. Table 1 summarizes the range of 32-bit fixed point number for different Q format representations.

In this format, the location of the binary point, or the integer word length, is determined by the statistical magnitude, or range of signal not to cause overflows. Since each signal can have a different value for the range, a unique integer word length can be assigned to each variable. For example, one sign bit, two integer bits and 29 fractional bits can be allocated for the representation of a signal having dynamic range of [−4, 3.999999998]. This means that the binary point is assumed to be located two bits below the sign bit. The format not only prevents overflows, but also has a small quantization level 2^{-29}.

Although the generalized fixed point format allows a much more flexible representation of data, it needs alignment of the binary point location for addition or subtraction of two data having different integer word lengths. However, the integer word length can be changed by using arithmetic shift. An arithmetic right shift of n-bit corresponds to increasing the integer word length by n. The output of multiplication has the integer word length which is sum of the two input integer word lengths, assuming that one superfluous sign bit generated in the two's complement multiplication is deleted by one left shift.

Table 1: Range of 32-bit fixed point number for different Q format representations

Type	Range		Type	Range	
	Min	Max		Min	Max
IQ30	−2	1.999 999 999	IQ15	−65536	65535.999 969 482
IQ29	−4	3.999 999 998	IQ14	−131072	131071.999 938 965
IQ28	−8	7.999 999 996	IQ13	−262144	262143.999 877 930
IQ27	−16	15.999 999 993	IQ12	−524288	524287.999 755 859
IQ26	−32	31.999 999 985	IQ11	−1048576	1048575.999 511 719
IQ25	−64	63.999 999 970	IQ10	−2097152	2097151.999 023 437
IQ24	−128	127.999 999 940	IQ9	−4194304	4194303.998 046 875
IQ23	−256	255.999 999 981	IQ8	−8388608	8388607.996 093 750
IQ22	−512	511.999 999 762	IQ7	−16777216	16777215.992 187 500
IQ21	−1024	1023.999 999 523	IQ6	−33554432	33554431.984 375 000
IQ20	−2048	2047.999 999 046	IQ5	−67108864	67108863.968 750 000
IQ19	−4096	4095.999 998 093	IQ4	−134217728	134217727.937 500 000
IQ18	−8192	8191.999 996 185	IQ3	−268435456	268435455.875 000 000
IQ17	−16384	16383.999 992 371	IQ2	−536870912	536870911.750 000 000
IQ16	−32768	32767.999 984 741	IQ1	−1073741824	1 073741823.500 000 000

For a bit-true and implementation independent specification of a fixed point operand, a three-tuple is necessary: the word length WL, the integer word length IWL, and the sign S. For every fixed point format, two of the three parameters WL, IWL, and FWL (fractional word length) are independent; the third parameter can always be calculated from the other two, WL = IWL + FWL. Note that a Q0 data type is merely a special case of a fixed point data type with an IWL that always equals WL—hence an integral data type can be described by two parameters only, the word length WL and the sign encoding S (an integral data type Q0 is not presented in Table 1).

LINEAR ALGEBRA ALGORITHM SELECTION

The vitality of the field of matrix computation stems from its importance to a wide area of scientific and engineering applications on the one hand, and the advances in computer technology on the other. An excellent, comprehensive reference on matrix computation is Golub and van Loan's text [16].

Commercial digital signal processing applications are constrained by the dictates of real-time implementations. Usually a big part of the DSP bandwidth is allocated for computationally intensive matrix factorizations [17, 18]. As the processing power of DSPs keeps increasing, more of these algorithms become practical for real-time implementation.

Five algorithms were investigated: Cholesky decomposition, LU decomposition with partial pivoting, QR decomposition, Jacobi singular-value decomposition, and GaussJordan algorithm.

These algorithms are well known and have been extensively studied, and efficient and accurate floating point implementations exist. We want to explore their implementation in fixed point and compare it to floating point.

PROCESS OF DYNAMIC RANGE ESTIMATION

Related Work

During conversion from floating point to fixed point, a range of selected variables is mapped from floating point space to fixed point space. Some published approaches for floating point to fixed point conversion use an analytic approach for range and error estimation [9, 13, 19–23], and others use a statistical approach [5, 11, 24, 25]. After obtaining models or statistics of range and error by analytic or statistical approaches, respectively, search algorithms can find an optimum word length. A useful survey and comparison of search algorithms for word length determination is presented in [26].

The advantages of analytic techniques are that they do not require simulation stimulus and can be faster. However, they tend to produce more conservative word length results. The advantage of statistical techniques is that they do not require a range or error model. However, they often need long simulation time and tend to be less accurate in determining word lengths. After obtaining models or statistics of range and error by analytic or statistical approaches, respectively, search algorithms can find an optimum word length.

Some analytical methods try to determine the range by calculating the L1 norm of a transfer function [27]. The range estimated using the L1 norm guarantees no overflow for any signal, but it is a very conservative estimate for most applications and it is also very difficult to obtain the L1 norm of adaptive or nonlinear systems. The range estimation based upon L1 norm analysis is applicable only to specific signal processing algorithms (e.g., adaptive lattice filters [28]). Optimum word length choices can be made by solving equations when propagated quantized errors [29] are expressed in an analytical form.

Other analytic approaches use a range and error model for integer word length and fractional word length design. Some use a worst-case error model for range estimation [19, 23], and some use forward and backward propagation for *IWL* design [21]. Still others use an error model for *FWL* [15, 19].

By profiling intermediate calculation results within expression trees-in addition to values assigned to explicit pro-gram variables, a more aggressive scaling is possible than those generated by the "worst case estimation" technique de-scribed in [9]. The latter techniques begin with range information for only the leaf operands of an expression tree and then combine range information in a bottom up fashion. A "worst-case estimation" analysis is carried out at each operation whereby the maximum and minimum result values are determined from the maximum and minimum values of the source operands. The process is tedious and requires the de-signer to bring in his knowledge about the system and specify a set of constraints.

Some statistical approaches use range monitoring for *IWL* estimation [11, 24], and some use error monitoring for *FWL* [22, *24]*. The work in [22] also uses an error model that has coefficients obtained through simulation.

In the "statistical" method presented in [11], the mean and standard deviation of the leaf operands are profiled as well as their maximum absolute value. Stimuli data is used to generate a scaling of program variables, and hence leaf operands, that avoid overflow by attempting to predict from the signal variances of leaf operands whether intermediate results will overflow.

During the conversion process of floating point numerical linear algebra algorithms to fixed point, the integer word length (*IWL*) part and the fractional word length (*FWL*) part are determined by different approaches while architecture word length (*WL*) is kept constant. In case when a fixed point DSP is target hardware, *WL* is constrained by the CPU architecture.

Float to fixed conversion method, used in this paper, originates in simulation-based, word length optimization for fixed point digital signal processing systems proposed by Kim and Sung [5] and Kim et al. [11]. The search algorithm at-tempts to find the cost-optimal solution by using "exhaustive" search. The technique presented in [11] requires moderate modification of the original floating point source code, and does not have standardized support for range estimation of multidimensional arrays.

The method presented here, unlike work in [5, 11], is minimally intrusive to the original floating point C/C++ code and has a uniform way to support multidimensional arrays and pointers which are frequently used in numerical linear algebra. The range estimation approach presented in the subsequent section offers the following features:

- minimum code intrusion to the original floating point C model. Only declarations of variables need to be modified. There is also no need to create a secondary *main()* function in order to output simulation results;
- support for pointers and uniform standardized sup-port for multidimensional arrays which are frequently used in numerical linear algebra;
- during simulation, key statistical information and value distribution of each variable are maintained. The distribution is kept in a 32-bin histogram where each bin corresponds to one *Q* format;
- output from the range-estimation tool is split in different text files on function by function basis. For each function, the range-estimation tool creates a separate text file. Statistical information for all tracked variables within one function is grouped together within a text file associated to the function. The output text files can be imported in Excel spreadsheet for review.

Dynamic Range Estimation Algorithm

The semiautomated approach proposed in this section utilizes simulation-based profiling to excite internal signals and obtain reliable range information. During the simulation, the statistical information is collected for variables specified for tracking. Those variables are usually the floating point variables which are to be converted to fixed point. The statistics collected is the dynamic range, the mean and standard deviation and the distribution histogram. Based on the collected statistic information *Q* point location is suggested.

The range estimation can be performed on function-by-function basis. For example, only a few of the most time consuming functions in a system can be converted to fixed point, while leaving the remaining of the system in floating point.

The method is minimally intrusive to the original floating point C/C++ code and has uniform way of support for multidimensional arrays and pointers. The only modification required to the existing C/C++ code is marking the variables whose fixed point behavior is to be examined with the range estimation directives. The range estimator then finds the statistics of internal signals throughout the floating point simulation using real inputs and determines scaling parameters.

To minimize intrusion to the original floating point C or C++ program for range estimation, the operator overloading characteristics of C++ are exploited. The new data class for tracing the signal statistics is named as *ti_float*. In order to prepare a range estimation model of a C or C++ digital signal processing program, it is only necessary to change the type of variables from *float* or *double* to *ti_float*, since the class in C++ is also a *type* of variable defined by users. The class not only computes the current value, but also keeps records of the variable in a linked list which is declared as its private static member. Thus, when the simulation is completed, the range of a variable declared as class is readily available from the records stored in the class.

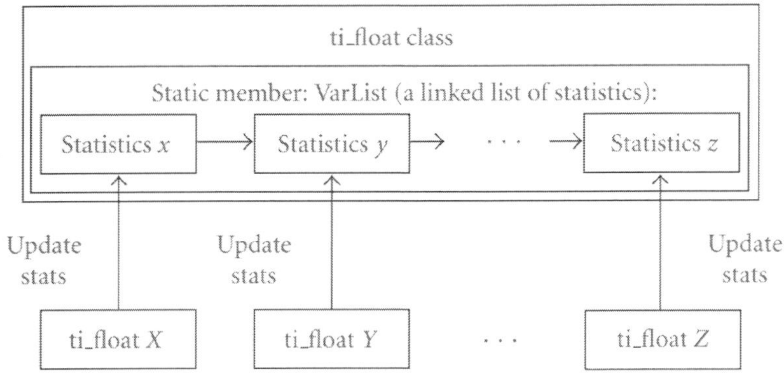

Figure 5: *ti_float* class composition.

Class *statistics* are used to keep track of the minimum, maximum, standard deviation, overflow, underflow and histogram of floating point variable associated with it. All in-stances of class *statistics* are stored in a linked-list class *VarList*. The linked list *VarList* is a static member of class *ti_float*. Every time a new variable is declared as a *ti_float*, a new object of class *statistics* is created. The new *statistics* object is linked to the last element in the linked list *VarList*, and associated with the variable. Statistics information for all floating point variables declared as *ti_float* is tracked and recorded in the *VarList* linked list. By declaring linked list of *statistics* objects as a static member of class *ti_float* we achieved that every instance of the object *ti_float* has access to the list. This approach minimizes intrusion to the origi-nal floating point C/C++ code. Structure of class *ti_float* is shown in Figure 5.

Every time a variable, declared as *ti_float*, is assigned a value during simulation, in order to update the variable statistics, the *ti_float* class searches through the linked list VarList for the *statistics* object which was associated with the variable.

The declaration of a variable as *ti_float* also creates association between the variable name and function name. This association is used to differentiate between variables with same names in different functions. Pointers and arrays, as frequently used in ANSI C, are supported as well.

Declaration syntax for *ti_float* is

ti_float <var name>("<funct name>,""'<var name>");

where <var name> is the name of floating point variable designated for dynamic range tracking, and <funct name> is the name of function where the variable is declared.

In case dynamic range of multidimensional array of float needs to be determined, the array declaration must be changed from

float <var name>[<M>][<N>]· · · [<Z>];

to

ti_float <var name>[<M>][<N>]· · · [<Z>]

={ ti_float ("<funct name>,""'<var name>,"

<M>*<N>* · · ·* <Z>)}.

Please note that declaration of multidimensional array of *ti_float* can be uniformly extended to any dimension. The declaration syntax keeps the same format for one, two, three, and *n* dimensional array of *ti_float*. In the declaration *<var name>* is the name of floating point array selected for dynamic range tracking. The *<func name>* is the name of function where the array is declared. The third element in the declaration of array of *ti_float* is size. Array size is defined by multiplying sizes of each array dimension.

In case of multidimensional *ti_float* arrays only one *statistics* object is created to keep track of statistics information of the whole array. In other words, *ti_float* class keeps statistic information for array at array level and not for each array element. Product defined as third element in the declaration defines the array size.

The *ti_float* class overloads arithmetic and relational operators. Hence, basic arithmetic operations such as addition, subtraction, multiplication, and division are conducted automatically for variables. This property is also applicable for relational operators, such as "==," ">," "<," ">=,""! =" and "<=." Therefore, any *ti_float* instance can be compared with floating point variables and constants. The contents, or private members, of a variable declared by the class are updated when the variable is assigned by one of the assignment operators, such as "=," "+ =," "− =," "∗ =," and "/ =." For example, *ti_float* is updated when the absolute of the present value is larger than the previously determined.

The floating point simulation model is prepared for range estimation by changing the variable declaration from *float* to *ti_float*. The simulation model code must be com-piled and linked with the overloaded operators of the *ti_float* class. The Microsoft Visual C++ compiler, version 6.0, is used throughout the floating point and range estimation development.

The dynamic range information is gathered during the simulation for each variable declared as *ti_float*. The statistical range of a variable is estimated by using histogram, standard deviation, minimum and maximum values. Finally, the integer word lengths of all signals declared as *ti_float* are suggested.

During floating point to fixed point conversion process we search for minimum integer word length (*IWL*) required for implementing algorithms effectively (therefore $FL = WL$

IWLmin). After completing the simulation *Q* point format in which the assigned value can be represented with mini-mum *IWL* is selected. The decision is made based on histogram data collected during simulation.

```
void choldc(float **a, int n, float p[])
{
        void nrerror(char error_text[]);
        int i,j,k;
        float sum;

        for (i=0;i<n;i++) {
                for (j=i;j<n;j++) {
                        for (sum=a[i][j],k=i-1;k>=1;k- -) sum -= a[i][k]*a[j][k];
                        if (i == j) {
                                if (sum <= 0.0)
                                        nrerror("choldc failed");
                                p[i]=sqrt(sum);
                        } else a[j][i]=sum/p[i];
                }
        }
}
```

Figure 6: Floating point code for Cholesky decomposition.

In this case, large floating point dynamic range is mapped to one of 31 possible fixed point formats from Table 1. To identify the best fixed point format the variable values are tracked by using a histogram with 32 bins. Each of these bins present one Q format. Every time during simulation, the tracked floating point variable is assigned a value, a corresponding Q format representation of the value is calculated and the value is binned to a corresponding Q point bin. In case floating point value is too large to be presented in 32-bit fixed point it is sorted in the Overflow bin. In case floating point value is too small to be presented in 32-bit fixed point it is sorted in the Underflow bin.

At the end of simulation, *ti_float* objects save collected statistics in a group of text files. Each text file corresponds to one function, and contains statistic information for variables declared as *ti_float* within that function.

Cholesky decomposition is used to illustrate porting from floating point to fixed point arithmetic. The overall procedure to estimate the ranges of internal variables can be summarized as follows.

- Implement Cholesky decomposition in floating point arithmetic C/C++ program. Floating point implementation of Cholesky decomposition is presented in Figure 6 [30].
- *Insert the range estimation directives.* In this case dynamic range is tracked for all floating point variables de-clared in *choldc()*

function. Dynamic range of *float* variable *sum*, two-dimensional array of floats *a*[][], and one-dimensional float array *p*[] are traced. Declarations for these variables are changed from float to *ti_float* as shown in lines (5), (7), and (8) shown in Figure 7. In line (7), a two-dimensional array of *ti_float* is declared. The declaration as-sociates the name of two-dimensional array "*a*" with function name "*choldc*."

Note that declaration of *ti_float* can be uniformly ex-tended for multidimensional arrays.

- *Rebuild model and run.* Code must be linked with library containing the *ti_float* implementation. During simulation, statistic data is collected for all variables declared as *ti_float*. After the simulation is complete, the collected data is saved in a group of text files. A text file is associated with each function. All variables declared as *ti_float* within a function are grouped and saved together. In this case, data associated to tracked variables from function *choldc()* are saved in text file named *choldc.txt*. Content of the *choldc.txt* is shown in Figure 8.

Statistics collected for each variable is presented in separate rows. In rows (7), (8), and (9) statistics for variables *p*, *a*, and *sum* are presented. The *Q* point information shown in column B presents *Q* format suggestion. For example, the tool suggests *Q*28 format for elements of two-dimensional array *a*. The count information, shown in column C, presents how many times particular variable was assigned a value during course of simulation. The information shown in columns D through I in Figure 8, respectively, present

- *Min*: smallest value of the selected variable during simulation;
- *Max*: largest value of the selected variable during simulation;
- *Abs Min*: absolute smallest value of the selected variable during simulation;
- *Abs Max*: absolute largest value of the selected variable during simulation;
- *Mean*: mean value of the selected variable during simulation;
- *Std dev:* standard deviation value of the selected variable during simulation.

In the remaining columns, histogram information is presented for each tracked variable. For example, during the course of simulation variable *sum* took twice values that can be represented in *Q*28 fixed point format, it took 100 times values that can be represented in *Q*29 fixed point format and it took 458 times values that can be represented in *Q*29 fixed point format. Overflow and Underflow bins track number of overflows and underflows, respectively.

```
choldc(float **ti_a, int n, float ti_p[])
{
        int i,j,k;

        ti_float sum("choldc," "sum");

        ti_float a[M][M] = {ti_float("choldc," "a," M*M)};
        ti_float p[M] = {ti_float("choldc," "p," M)};

        for (i=0; i<n; i++)
        {

                for (j=0; j<n; j++) a[i][j] = ti_a[i][j];
        }

        for (i=0;i<n;i++) {
                for (j=i;j<n;j++) {
                        for (sum=a[i][j],k=i-1;k>=0;k--) sum -= a[i][k]*a[j][k];
                        if (i == j) {
                                if (sum <= 0.0)
                                        nrerror("choldc failed");
                                p[i]=sqrt(sum);
                        } else a[j][i]=sum/p[i];
                }
        }

        for (i=0; i<n; i++)
        {
                ti_p[i] = p[i];
                for (j=0; j<n; j++) ti_a[i][j] = a[i][j];
        }
}
```

Figure 7: Floating point code for Cholesky decomposition prepared for range estimation.

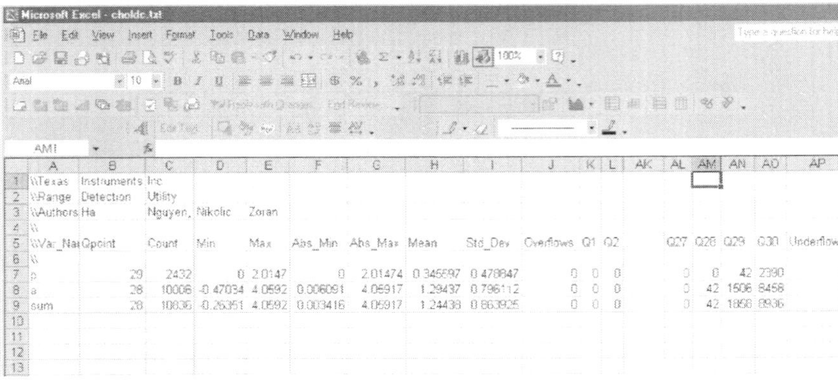

Figure 8: Output from range estimation tool imported in excel spreadsheet.

BIT-TRUE FIXED POINT SIMULATION

Once Q point position is determined, fixed point system simulation is required to validate if achieved fixed point performance is satisfactory. This intermediate fixed point simulation step on PC or workstation is required before porting the fixed point code to a DSP platform. Cosimulating this fixed point algorithm with the original floating point code will give an accuracy evaluation.

Since ANSI C or C++ offers no efficient support for fixed point data types, it is not possible to easily carry the fixed point simulation in pure ANSI C or C++. Several library ex-tensions to C++ have been proposed in the past to compensate for this deficiency [7, 31]. These fixed point language extensions are implemented as libraries in C++ and offer a high modeling efficiency. They supply generic fixed point data types and various casting modes for overflow and quantization handling. The simulation speed of these libraries on the other hand is rather poor.

The SystemC fixed point data types and cast operators are utilized in proposed design flow [7]. Since ANSI C is a subset of SystemC, the additional fixed point constructs can be used as bit-true annotations to dedicated operands of the original floating point ANSI C file, resulting in a *hybrid* specification. This partially fixed point code is used for simulation.

In the following paragraphs, a short overview of the most frequently used fixed point data types and functions in SystemC is provided. A more detailed description can be found in the SystemC user's manual [7].

The data types *sc_fixed* and *sc ufixed* are the data types of choice. The two's complement data type *sc_fixed* and the unsigned data type *sc ufixed* receive their format when they are declared, that is, the fixed point attributes must be known at compile time (static arguments). Thus they behave according to these fixed point parameters throughout their lifetime. Pointers and arrays, as frequently used in ANSI C, are sup-ported as well.

For a cast operation to a fixed point format *<WL, IWL, SIGN>*, it is also important to specify the overflow and precision reduction in case the target data type cannot hold the original value. The most important casting modes are listed below. SystemC also specifies many additional cast modes to model target specific behavior.

i. *Quantization modes*

a) *Truncation* (SC_TRN). The bits below the specified LSB are cut off. This quantization mode is the default for SystemC fixedpoint types and will be used if no other value is specified.

b) *Rounding* (SC_RND). Adds LSB/2 first, before cutting off the bits below the LSB.

ii. *Overflow modes*

a) *Wrap-around* (SC_WRAP). In case of an overflow the MSB carry bit is ignored. This overflow mode is the default for SystemC fixed point types and will be used if no other value is specified.

b) *Saturation* (SC SAT). In case the minimum or maximum values are exceeded, the result is set to the minimum or maximum values, respectively.

Described above are the *algorithmic level* transformations as illustrated in Figure 9, that change the behavior or accuracy of an algorithm.

Transformation starts from a floating point program, where the designer abstracts from the fixed point problems and does not think of a variable as finite length register.

Fixed point formats are suggested by range estimation tool. Based on this advice, when migrating from floating point C to bit-true fixed point C code, the floating point variables should be converted to variables with appropriate fixed point range.

To illustrate this step, *choldc()* function from Figure 6 is converted to fixed point based on advice from range estimation tool. It is assumed that function *choldc()* accepts floating point inputs, performs all calculations in fixed point, and then converts the results back to floating point. Based on data collected during range estimation step, floating point variables in *choldc()* should be converted to appropriate fixed point formats. The output from the range estimation tool (Figure 8) recommends that floating point variables *sum*, *p*[] and *a*[][] should have *Q*28, *Q*29, and *Q*28 fixed point formats, respectively. In listing shown in Figure 9, in line (5), variable *sum* is declared as *Q*28 (*IWL = 4*). Variables *a*[][], and *p*[] are declared in lines (7) and (8) as *Q*28 and *Q*29, respectively. Note that lines (16)–(27) from listing in Figure 9 are equivalent to lines (7)–(16) from Figure 6. Since variables *ti a[][]* and *ti p[]* passed from calling function to *choldc()* are floating point variables, it is required to convert them to fixed point variables (lines (10)–(14) in Figure 9). The *choldc()* function should return floating point results therefore before returning the fixed point results must be converted back to floating point (lines (28)–(32) in Figure 9).

The resulting completely bit-true algorithm in *SystemC* is not directly suited for implementation on a DSP. The algorithm needs to be mapped to a DSP target. This is an *implementation level* transformation, where the bit-true behavior normally remains unchanged.

ALGORITHM PORTING TO A TARGET DSP

Selecting a target DSP, and porting the bit-true fixed point numerical linear algebra algorithm to its architecture is not a trivial task. The internal DSP architecture plays a significant role in how efficiently the algorithm runs in real time. The internal architecture, number and size of the internal data paths, type and bandwidth of the external memory interface, number and precision of functional units, and cache architecture all play important role in how well numerical algebra tasks will be carried in real time.

Programming modern DSP processors manually utilizing assembly language is a very tedious task. In awareness of this problem, the modern DSP architectures have been developed using a processor/compiler codesign methodology which led to compiler-efficient processor designs.

Despite improvements in development tools, a significant gap in the system design flow is still evident. Today there is no direct path from a floating point system level simulation to an optimized fixed point implementation on a DSP. While multiplication is supported directly on the fixed point DSPs, division and square root are not; hence they must be computed iteratively. Many numerical linear algebra algorithms require "square root" and "reciprocal square root" operation. By standardizing these building blocks, we are minimizing manual implementation and necessary optimization of target specific code for the DSP. This will decrease time-to-market and make design changes less tedious, error prone and costly.

```
choldc(float **ti_a, int n, float ti_p[])
{
        int i,j,k;

        sc_fixed<32,4> sum;

        sc_fixed<32,4> a[M][M];
        sc_fixed<32,3> p[M];

        for (i=0; i<n; i++)
        {

                for (j=0; j<n; j++) a[i][j] = ti_a[i][j];
        }

        for (i=0;i<n;i++) {
                for (j=i;j<n;j++) {
                        sum=a[i][j];
                        for (k=i-1;k>=0;k- -) sum -= a[i][k]*a[j][k];
                        if (i == j) {
                                if (sum <= 0.0)
                                        nrerror("choldc failed");
                                p[i]=sqrt(sum);
                        } else a[j][i]=sum/p[i];
                }
        }

        for (i=0; i<n; i++)
        {
                ti_p[i] = p[i];
                for (j=0; j<n; j++) ti_a[i][j] = a[i][j];
        }
}
```

Figure 9: Fixed point implementation of Cholesky decomposition algorithm in SystemC.

DSP Architecture Overview

In this paper, we selected TMS320C6000 DSP family as an implementation target for numerical linear algebra algorithms. The TMS320C6000 family consists of fixed point DSPs [32], and floating point DSPs [33]. TMS320C6000 DSPs have an architecture designed specifically for real-time signal processing [34].

To achieve high performance through increased instruction-level parallelism, the architecture of the C6000 platform use advanced Very Long Instruction Word (VLIW). A traditional VLIW architecture consists of multiple execution units running in parallel, performing multiple instructions during a single clock cycle. Parallelism is the key to high performance, taking these DSPs well beyond the performance capabilities of traditional superscalar designs. The TMS320C6000 DSPs have a highly deterministic architecture, having few restrictions on how or when instructions are fetched, executed, or stored. This architectural flexibility enables high-efficiency levels of the TMS320C6000 optimizing C compiler. Features of the C6000 devices include

- advanced (VLIW) CPU with eight functional units, including two multipliers and six arithmetic units. The CPU can execute up to eight 32-bit instructions per cycle;
- these eight functional units contain: two multipliers and six ALUs instruction packing: reduced code size;
- all instructions can operate conditionally: flexibility of code;
- variable-width instructions: flexibility of data types;
- fully pipelined branches: zero-overhead branching.

An important attribute of a real-time implementation of a matrix algorithm concerns the actual volume of data that has to be moved around during execution. Matrices sit in memory but the computations that involve their entries take place in functional units. The control of memory traffic is crucial to performance. We need to keep the fast arithmetic units busy with enough deliveries of matrix data and we have to ship the result back to memory fast enough to avoid backlog.

Customization of Bit-True Fixed Point Algorithm to a Fixed Point DSP Target

Compiling the bit-true fixed point model, developed in Section 4, by using a target DSP compiler does not give optimum performance. The C64x+ DSP compilers support C++ language constructs, but compiling the fixed point libraries for the DSP is no viable alternative as the implementation of the generic data types makes extensive use of operator over-loading, templates and dynamic memory management. This will render fixed point operations rather inefficient com-pared to integer arithmetic performed on a DSP. Therefore, target specific code generation is necessary.

In this study, we have chosen the TMS320C64x+ fixed point CPU and its C compiler as an implementation target [32, 35, 36]. We had to develop a target-optimized DSP C code for C64x+ CPU core. The most frequently used routines in numerical linear algebra are optimized in fixed point to C64x+ CPU.

Texas Instruments has developed IQmath library for TI's TMS320C28x processor [37]. The C28x IQmath library was used as a starting point to create a similar library for C64x+ CPU. The C64x+ IQmath library is a highly optimized and high-precision mathematical function library for C/C++ programmers to seamlessly port the bit-true fixed point algorithm into fixed point code on the C64x+ family of DSP devices. These routines are intended for use in computation-ally intensive real-time applications where optimal execution speed and high accuracy are critical. By using these routines, execution speeds are considerably faster than equivalent code written in standard ANSI C language can be achieved.

The resulting system enables automated conversion of the most frequently used ANSI floating point math functions such as sqrt(), isqrt(), div(), sin(), cos(), atan(), log(), and exp() by replacing these calls with their fixed point equivalents coded using portable ANSI C. This substitution of function calls is part of the floating point to fixed point conversion process.

Numerical precision and dynamic range requirement will vary considerably from one application to the other. IQmath Library facilitates the application programming in fixed point arithmetic, without fixing the numerical precision up-front. This allows the system engineer to check the application performance with different numerical precision and finally fix the numerical resolution.

Typically C64x+ IQmath function supports $Q0$ to $Q30$ format. In other words, Q point can be placed anywhere assuming 32-bit word length (WL). Nevertheless some functions like IQNsin, IQNcos, IQNatan2, IQNatan2PU, IQatan do not support $Q30$ format, due to the fact that these functions input or output need to vary between $-\pi$ to π radians. For definition of $Q0$ to $Q30$ format please refer to Table 1.

A subset of IQmath functions used in this paper is presented in Table 2.

Table 2: List of relevant functions from IQmath library

Function name	Remarks
IQabs	Absolute value of IQ number
IQdiv	Fixed point division
IQXtoY	Conversion between two different IQ formats
IQisqrt	High-precision inverse square root
IQmag	Magnitude square: sqrt(A^2 + B^2)
IQmpy	IQ multiplication
IQmpyIQx	Multiply two different IQ numbers
IQrmpy	IQ multiplication with rounding
IQrsmpy	IQ multiplication with rounding & saturation
IQsqrt	High-precision square root
IQtoF	IQ to floating point
FtoIQ	Convert float to IQ

In order to include an IQmath function in C code the following steps must be followed:

- include the *IQmathLib.h* include file;
- link your code with the IQmath object code library, *IQmath.lib*
- use a correct linker command file to place “IQmath” section in program memory;
- the section “IQmathTables” contains lookup tables for IQmath functions.

The C code functions from IQmath library compile into efficient C64x+ assembly code. The IQmath functions are implemented by using C64x+ specific C language extensions (intrinsics) and compiler directives to restructure the off-the-shelf C code while maintaining functional equivalence to the original code [36]. The intrinsics are built in functions that usually map to specific assembly instructions. The C64x+ instruction such as multiplication of two 32-bit numbers to 64-bit result is utilized to have higher precision multiplication [32].

To illustrate this step, a bit-true fixed point version of function *choldc()* shown in Figure 9 is ported to fixed point DSP.

The process of porting to a fixed point target starts with a bit-true fixed point model (Figure 9). The fixed point variables from listing shown in Figure 9 are converted to corresponding fixed point formats supported by IQmath library. In listing presented in Figure 10, lines (11)–(14) and (33)–(38) convert between floating point and fixed point formats. Lines (16)–(30) from listing in Figure 10 are equivalent to lines (16)–(26) from listing in Figure 9. Note that fixed point multiplication and square root operations are replaced with the equivalents from IQmath library. These functions are optimized for maximum performance on target fixed point DSP architecture.

Note that *Q*28 fixed point format is used for array *a*[][] (*a*[][] is declared as iq28 in line (5) in Figure 10). In line (19), two elements of the array are multiplied by using the IQmath function *IQ28mpy()*. In line (26), variable *p*[*i*] is converted from *iq29* to *iq28* fixed point format. Although range estimation tool suggested *Q*29 format for variable *p*[],a few CPU cycles can be saved if the variable is in *Q*28 fixed point format. If all fixed point variables in the function were in *Q*28 fixed point format, the function would execute slightly faster since it would not be necessary to spend CPU cycles for conversion between different formats (lines (23) and (26) in Figure 10).

```
(1) void choldc(float **aa, int n, float pp[])
(2) {
(3)         void nrerror(char error_text[]);
(4)         int i,j,k,ip,iq;
(5)         _iq28 a[M][M];
(6)         _iq29 p[M];
(7)         _iq28 sum;
(8)
(9)        a=imatrix(1,n,1,n);
(10)       p=ivector(1,n);
(11) //convert input matrix to IQ format
(12)         for (i=0;i<n;i++) {
(13)                   for (j=0;j<n;j++) a[i][j]= _FtoIQ28(aa[i][j]);
```

```
    }
    for (i=0;i<n;i++) {
        for (j=i;j<n;j++) {
            for (sum=a[i][j],k=i-1;k>=0;k- -)
             sum -= _IQ28mpy(a[i][k],a[j][k]);
            if (i == j) {
                if (sum <= 0.0)
                    nrerror("choldc failed");
                p[i]=_IQXtoIQY(_IQ28sqrt(sum),28,29);
            } else {
                _iq28 tmp;
                  tmp = _IQXtoIQY(p[i],29,28);
                  a[j][i]=_IQ28div(sum,tmp);
                  }
        }
    }
//convert back to floating point
    for (i=0;i<n;i++) {
        pp[i]= _IQ29toF(p[i]);
            for (j=0;j<n;j++)
              aa[i][j]= _IQ28toF(a[i][j]);
    }
}
```

Figure 10: Fixed point implementation of Cholesky decomposition algorithm in IQmath.

Usually, the implementation optimized for a target DSP must not only run on the DSP but it should also run and simulate on a work station or a PC. Since the IQmath library functions are implemented in C, it is possible to recompile and run fixed point target DSP code on a PC or workstation providing that DSP intrinsics library for the host exists.

RESULTS

Real-time performance of selected numerical linear algebra algorithms is compared between their implementations on fixed point DSP and floating point DSP platforms. Implementation of the numerical linear algebra algorithms on a floating point DSP target was straight forward since their original implementation was in floating point C/C++ [18]. On the other hand, in order to run on a fixed point DSP target, the numerical linear algebra algorithms described in Section 2 had to be ported to fixed point arithmetic.

Conversion from floating point to fixed point arithmetic was completed based on the flow described in Sections 3, 4, and 5. Dynamic range of floating point variables is estimated by the range-estimation tool presented in Section 3. Based on the recommendation from the range-estimation tool, we created a bit-true fixed point SystemC model as described in Section 4. Performance of the bit-true SystemC fixed point algorithm is first validated. After performance validation, the bit-true fixed point algorithm is ported to a target DSP as described in Section 5.

The flow presented in this paper significantly shrinks the time spent on algorithm conversion from a floating point to fixed point DSP target. The conversion process turns out to be at least four times faster than trial-and-error determination of the fixed point formats by hand.

The simulation speed of bit-true fixed point implementation in SystemC is rather slow compared to the original floating point C program. The SystemC simulation runs approximately one thousand times slower. The simulation time required for the range estimation process is 5–20 times shorter than bit-true fixed point model simulation in SystemC.

Optimization for different design criteria, like through-put, chip size, memory size, or accuracy, are in general mutually exclusive goals and result in a complex design. We use three points to compare performance between fixed point and floating DSP platforms for running the numerical linear algebra algorithms:

- speed which translates to number of CPU cycles required to run the algorithm, and CPU frequency;
- code size;
- accuracy.

To optimize the speed performance of the algorithms, only compiler-driven optimization is used. We wanted to pre-serve connection to the original floating point C algorithm throughout different stages of the conversion flow described in Sections 3, 4, and 5. In order to keep simple mapping between the different stages of the float-to-fixed conversion flow we did not change the original algorithms. In order to maintain portability between different platforms (work station/target DSP) the algorithm implementation is kept in C/C++. Although better performance can be achieved by implementation of critical functions (such as square root) in assembly this was not exploited to maintain code portability. For the occasional cases where additional CPU performance is needed, additional techniques are available to improve performance of C/C++ applications [38].

In the following three sections each aspect will be dis-cussed separately.

The selected numerical linear algebra algorithms are implemented on the TMS320C6000 DSP family from Texas Instruments.

The algorithm performance in floating point was evaluated on TMS320C6713 (C67x CPU core) and TMS320C6727 DSPs (C67x+ CPU core). Compiler used for both cases was v5.3.0.

The performance of the numerical algebra algorithms on the fixed point DSP is evaluated on C64x+ CPU core. To evaluate algorithm performance in fixed point, we used TMS320C6455 DSP (C64x+ CPU core). Compiler used was v6.0.1.

Number of CPU Cycles/Speed

Code/compiler Optimizations

Pivoting is nothing more than interchange of rows (partial pivoting) or rows and columns (full pivoting) so as to put a particularly desirable element in the diagonal position from which the pivot is about to be selected. Pivoting operation can be separated to (1) search for pivot element, and (2) interchange rows, if needed. Search for pivot element adds a slight overhead on a DSP since conditional branch pre-vents efficient pipelining. The computational overhead of row swapping (permutation operation) is significantly reduced on TMS320C6000 DSPs, since, interchange of rows (once the pivot element is found), is fully pipelined by the compiler.

The Gauss-Jordan algorithm requires row operations and pivoting (swapping rows) for numerical stability. The compiler successfully pipelines row swapping loops, and scaling loops in Gauss-Jordan algorithm.

The LU factorization algorithm uses Crout's method with partial pivoting. Crout's algorithm solves a set of equations by arranging the equations in a certain order. Pivoting is essential for stability of Crout's algorithm. In LU decomposition the compiler is successfully pipelining five inner loops: loop over row elements to get the implicit scaling information, the inner loop over columns of Crout's method, the inner loop in search for largest pivot element, the row inter-change loop, and pivot divide loop.

Performing the pivoting by interchange of row indexes significantly speeds up decomposition of large matrices. In case of small matrices the pivoting by interchange of row indexes is only slightly faster. It takes ~ 30 CPU cycles to inter-change two rows in 5 × 5 matrix which is less than 1.5% of total number of cycles required for *LU* decomposition. The accuracy of decomposition is not affected by either pivoting implementation. In our implementation of LU decomposition we perform pivoting by really interchanging rows.

Cholesky decomposition is extremely stable without any pivoting. Cholesky decomposition requires multiplication, division, and the square root computation. In the fixed point implementation of Cholesky decomposition the square root, division, and multiplication are replaced by IQmath C functions optimized for C64x+ CPU architecture. The numerical linear algebra algorithms usually contain double- or triple nested loops. The compiler is the most aggressive on innermost loops. The inner loop of block dot product implementation of Cholesky decomposition (lines (18)-(19) in Figure 10) is successfully pipelined by the compiler. The compiler extracts an impressive amount of parallelism out of the application. Optimized with the appropriate flags the inner loop is unrolled so that a multiple of 2 elements are computed at once.

Givens and Householder transformations are frequently used in matrix factorizations [16]. When Givens rotations are used to diagonalize a matrix, the method is known as a Jacobi transformation. For this reason, Givens rotations are also known as Jacobi rotations. In numerical terms, both Givens and Housholder are very stable and accurate methods of introducing zeros to a matrix. Backward error analysis reveals that error introduced by limited precision computation is on order of machine precision, which is an important fact given that we have limited number of bits on fixed point.

Jacobi methods are suitable for fixed point implementation because they provide more control over scaling of values when compared to most other methods, for example, QR iteration. The exact Jacobi Algorithm [16] involves the calculation of a square root, the calculation of a reciprocal of a square root, and multiple divisions. We implement Jacobi rotations in which division, the square root computation, and the reciprocal of the square root are replaced by IQ-math C functions optimized for C64x+ CPU architecture. Algorithms to compute the Jacobi SVD are computationally intensive when compared to the traditional factorizations. Unlike Cholesky, the Jacobi SVD algorithm is iterative. We demonstrate here that Jacobi SVD algorithm translates well to fixed point DSPs; and that the convergence property of the algorithms is not jeopardized by fixed point computations. The compiler successfully pipelines four rotation loops.

Table 3: Cycle count and code size for floating point emulation of the key operations for numerical linear algebra (fixed point C64x+ CPU).

Floating point emulation on C64x+ CPU core	C64x+ [CPU clocks]	Code size [bytes]
Addition	66	384
Multiplication	69	352
Square root	3246	512
Division	114	320

In QR decomposition, we use Householder reflection algorithm. In practice, using Givens rotations is slightly more expensive than reflections. Givens rotations are slower but they are easier to implement and debug, and they only re-quire four temporary variables when calculating the orthogonal operation compared with number of reflections, they are slightly more accurate than Householder method. All of these effects stem from the fact that Givens examines only two elements on the top row of a matrix at a time, whereas Householder needs to examine all the elements at once. The compiler is successfully pipelining two inner loops of successive Householder transformations.

Target Customization of Critical Functions

Square root, inverse square root, multiplication and division are by far the most expensive real floating point operations. These operations are necessary to compute Jacobi SVD, Cholesky decomposition, QR decomposition, and LU decomposition. Their efficient implementation is crucial for overall system performance. In Tables 3 and 4, we compare performance of these functions between two implementations: floating point emulation and pure fixed point implementation on fixed point C64x+ CPU. Table 3 presents cycle count and memory footprint when these functions are implemented by emulating floating point on fixed point C64x+ CPU.

In Table 4, code size and cycle count for IQmath implementation, on C64x+ CPU core, of these four critical functions are presented.

Table 4: Cycle count and code size for IQmath implementation of the key operations for numerical linear algebra (fixed point C64x+ CPU)

Fixed point implementation IQmath on C64x+ CPU core	C64x+ [CPU clocks]	Code size [bytes]
Addition	8	24
Multiplication	15	32
Square root	64	320
Inverse square root	75	384
Division	80	288

The IQmath division, square root, and inverse square root functions are computed using two iterations of the Newton-Raphson method. Each Newton-Raphson iteration doubles number of significant bits. First iteration gives 16-bit accuracy, and second iteration gives 32-bit accuracy. To initialize the iterations a 512 byte lookup table is used for square root and inverse square root, and 1024 byte lookup table is used for division. The serial nature of Newton iterations does not allow compiler to use pipelining.

CPU Cycle Count for Different Algorithm Realizations

In Table 5, CPU cycle counts are presented for the selected numerical linear algebra algorithms. Floating point section of Table 5 presents results for the following floating point DSP realizations:

- algorithm performance in CPU clocks for implementation on TMS320C6711 (C67x CPU core);
- algorithm performance in CPU clocks for implementation on TMS320C6727 (C67x+ CPU core);
- algorithm performance in CPU clocks for inline implementation on TMS320C6727 (C67x+ CPU core).

Fixed point section of Table 5 presents results for the following fixed point DSP realizations:

- algorithm performance in CPU clocks for implementation using floating point emulation on C64x+ CPU core;
- algorithm performance in CPU clocks for fixed point implementation using IQmath library on C64x+ CPU core;
- algorithm performance in CPU clocks for fixed point implementation using inline functions from IQmath on C64x+ CPU core.

The floating point implementation of the numerical linear algebra algorithms takes minimum effort and results in a reasonable performance. On the other hand, it turns out that floating point emulation on fixed point DSP costs many CPU cycles. On average, floating point emulation on C64x+ CPU takes 10–20 times more CPU cycles than running floating point code on C67x+ CPU core (Table 5). The fixed point CPU cores are capable of running at higher frequencies than floating point CPU cores. A practical clock ratio between fixed point and floating point CPUs is close to three. Even at clock rates that are three times higher than clock rates of the floating point DSP, the performance of floating point emulation on fixed point DSP is still inferior. The floating point emulation performance is satisfactory only if there are no big real-time implementation restrictions. To get the maximum performance from the fixed point DSP the algorithms must be converted to fixed point arithmetic.

Table 5: Cycle count relative to selected numerical linear algebra implementations

CPU cycles	Floating point DSP realizations			Fixed point DSP realizations		
	C67x -pm -o3	C67x+ -pm -o3	Inlined C67x+ -pm -o3	Floating point emulation C64x+ -pm -o3	IQMath C64x+ -pm -o3	Inlined IQMath C64x+ -pm -o3
Jacobi SVD (5 × 5)	99433	89238	24796	514386	80000	43753
Cholesky (5 × 5)	4350	4130	1597	21961	2751	1859
LU (5 × 5)	6061	5536	2288	15552	4988	2687
QR (5 × 5)	8006	7357	3201	34418	8570	5893
Gauss-Jordan (5 × 5)	8280	7550	4921	35681	14020	6308

The range-estimation step (Section 3) is carried in or-der to create a bit-true fixed point model (Section 4). Speed performance of numerical linear algebra algorithms on fixed point DSP becomes comparable to floating point DSP only if steps outlined in Section 5 are taken. The bit-true fixed point model is adapted to a fixed point DSP target by using a library of C functions optimized for C64x+ architecture (Section 5).

The two leftmost columns in the "*floating point realization*" part of Table 5 represent cycle counts for the algorithms executed on C67x and C67x+ floating point cores. In these cases, the floating point algorithms are calling square root, inverse square root, and division functions from an external library. The middle column of the "*fixed point realization*" part of Table 5 represents cycle counts for the algorithms executed on C64x+ fixed point core. In this case, the fixed point algorithms are calling fixed point implementation of square root, inverse square root, and division functions from an external IQmath library. Note that if external libraries are used, algorithm realization on floating point DSP takes roughly the same amount of cycles as implementation in fixed point running on a fixed point DSP. Since floating point DSPs usually run at lower clock rates, the overall execution time is much shorter on fixed point DSPs.

The maximum performance can be achieved only when inline function expansion is used (Table 5). In this case, the C/C++ source code for the functions such as square root, in-verse square root, and division is inserted at the point of the call. Inline function expansion is advantageous in short functions for the following reasons:

- it saves the overhead of a function call;
- once inlined, the optimizer is free to optimize the function in context with the surrounding code.

Speed performance improvement was also achieved by helping the compiler determine memory dependencies by using restrict keyword. The restrict keyword is a type qualifier that may be applied to pointers, references, and arrays. Its use rep-resents a guarantee by the programmer that, within the scope of the pointer declaration, the object pointed to can be accessed only by that pointer. This practice helps the compiler optimize certain sections of code because aliasing information can be more easily determined.

Table 6: Jacobi SVD algorithm: number of Jacobi rotations for different matrix sizes

Matrix dimension	Number of Jacobi rotations
5×5	40
10×10	196
15×15	536
20×20	978
25×25	1622
30×30	2532

By using the above optimization techniques and by using the highest level of compiler optimizations, speed performance of the fixed point implementation can be up to 10 times improved over floating point emulation. By using the above optimization, the fixed point implementation gets close in cycle counts to floating point DSP implementation.

Figure 11 presents number of CPU cycles required to calculate the selected linear algebra algorithms in fixed point arithmetic for different matrix sizes $n \times n$ on a fixed point C64x+ CPU. The fixed point algorithms are implemented in pure C language, and to collect CPU cycle numbers presented in Figure 11 inline function expansion and the highest compiler optimization are used.

Due to its iterative nature the most time consuming algorithm is Jacobi SVD. Algorithm that computes Jacobi SVD and Cholesky factorization algorithm are both $O(n^3)$, but the constant involved for Jacobi SVD is typically ten times the size of the Cholesky constant (Figure 11). The classic Jacobi procedure converges at a linear rate and the asymptotic convergence rate of the method is considerably better than linear [16]. It is customary to refer to N Jacobi updates as a sweep (where N is matrix rank). There is no rigorous theory that enables one to predict the number of sweeps that are required for the algorithm to converge. However, in practice, the number of sweeps is proportional to $\log(n)$. The number of rotations on fixed point C64x+ CPU core for different matrix sizes is presented in Table 6.

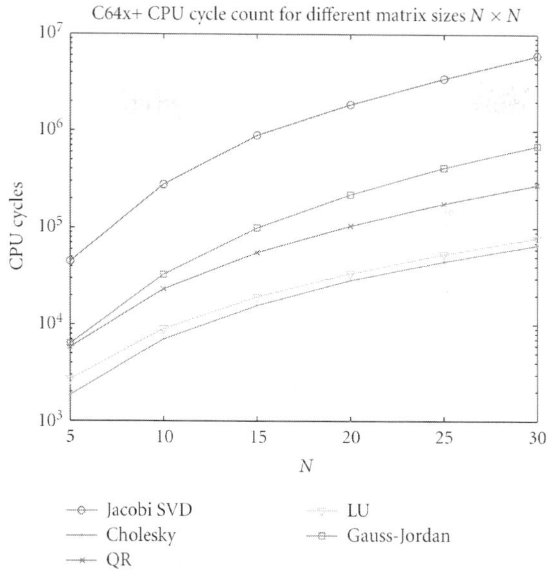

Figure 11: Number of C64x+ CPU cycles required to calculate selected linear algebra algorithms in fixed point arithmetic for different size matrices.

The fixed point CPU cores are capable of running at higher frequencies than floating point CPU cores, and optimized fixed point implementation will usually execute faster. In Figure 12, execution time of the numerical linear algebra algorithms is compared between floating point and fixed point DSPs. In both cases the highest compiler optimization and inline function expansion is used to achieve the lowest cycle count.

The floating point implementation takes slightly less CPU cycles (comparing third column in "*floating point realization*" to third column in "*fixed point realization*" part of Table 5). On the other hand, the fixed point realization exe-cutes faster since the C64x+ CPU core is capable of running at higher clock rates than the C67x+ CPU core. In case when the fixed point DSP runs at 1 GHz and the floating point DSP runs at 300 MHz, fixed point algorithm realization usually executes on average 2.4 times faster.

For a symmetric matrix whose dimension is 30 × 30, the fixed point CPU running at 700 MHz can calculate over 167 Jacobi SVD decompositions per second.

Further performance improvement of the fixed point realization of the selected numerical algorithms can be achieved by hand-optimized implementation in assembly language. Since writing hand optimized assembly is a tedious and time-consuming task, this step is recommended only in cases when C compiler optimizations are not sufficient and an absolute maximum performance is required.

By hard coding *Q* format and implementing Cholesky factorization in hand-optimized assembly, speed performance can be more than doubled for large matrix sizes. The best achievable total cycle count for hand-optimized assembly implementation of Cholesky decomposition of 8×8 matrix is about 2400 cycles using all assembly. Total cycle count for IQmath inline implementation of Cholesky decomposition of 8×8 matrix is about 3500 cycles.

The algorithm realization in C language offers portability. Portability enables designer to run and verify the code on different platforms. This is typically a very important as-pect of system design. The portability is lost in case of hand-optimized assembly implementations. Therefore, hand optimized assembly has the advantage of increasing algorithm speed performance but, on the other side, the implementation process, is time-demanding and offers no code portability. The code modification and maintenance is also much easier if the implementation is kept in C language.

Memory Requirements

The design of efficient matrix algorithm requires careful thinking about the flow of data between the various levels of storage. The vector touch and data reuse issues are important in this regard. In this study both levels of the CPU cache were enabled. In the case of TMS320C6727 DSP, which has a flat memory model, all data and program were kept in the internal memory.

DSPs with cache memory accesses that are localized have less overhead than those with wider ranging access. A matrix algorithm which has mostly row operations, or mostly column operations, can be optimized to take advantage of pat-tern of memory accesses. The Cholesky factorization used for solving normal equations (or any equivalent method such as Gaussian elimination) mixes both row and column operations and is therefore difficult to optimize. QR factorization can be easily arranged to do exclusively row operations [39].

Code size for different algorithm realizations is shown in Table 7.

Increase in speed performance by expanding functions inline increases code size. Function inline expansion is optimal for functions that are called only from a small number of places and for small functions.

If no inline function expansion is used, floating point DSP code size is roughly equivalent to fixed point DSP code (Table 7). For floating point DSP, in the cases of Jacobi SVD, LU, and Gauss-Jordan by expanding functions inline code size decreases. Program level optimization (specified by using the -pm option with the -o3 option) with inline function expansion can sometimes lead to overall code size reduction since compiler can see the entire program, and perform several optimizations that are rarely applied during file-level optimization. Once expanded inline, the functions are optimized in context with the surrounding code.

Accuracy of Fixed Point Implementation

Precision of a number indicates the exactness of the quantity, which is expressed by the number of significant digits. A ma-chine number has limited precision, and as a result, it may be only an approximation of the value it intends to represent. It is difficult to know how much precision is enough. The number of significant digits necessary for one computation will not be adequate for another. Greater precision costs more computation time, so designers must consider the tradeoff carefully.

Table 7: Algorithm code size relative to various numerical linear algebra implementations

Algorithm code size footprint [bytes]	Floating point DSP realizations			Fixed point DSP realizations		
	6711 -pm -o3	6727 -pm -o3	Inlined 6727 -pm -o3	Floating point emulation C64x+ -pm -o3	IQMath C64x+ -pm -o3	Inlined IQMath C64x+ -pm -o3
Jacobi SVD	3200	3008	2976	2688	2528	7072
Cholesky	544	512	676	448	832	1472
LU	1440	1440	1328	1152	1536	2560
QR	1376	1312	1756	1024	1472	3232
Gauss-Jordan	2112	2048	1888	1344	2048	2496

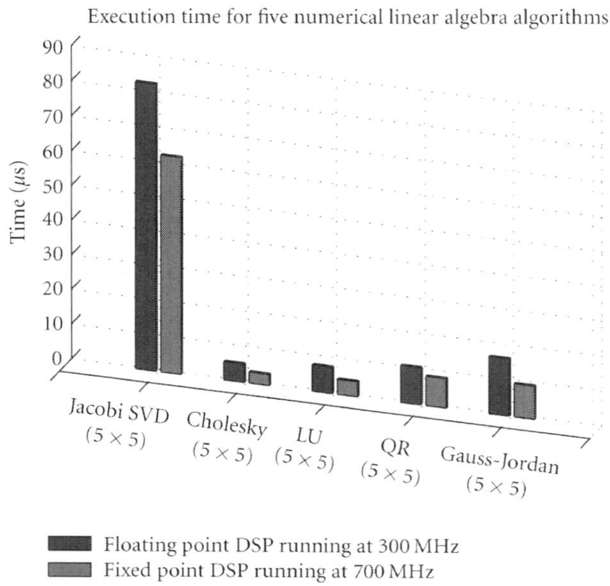

Figure 12: Execution time on floating point C67x+ CPU running at 300 MHz and on C64x+ CPU running at 700 MHz.

The main advantage of floating point over fixed point is its constant relative accuracy. The quantization error gets compounded through error propagation as more arithmetic operations are performed on approximated values. The error can grow with each arithmetic operation until the result no longer represents the true value.

With floating point data types, precision remains approximately constant over most of the dynamic range while with fixed point types, in contrast, the signal to quantization noise ratio increases as the signal decreases in amplitude. To maintain high levels of precision, the signal must be kept within a certain range, large enough to maintain a high signal to quantization noise ratio, but small enough to remain within the dynamic range supported by the fixed point data type. This provides motivation for defining optimal fixed point data types for algorithm variables.

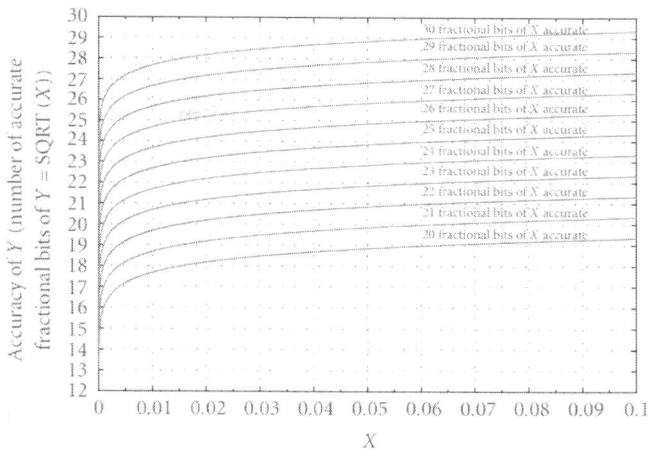

Figure 13: Accuracy of square root calculation SQRT(x) depends on accuracy of the operand x.

Fixed point number formats use tradeoff between dynamic range and accuracy (Table 1). In this implementation, 32-bit target DSP architecture forces tradeoffs between dynamic range and precision. The 32-bits are divided to integer part (characterize dynamic range) and fractional part (define precision). To perform an arithmetic operation between two fixed point numbers, they must be converted to the same fixed point format. Since *WL* of the DSP architecture is 32-bit long, conversion between different fixed point formats is associated with lost of accuracy. For example, to calculate sum of two fixed point variables $a + b$, where a is presented in $Q16$ format and b is presented in $Q22$ format, variable b must be converted to $Q16$ format. The conversion between two for-mats is done by right shifting variable b six times. During the conversion from $Q22$ to $Q16$ format six fractional digits of variable b are lost.

The basic operations such as square root, and division can be very sensitive to the operand noise. In the case of square root accuracy, the result depends on value and ac-curacy of the input (Figure 13). For small operand values, square root operation amplifies inaccuracy of the input variable. Figure 13 presents noise sensitivity of the square root operation SQRT(x) for x < 0.1. Each of the curves corresponds to different accuracy of variable x. As shown in Figure 13, accuracy of SQRT(X) depends on both: value x and accuracy of x. For example, calculating a square root of 0.015 represented with 22 accurate fractional bits gives result with only 20 accurate fractional bits. Therefore, in this case by calculating square root two precision bits are lost.

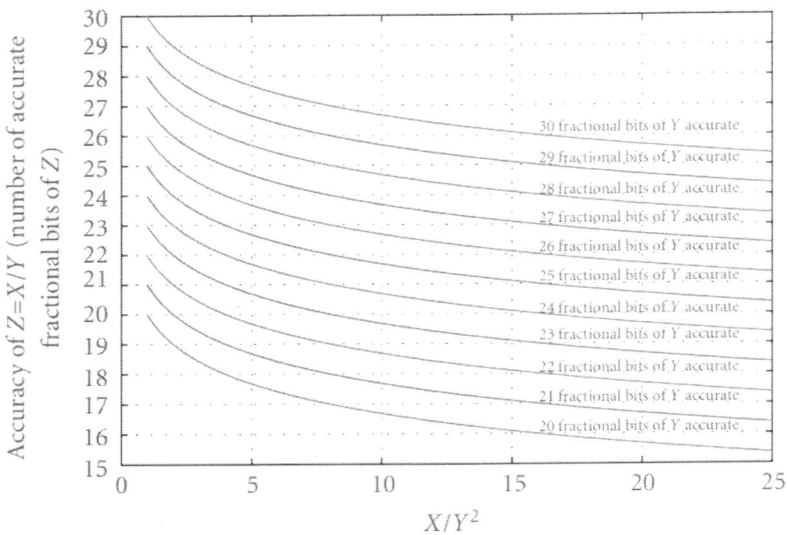

Figure 14: Accuracy of division Z = X/Y depends on accuracy of the operand Y and ratio X/Y^2 (assuming that Y value is much larger than inaccuracy of X).

Division operation exhibits similar behavior to square root. In case of division (Z = X/Y) accuracy of the result depends on value and accuracy of the operands (Figure 14). Assumption taken here is that value Y is much larger than inaccuracy of X. In most cases this assumption is valid. In cases when Y2 X division operation amplifies inaccuracy of the operand Y. Figure 14 presents noise sensitivity of division operation Z = X/Y for $X/Y^2 < 25$. Each of the curves corresponds to different accuracy of variable Y. As shown in Figure 14, accuracy of X/Y depends on: ratio X/Y^2 and accuracy of Y. For example, if X = 1, and Y = 0.25, and Y has 22 accurate fractional bits calculating, Z = X/Y will give only 18.5 accurate fractional bits. Therefore, in this case by calculating division 3.5 precision bits are lost.

In order to determine accuracy of the fixed point arithmetic implementation of numerical linear algebra algorithm we compare the results obtained from our fixed point algorithm to the ones obtained from a floating point implementation. The accuracy of the fixed point implementation is quantified by the number of accurate fractional bits. The number of accurate fractional bits is defined by

$$\text{Number\_of\_Accurate\_Fractional\_Bits} = -\log_2 \left| \max\left(f_{xp} - f_p\right) \right|, \quad (1)$$

where | max(fxp− f p)| represents maximum absolute error between floating point and fixed point representations. The value obtained from the fixed point algorithm is represented by fxp, while f p is the (reference) value obtained from the exact floating point implementation.

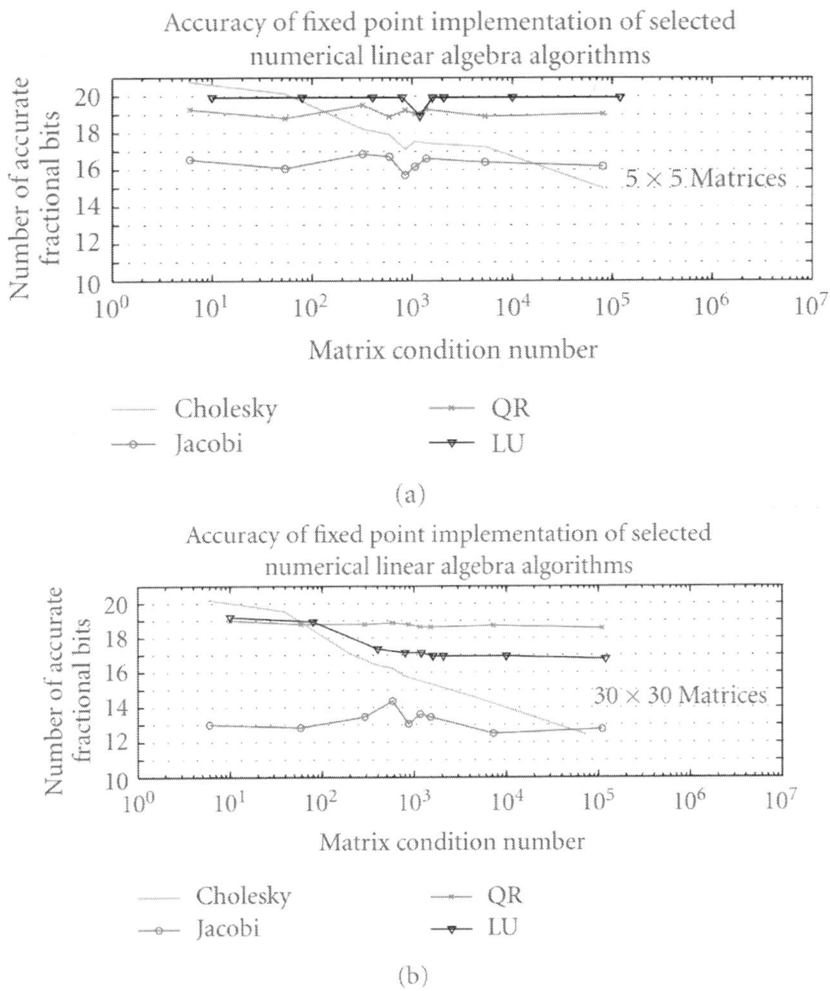

Figure 15: Number of accurate fractional bits. Fixed point implementation of selected numerical linear algebra algorithms (results are for 5×5 and for 30×30 matrices with different condition numbers).

The Q28 fixed point format is used for Cholesky, QR and LU factors. Number of accurate fractional bits for Cholesky, QR, LU factors, and Jacobi SVD eigenvalues for matrices with different condition numbers is presented in Figure 15.

For 5 × 5 matrices, even for relatively high matrix condition number (1.28e5), accuracy of LU, QR, and Jacobi SVD eigenvalues stays unchanged (Figure 15). Number of accurate fractional bits for Cholesky factors declines with large matrix condition numbers. The reasons for decline of accuracy of Cholesky factors (Figure 15) are the following:

- inaccuracy of fixed point operations due to limited word length of the DSP architecture (WL = 32);
- error sensitivity of square root operation when the operand is a small number.

For matrix dimensions of 5 × 5, the fixed point variable sum (calculated in lines (18)-(19), Figure 10) has approximately (24)-(25) accurate fractional bits. The primary sources for inaccuracy of this loop are arithmetic operations and truncation of multiplication result from 64 to 32-bits.

Taking a square root of the variable sum (line (23), in Figure 10) amplifies the inaccurate fractional bits in case when sum is much smaller than one. For example, when value of the variable *sum* is close to 0.06, square root calculation doubles the inaccuracy. Calculating square root of variable *sum,* in case when its value is equal to 0.06 with 25 accurate fractional bits, gives a result with 24 accurate bits (Figure 13). The value of variable *sum* gets small for matrices with large condition numbers, which is causing error to increase (Figure 15).

According to Figure 15, for 5 × 5 matrices with condition number lower than 100, Cholesky factors have 24.13 accurate bits (20.13 accurate fractional bit and four integer bits since $IWL = 4$). For matrix with condition number of 1.28e5, Cholesky factors have 18.99 accurate bits (14.99 accurate fractional bits and four integer bits since $IWL = 4$).

According to Figure 15, for 5×5 matrices *LU* factors have 22.93 accurate bits (18.93 accurate fractional bits and four integer bits since $IWL = 4$).

For LU decomposition we used Crout's algorithm. Pivoting is absolutely essential for the stability of Crout's method. Only partial pivoting (interchange of rows) is implemented. However, this is enough to make the method stable for the proposed tests. Without pivoting division by small numbers can lead to a severe loss of accuracy during LU decomposition.

According to Figure 15, for 5×5 matrices *QR* factors have 22.79 accurate (18.79 accurate fractional bits and four integer bits since $IWL = 4$).

In case of Jacobi SVD, eigenvalues and eigenvectors are presented in *Q*28 fixed point format ($IWL = 4$). In order to calculate Jacobi SVD a number of intermediate variables with different fixed point formats are used. The maximum number of fractional bits is utilized for most of the internal variables. In order to accommodate large intermediate results the $Q16$ fixed point format is used for some internal variables. Conversion between different fixed point formats is associated with lost of accuracy, so not all 28 fractional bits of the result are accurate.

For the considered tests the eigenvalue problem is always well conditioned, also for ill conditioned matrices, since the involved matrices are symmetric positive definite.

In the case of 30×30 matrices computational accuracy decreases due to the increase in number of arithmetic operations required to calculate matrix decompositions (lower panel in Figure 15). For 30×30 matrices, Jacobi SVD method is 3-bit less accurate than in case of 5×5 matrices. For 5×5 matrices accuracies of *LU* and *QR* factorization are similar (accumulation of computational inaccuracy is not big enough to affect overall accuracy of *LU* decomposition). Large number of computations takes its toll on *LU* decomposition in case of 30×30 matrices. During *LU* decomposition calculations of the elements of L matrix require division by the elements on main diagonal of U matrix. For large matrix condition numbers the lower right diagonal element of matrix U becomes smaller, and due to in-creased number of operations less accurate. Division by small and less accurate numbers amplifies inaccuracy (Figure 14). Therefore, with the increase of the matrix condition number, *LU* decomposition accuracy decreases for 30×30 matrices.

Accuracy of the fixed point implementation of linear algebra algorithms relies on IQmath functions. IQmath functions are optimized for the C64x+ architecture and use 64-bit precision wherever possible (IQmath functions employ the CPU intrinsic operation that multiplies two 32-bit values in a 64-bit result).

CONCLUSION

The primary goal of this paper is to address implementententional aspects of the numerical linear algebra for real-time applications on fixed point DSPs. In this paper, we compared performance (accuracy, memory requirements, and speed) between floating point and fixed point implementations for five linear algebra algorithms. Numerical linear algebra algorithms are defined in terms of the real number system, which has infinite precision. These algorithms are implemented on DSPs with finite precision. Computer roundoff errors can and do cause numerical linear algebra algorithms to diverge. The algorithms considered here proved to be numerically stable in fixed point arithmetic for the proposed tests.

Most floating point software routines are very slow with-out considerable hardware support. This can make floating point algorithms costly. The best way to write code for target hardware that does not support floating point is to not use floating point. Advantages of implementation in fixed point are the following:

- fractional arithmetic can be performed on fixed point numbers using integer hardware which is considerably faster than floating point hardware;
- less hardware implies low power consumption for battery powered devices;
- a fixed point algorithm can use less data memory com-pared to its floating point implementation.

In fixed point representation of fractional numbers, dynamic range and fractional accuracy are complementary to each other. This poses a unique problem during arithmetic operations. Some of the common problems with fixed point numbers are the following:

- a fixed point number has limited integer range of values and does not support automatic scaling as in floating point. It is not possible to represent very large and very small numbers with this representation;
- conversion between different fixed point formats is associated with lost of accuracy;
- drastic change in value results if intermediate result exceeds maximum allowed. It is easy for an arithmetic operation to produce an "overflow" or "under-flow." Thus the choice of the fixed point representation should be made very carefully and it should best suit the algorithms need. Most DSPs support saturation arithmetic to handle this problem.

In this paper, we introduced a flow analysis that is necessary for the transformation from floating point arithmetic to fixed point. The software tools presented in this paper semiautomatically convert floating point DSP algorithms implemented in C/C+ to fixed point algorithms that achieve maximum accuracy. In our approach, a simulation-based method is adopted to estimate dynamic ranges, where the range of each signal is measured during the floating point simulation using realistic input signal files. The range estimator finds the statistics of internal signals throughout the floating point simulation using real inputs and determines scaling parameters. This method is applicable to both non-linear and linear systems since it derives an adequate estimation of the range from a finite length of simulation results.

We also introduce a direct link to DSP implementation by processor specific C code generation and advanced code optimization. The fixed point algorithm implementation heavily relies on the IQmath library. The IQmath library provides blocks that perform C64x+ processor-optimized, fixed point mathematical operations. The IQmath library functions generally input and output fixed point data types and use numbers in *Q* format. The fixed point DSP target code yields bit-by-bit the same results as the bit-true SystemC code from host simulation. This enables comparative simulation to the reference model. The main bottleneck of the float to fixed point conversion flow is simulation speed of bit-true fixed point model in SystemC. By implementation in fixed point a speedup by a factor of 10 can be achieved compared to floating point emulation.

The numerical linear algebra algorithms require slightly less CPU cycles on a floating point DSP, but since the DSPs run at slower clock rates the algorithms can still execute faster on a fixed point DSP. On the other hand, accuracy of the fixed point implementation is not as good as in floating point. It is the accuracy of a floating point number that is so expensive. By implementing the algorithms in fixed point the correct-ness of the result is compromised. For some applications, a fast but possibly inexact solution is more acceptable than a slow but correct solution. Floating point representation al-ready approximates values. Approach presented in this paper is another approximation which is less accurate than floating point but provides for an increase in speed. Speed for ac-curacy is an important tradeoff, and its applicability should be examined at each level that abstracts floating point arithmetic.

For the numerical linear algebra algorithms considered, the fixed point DSP and its optimizing compiler make an efficient combination. These optimizations lead to a consider-able improvement in performance in many cases as the compiler was able to utilize software pipelining and instruction level parallelism to speed up the code. It has turned out that software pipelining and inline function expansion is the key to achieving high performance. The high performance was achieved by using only compiler optimization techniques. It is possible to achieve even further performance improvement by careful analysis and code restructuring.

All phases of the fixed point design flow discussed in the paper are based on C/C++ language implementation which makes it maintainable, readable, and applicable to a number of different platforms on which the flow can execute correctly and reliably.

REFERENCES

1. P. van Dooren, "Numerical aspects of system and control algorithms," Journal A, vol. 30, no. 1, pp. 25–32, 1989.
2. I. Jollife, Principal Component Analysis, Springer, New York, NY, USA, 1986.
3. M. S. Grewal and A. P. Andrews, Kalman Filtering Theory and Practice, Prentice Hall Information and Systems Sciences Series, Prentice-Hall, Upper Saddle River, NJ, USA, 1993.
4. G. Frantz and R. Simar, "Comparing Fixed and Floating Point DSPs," SPRY061, Texas Instruments, 2004.
5. S. Kim and W. Sung, "A floating-point to fixed-point assembly program translator for the TMS 320C25," IEEE Transactions on Circuits and Systems II: Analog and Digital Signal Processing, vol. 41, no. 11, pp. 730–739, 1994.
6. Simulink, "Simulation and Model Based Design," Simulink Reference, Version 6, The Mathworks 2006.
7. IEEE Std 1666-2005 IEEE Standard SystemC Language Reference Manual, http://standards.ieee.org/getieee/1666/ download/1666-2005.pdf.
8. "Matlab The Language of Technical Computing," Function Reference, Version 7, The Mathworks 2006.
9. M. Coors, H. Keding, O. Luthje, and H. Meyr, "Design and ¨ DSP implementation of fixed-point systems," EURASIP Journal on Applied Signal Processing, vol. 2002, no. 9, pp. 908–925, 2002.
10. B. Liu, "Effect of finite word length on the accuracy of digital filters—a review," IEEE Transactions on Circuit Theory, vol. 18, no. 6, pp. 670–677, 1971.
11. S. Kim, K.-I. I. Kum, and W. Sung, "Fixed-point optimization utility for C and C++ based digital signal processing programs," IEEE Transactions on Circuits and Systems II: Analog and Digital Signal Processing, vol. 45, no. 11, pp. 1455–1464, 1998.
12. A. Gelb, Applied Optimal Estimation, The MIT Press, Cambridge, Mass, USA, 1992.
13. T. Aamodt and P. Chow, "Numerical error minimizing floating point to fixed-point ANSI C compilation," in The 1st Workshop on Media Processors and Digital Signal Processing (MPDSP '99), pp. 3–12, Haifa, Israel, November 1999.

14. K. Han and B. L. Evans, "Optimum word length search using sensitivity information," EURASIP Journal on Applied Signal Processing, vol. 2006, Article ID 92849, 14 pages, 2006.
15. C. Shi and R. W. Brodersen, "Floating-point to fixed-point conversion with decision errors due to quantization," in Proceedings of IEEE International Conference on Acoustics, Speech, and Signal Processing (ICASSP '04), vol. 5, pp. 41–44, Montreal, Que, Canada, May 2004.
16. G. Golub and C. van Loan, Matrix Computations, Johns Hopkins University Press, Baltimore, Md, USA, 1996.
17. G. Golub and I. Mitchell, "Matrix factorizations in fixed point on the C6x VLIW architecture," Stanford University, Stanford, Calif, USA, 1998.
18. G. A. Hedayat, "Numerical linear algebra and computer architecture: an evolving interaction," Tech. Rep. UMCS-93-1-5, Department of Computer Science, University of Manchester, Manchester, UK, 1993.
19. S. A. Wadekar and A. C. Parker, "Accuracy sensitive word length selection for algorithm optimization," in Proceedings of IEEE International Conference on Computer Design: VLSI in Computers and Processors (ICCD '98), pp. 54–61, Austin, Tex, USA, October 1998.
20. G. A. Constantinides, P. Y. K. Cheung, and W. Luk, "Word length optimization for linear digital signal processing," IEEE Transactions on Computer-Aided Design of Integrated Circuits and Systems, vol. 22, no. 10, pp. 1432–1442, 2003.
21. M. Stephenson, J. Babb, and S. Amarasinghe, "Bit width analysis with application to silicon compilation," in Proceedings of the ACM SIGPLAN Conference on Programming Language Design and Implementation, pp. 108–120, Vancouver, BC, Canada, June 2000.
22. C. Shi and R. W. Brodersen, "Automated fixed-point data-type optimization tool for signal processing and communication systems," in Proceedings of 41st Annual Conference on Design Automation, pp. 478–483, San Diego, Calif, USA, June 2004.
23. A. Nayak, M. Haldar, A. Choudhary, and P. Banerjee, "Precision and error analysis of MATLAB applications during automated hardware synthesis for FPGAs," in Proceedings of Design, Automation and Test in Europe, Conference and Exhibition (DATE '01), pp. 722–728, Munich, Germany, March 2001.

24. R. Cmar, L. Rijnders, P. Schaumont, S. Vernalde, and I. Bolsens, "A methodology and design environment for DSP ASIC fixed point refinement," in Proceedings of Design, Automation and Test in Europe, Conference and Exhibition (DATE '99), pp. 271–276, Munich, Germany, March 1999.
25. S. Kamath, N. Magotra, and A. Shrivastava, "Quantization analysis tool for fixed-point implementation of real time algorithms on the TMS320C5000," in Proceedings of IEEE International Conference on Acoustics, Speech, and Signal Processing (ICASSP '02), vol. 4, pp. 3784–3787, Orlando, Fla, USA, May 2002.
26. K. Han and B. L. Evans, "Word length optimization with complexity-and-distortion measure and its application to broadband wireless demodulator design," in Proceedings of IEEE International Conference on Acoustics, Speech, and Signal Processing (ICASSP '04), vol. 5, pp. 37–40, Montreal, Que, Canada, May 2004.
27. L. B. Jackson, "On the interaction of the round-off noise and dynamic range in digital filters," The Bell System Technical Journal, vol. 49, no. 2, pp. 159–184, 1970.
28. V. J. Mathews and Z. Xie, "Fixed-point error analysis of stochastic gradient adaptive lattice filters," IEEE Transactions on Acoustics, Speech, and Signal Processing, vol. 38, no. 1, pp. 70–80, 1990.
29. A. V. Oppenheim, R. W. Schafer, and J. R. Buck, Discrete-Time Signal Processing, Prentice-Hall, Upper Saddle River, NJ, USA, 1998.
30. W. H. Press, B. P. Flannery, S. A. Teukolsky, and W. T. Vetterling, Numerical Recipes in C: The Art of Scientific Computing, Cambridge University Press, Cambridge, UK, 1992.
31. W. Cammack and M. Paley, "Fixpt: a C++ method for development of fixed point digital signal processing algorithms," in Proceedings of the 27th Annual Hawaii International Conference on System Sciences (HICSS '94), vol. 1, pp. 87–95, Maui, Hawaii, USA, January 1994.
32. "TMS320C64/C64x+ DSP CPU and Instruction Set Reference Guide," SPRU732C, Texas Instruments, August 2006, http://focus.ti.com/lit/ug/spru732c/spru732c.pdf.
33. "TMS320C67x/C67x+ DSP CPU and Instruction Set Reference Guide," SPRU733, Texas Instruments, May 2005.
34. N. Seshan, T. Hiers, G. Martinez, A. Seely, and Z. Nikolic,́ "Digital signal processors for communications, video infrastructure, and audio,"

in Proceedings of IEEE International SOC Conference (SOCC '05), pp. 319–321, Herndon, Va, USA, September 2005.

35. "TMS320C64x+ DSP Mega-module Reference Guide," SPRU871, Texas Instruments, June 2007, http://focus.ti.com/lit/ug/ spru871g/ spru871g.pdf.
36. "TMS320C6000 Programer's Guide," SPRU198i, Texas Instruments, March 2006, http://focus.ti.com/lit/ug/spru198i/ spru198i.pdf.
37. IQmath Librar, A Virtual Floating Point Engine, Module User's Guide C28x Foundation Software, version 1.4.1, Texas Instruments, 2002.
38. E. Granston, "Hand tuning loops and control code on the TMS320C6000," Application Report SPRA666, Texas Instruments, Stafford, Tex, USA, August 2006.
39. J. Halleck, "Least squares network adjustments via QR factorization," Surveying and Land Information Systems, vol. 61, no. 2, pp. 113–122, 2001.

CHAPTER 9

PERFORMANCE VERSUS ENERGY CONSUMPTION OF HYPERSPECTRAL UNMIXING ALGORITHMS ON MULTI-CORE PLATFORMS

Alfredo Remon[1], Sergio Sanchez [2], Sergio Bernabe[2], Enrique S Quintana-Ort´ı[1] and Antonio Plaza[2]

[1]Department of Engineering and Computer Sciences, University Jaime I, Castellon, Spain

[2]Hyperspectral Computing Laboratory (HyperComp), Department of Technology of Computers and Communications, University of Extremadura, Caceres, Spain.

ABSTRACT

Hyperspectral imaging is a growing area in remote sensing in which an imaging spectrometer collects hundreds of images (at different wavelength

Citation (APA): Remón, A., Sánchez, S., Bernabé, S., Quintana-Ortí, E. S., & Plaza, A. (2013). Performance versus energy consumption of hyperspectral unmixing algorithms on multi-core platforms. EURASIP Journal on Advances in Signal Processing, 2013(1), 68. (15 pages).

channels) for the same area on the surface of the Earth. Hyperspectral images are extremely high-dimensional, and require on-board processing algorithms able to satisfy near real-time constraints in applications such as wildland fire monitoring, mapping of oil spills and chemical contamination, etc. One of the most widely used techniques for analyzing hyperspectral images is spectral unmixing, which allows for sub-pixel data characterization. This is particularly important since the available spatial resolution in hyperspectral images is typically of several meters, and therefore it is reasonable to assume that several spectrally pure substances (called *endmembers* in hyperspectral imaging terminology) can be found within each imaged pixel. There have been several efforts towards the efficient implementation of hyperspectral unmixing algorithms on architectures susceptible of being mounted onboard imaging instruments, including field programmable gate arrays (FPGAs) and graphics processing units (GPUs). While FPGAs are generally difficult to program, GPUs are difficult to adapt to onboard processing requirements in spaceborne missions due to its extremely high power consumption. In turn, with the increase in the number of cores, multi-core platforms have recently emerged as an easier to program platform compared to FPGAs, and also more tolerable radiation and power consumption requirements. However, a detailed assessment of the performance versus energy consumption of these architectures has not been conducted as of yet in the field of hyperspectral imaging, in which it is particularly important to achieve processing results in real-time. In this article, we provide a thoughtful perspective on this relevant issue and further analyze the performance versus energy consumption ratio of different processing chains for spectral unmixing when implemented on multi-core platforms.

Keywords: Hyperspectral Image, Hyperspectral Data, Spectral Unmixing, Energy Consumption Ratio, Principal Component Transform

INTRODUCTION

Hyperspectral imaging instruments are capable of collecting hundreds of images, corresponding to different wavelength channels, for the same area on the surface of the Earth [1]. For instance, NASA is continuously gathering imagery data with instruments such as the Airborne Visible Infra-Red Imaging Spectrometer (AVIRIS), which operates in the 0.4–2.5 μ m spectral range, with 10 nm spectral resolution and 30 m spatial resolution [2]. As a follow-up to the success of AVIRIS (an airborne instrument), a new generation of satellite instruments for Earth observation are operating or under development (see Table 1). As indicated there, most current hyperspectral missions are spaceborne in nature [3].

Table 1: Overview of some present and future remote sensing missions including hyperspectral sensors

	EO-1 Hyperion[a]	Prisma[b]	EnMAP[c]	HyspIRI[d]
Country of origin	USA	Italy	Germany	USA
Spatial resolution (m)	30	5–30	30	60
Revisit time (days)	16	3/7	4	18
Spectral range (nm)	400–2,500	400–2,500	420–2,450	380–2,500
Spectral resolution (nm)	10	10	6.5–10	10
Swath width (km)	7.7	30	30	120
Earth coverage	Partial	Full	Full	Full
Launch	2000	2012	2015	2018

[a]http://eo1.gsfc.nasa.gov.

[b] http://www.asi.it/en/flash_en/observing/prisma

[c]http://www.enmap.org.

[d]http://hyspiri.jpl.nasa.gov.

One of the main problems in the analysis of hyperspectral data cubes is the presence of mixed pixels [4, 5], which arise when the spatial resolution of the sensor is not fine enough to separate spectrally distinct materials (see Figure 1). Spectral unmixing [6, 7, 8] is one of the most popular techniques to analyze hyperspectral data. It involves the separation of a pixel spectrum into its pure component (endmember) spectra [9, 10], and the estimation of the abundance value for each endmember [11, 12]. The linear mixture model assumes single scattering between the endmember substances resulting from the fact that they are sitting side-by-side within the field of view of the imaging instrument (see Figure 2a). On the other hand, the nonlinear mixture model [13, 14, 15] assumes nonlinear interactions and multiple scattering between endmember substances (see Figure 2b). In practice, the linear model is more flexible and can be easily adapted to different analysis scenarios [16]. It can be simply defined as follows [17]:

$$\mathbf{y} = \mathbf{E}\mathbf{a} + \mathbf{n} = \sum_{i=1}^{p} \mathbf{e}_i a_i + \mathbf{n}, \qquad (1)$$

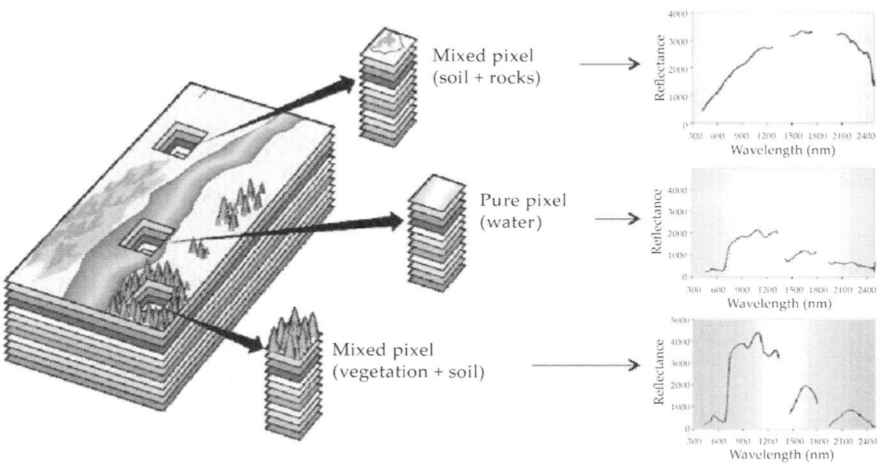

Figure 1: Presence of mixed pixels in remotely sensed hyperspectral images.

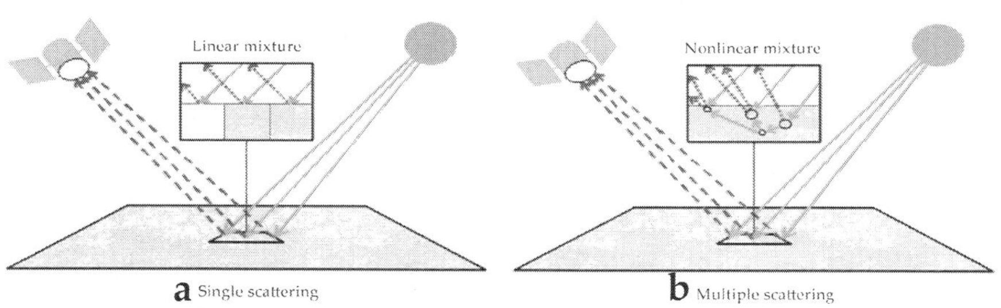

Figure 2: Mixture models in remotely sensed hyperspectral imaging. (a) Linear and (b) nonlinear.

where **y** is an *n*-dimensional pixel vector given by a collection of values at different wavelengths, $E=\{ei\}^{p}_{i=1}$ is a matrix containing *p* endmembers, $a=[a_1,a_2,\ldots,a_p]$ is a *p*-dimensional vector containing the abundance fractions for each of the *p* endmembers in **y**, and **n** is a noise term. Generalizing this expression for all the hyperspectral pixels in the scene (in compact matrix notation) yields $\mathbf{Y}=\mathbf{E}\ \mathbf{A}+\mathbf{N}$, where **Y** is the full hyperspectral image with *m*pixels, each with *n* bands, **E** is the endmember matrix with dimensions $n \times p$, **A** is an $p \times m$matrix containing the endmember abundances for each pixel of the scene, and **N** is a $n \times m$ noise matrix. With the aforementioned notation in mind, solving the linear mixture model involves: (1) estimating the number of endmembers, *p*, in the hyperspectral scene; (2) identifying a collection of $E=\{ei\}^{p}_{i=1}$ endmembers; and (3) estimating the fractional abundances of the *p*endmembers for each pixel in the hyperspectral data set.

Several techniques have been proposed to solve this problem under the linear mixture model assumption in recent years (see [18, 19, 20, 21, 22, 23, 24, 25, 26, 27, 28, 29, 30, 31, 32, 33, 34, 35, 36, 37, 38, 39, 40, 41, 42, 43], among several others), but all of them are quite expensive in computational terms. Although these techniques map nicely to high performance computing platforms such as commodity clusters [44], these systems are difficult to adapt to on-board processing requirements introduced by applications with real-time constraints such as wild land fire tracking, biological threat detection, monitoring of oil spills and other types of chemical contamination [45, 46, 47]. In those cases, low-weight integrated components such as field programmable gate arrays (FPGAs) [48, 49, 50] and graphics processing units (GPUs) [51, 52] have the potential to reduce payload in current and future Earth observation missions, which are mainly spaceborne in nature as indicated by Table 1. Furthermore, GPUs offer fast processing at low cost (in the literature, so far GPUs have been the only platform shown to be able to process hyperspectral images in real-time) and easy programmability which are very appealing for future remote sensing missions [53, 54, 55, 56, 57, 58]. On the negative side, FPGAs are difficult to program and there is currently a lack of efficient implementations of a full spectral unmixing chain for hyperspectral image processing in these type of architectures. Besides, GPUs are currently not suitable for spaceborne missions due to their high power consumption and the lack of radiation tolerance. These aspects are critical for the definition of the mission (payload) and its overall success and lifetime. In turn, with the increase in the number of cores, multi-core platforms have recently emerged as an easier to program platform as compared to FPGAs, and also more flexible to accomodate to radiation tolerance and power consumption requirements. Nevertheless, a detailed assessment of the performance versus energy consumption of these architectures has not been conducted as-of-yet in the field of hyperspectral imaging, in which it is particularly important to achieve processing results in real-time with low energy cost.

In this article, we provide a thoughtful perspective on this relevant issue and further analyze the performance versus energy consumption ratio of different processing chains for spectral unmixing when implemented on multi-core platforms. This kind of analysis has not been previously conducted in the literature, and in our opinion it is very important in order to really calibrate the possibility of using multi-core platforms for efficient hyperspectral image processing in real remote sensing missions.

The remainder of the article is organized as follows. Section 2 reviews the different modules that conform the considered unmixing chains discussed in this work. Section 3 describes the parallel implementation of these modules on multi-core platforms. Section 4 presents an experimental evaluation of the proposed implementations in terms of unmixing accuracy, parallel performance and energy consumption, reporting several multi-core implementations able to provide real-time analysis performance and discussing their energy consumption requirements. Section 5 concludes the article with some remarks and hints at plausible future research lines.

SPECTRAL UNMIXING MODULES

In this section, we describe different modules for spectral unmixing of hyperspectral data. These modules will be then used to define spectral unmixing chains, which are composed of three main stages: (1) estimating the number of endmembers in the original hyperspectral scene; (2) identifying a collection of endmembers in the scene; and (3) estimating the fractional abundances of endmembers in each pixel of the scene. In the following, we describe different methods in each category, offering a few remarks that describe computational aspects such as the operations that are involved and the cost of the algorithm in terms of floating-point arithmetic operations (flops). In the following cost expressions, for simplicity, we neglect lower order terms, taking into account that in practice p, $n \ll m$.

Methods for Estimating the Number of Endmembers

This section introduces two different methods for estimating the number of endmembers: virtual dimensionality (VD) [59] and hyperspectral signal identification by minimum error (HySime) [60].

Virtual Dimensionality (VD)

The VD method first calculates the eigenvalues of the covariance matrix $K_{L\times L} = 1/N(\mathbf{Y} - \overline{\mathbf{Y}})^T(\mathbf{Y} - \overline{\mathbf{Y}})$ and the correlation matrix $R_{L\times L} = \mathbf{K}_{L\times L} + \overline{\mathbf{Y}\mathbf{Y}}^T$, referred to as covariance-eigenvalues and correlation-eigenvalues, for each of the spectral bands in the original hyperspectral image **Y**. The VD concept follows the "pigeon-hole principle". If we represent a signal source by a pigeon and a spectral band by a hole, we can use a spectral band to accommodate one source. Thus, if a distinct spectral signature makes a contribution to the eigenvalue-represented signal energy in one spectral band, then its associated correlation eigenvalue will be greater than its corresponding covariance eigenvalue in this particular band. Otherwise, the correlation eigenvalue would be very close to the covariance eigenvalue, in which case only noise would be present in this particular band. By applying this concept, a Neyman–Pearson detector [59] is introduced to formulate the issue of whether a distinct signature is present or not in each of the spectral bands of **Y** as a binary hypothesis testing problem. Here, the decision is made based on an input parameter of the algorithm that is called the false alarm probability or P_F, which is used to establish the sensitivity of the algorithm in terms of how much error can be tolerated in the identification of the actual number of endmembers in the image data. With this interpretation in mind, the issue of determining an estimate p^ for the number of endmembers is further simplified and reduced to a specific value of P_F that is preset by the Neyman–Pearson detector.

From the computational point of view, the most complex operation in this algorithm is related with the calculation of the covariance and correlation matrices which need to be compared in order to determine the number of endmmebers. If we recall that the number of bands of the hyperspectral image is denoted by n, the total cost of each calculation is given by n^2 flops.

HySime Method

The HySime method consists of two parts. Algorithm 1 describes the noise estimation part, which obtains an $N \times L$ matrix ξ^ containing an estimation of the noise present in the original hyperspectral image **Y**[60]. This algorithm follows an approach which addresses the high correlation exhibited by close spectral bands. The main advantage of Algorithm 1 is that the computational complexity is substantially lower than that of other algorithms for noise estimation in hyperspectral data in the literature. Additional details about Algorithm 1 can be found in [60] and we do not repeat them for space considerations. On the other hand, Algorithm 2 describes the signal subspace identification part of the algorithm, which first computes the noise correlation matrix $R\hat{}_n$ and then computes the signal correlation matrix $R\hat{}_x$. Next, the eigenvectors of the signal correlation matrix are obtained and sorted in ascending order. Finally, a minimization function is applied to obtain an estimate p^ of the number of endmembers in the subspace X^. The main purpose of this algorithm is to select the subset of eigenvectors that best represents the signal subspace in the minimum mean squared error sense. As in the case of the previous algorithm, the most complex operations are due to the calculation of the covariance and correlation matrices. Again, the total cost of each calculation is given by n^2 flops, where n is the number of bands of the hyperspectral image.

Algorithm 1 Noise estimation

INPUT: $\mathbf{Y} \equiv [\mathbf{y}_1, \mathbf{y}_2, \ldots, \mathbf{y}_N]$
$\mathbf{Z} := \mathbf{Y}^T, \widehat{\mathbf{R}} := (\mathbf{Z}^T\mathbf{Z})$
$\mathbf{R}' := \widehat{\mathbf{R}}^{-1}$
for i:=1 to L **do**
$\quad \widehat{\beta_i} := ([\mathbf{R}']_{\alpha_i,\alpha_i} - [\mathbf{R}']_{\alpha_i,i}[\mathbf{R}']_{i,\alpha_i}/[\mathbf{R}']_{i,i})[\widehat{\mathbf{R}}]_{\alpha_i,i}$
{Note that $\alpha_i = 1, \ldots, i-1, i+1, \ldots, L$ }
$\quad \widehat{\xi_i} := \mathbf{z}_i - \mathbf{Z}_{\alpha_i}\widehat{\beta_i}$
end for
OUTPUT: $\widehat{\xi}$ {$N \times L$ matrix with the estimated noise}

Algorithm 2 Signal subspace estimation

INPUTS: $\mathbf{Y} \equiv [\mathbf{y}_1, \mathbf{y}_2, \ldots, \mathbf{y}_N]$, $\widehat{\mathbf{R}}_y \equiv, (\mathbf{Y}\mathbf{Y}^T)/N, \widehat{\xi}$
$\widehat{\mathbf{R}}_n := 1/N \sum_i (\widehat{\xi}_i \widehat{\xi}_i^T)$
$\widehat{\mathbf{R}}_x := 1/N \sum_i ((\mathbf{y}_i - \widehat{\xi}_i)(\mathbf{y}_i - \widehat{\xi}_i^T))$ {estimate of $\widehat{\mathbf{R}}_x$}
$\mathbf{E} := [\mathbf{e}_1, \ldots, \mathbf{e}_L]$ {$\mathbf{e}_i$ are eigenvectors of $\widehat{\mathbf{R}}_x$}
$\delta := [\delta_1, ..., \delta_L]$
$(\widehat{\delta}, \widehat{\pi}) :=$ sort(δ) {sort δ_i in ascending order of the corresponding eigenvalues and save the permutation $\widehat{\pi}$}
$\widehat{p} :=$ number of terms $\widehat{\delta_i} < 0$ {This is a simplified description of the minimization function, which is fully described in [60] and ommitted here for space considerations.}
OUTPUT: $\widehat{\mathbf{X}} = \langle[\widehat{\mathbf{e}}_{i1}, \cdots, \widehat{\mathbf{e}}_{ip}]\rangle$ {Estimated subspace}

Methods for Endmember Identification

This section introduces two different methods for identifying the endmember signatures in the hyperspectral data: orthogonal subspace projection with Gram-Schmidt orthogonalization (OSP-GS) [18] and N-FINDR [23].

Orthogonal Subspace Projection with Gram-Schmidt Orthogonalization (OSP-GS)

The OSP algorithm [18] was originally developed to find spectrally distinct signatures using orthogonal projections. For this work, we have used an optimization of this algorithm (see [61, 62]) which allows calculating the OSP without requiring the computation of the inverse of the matrix that contains the endmembers already identified in the image. This operation, which is difficult to implement in parallel, is accomplished using the Gram-Schmidt method for orthogonalization. This process selects a finite set of linearly independent vectors $\mathbf{A} = \{\mathbf{a}_1, \ldots, \mathbf{a}_p\}$ in the inner product space $\mathbf{R}^n$ in which the original hyperspectral image is defined, and generates an orthogonal set of vectors $\mathbf{B} = \{\mathbf{b}_1, \ldots, \mathbf{b}_p\}$ which spans the same *p-dimensional* subspace of $\mathbf{R}^n$ $(p \leq n)$ as $\mathbf{A}$. In particular, $\mathbf{B}$ is obtained as follows:

$$
\begin{aligned}
&\mathbf{b}_1 = \mathbf{a}_1, && \mathbf{e}_1 = \frac{\mathbf{b}_1}{\|\mathbf{b}_1\|} \\
&\mathbf{b}_2 = \mathbf{a}_2 - \mathrm{proj}_{\mathbf{b}_1}(\mathbf{a}_2), && \mathbf{e}_2 = \frac{\mathbf{b}_2}{\|\mathbf{b}_2\|} \\
&\mathbf{b}_3 = \mathbf{a}_3 - \mathrm{proj}_{\mathbf{b}_1}(\mathbf{a}_3) - \mathrm{proj}_{\mathbf{b}_2}(\mathbf{a}_3), && \mathbf{e}_3 = \frac{\mathbf{b}_3}{\|\mathbf{b}_3\|} \\
&\mathbf{b}_4 = \mathbf{a}_4 - \mathrm{proj}_{\mathbf{b}_1}(\mathbf{a}_4) - \mathrm{proj}_{\mathbf{b}_2}(\mathbf{a}_4) - \mathrm{proj}_{\mathbf{b}_3}(\mathbf{a}_4), && \mathbf{e}_4 = \frac{\mathbf{b}_4}{\|\mathbf{b}_4\|} \\
&\vdots && \vdots \\
&\mathbf{b}_p = \mathbf{a}_p - \sum_{j=1}^{p-1} \mathrm{proj}_{\mathbf{b}_j}(\mathbf{a}_p), && \mathbf{e}_p = \frac{\mathbf{b}_p}{\|\mathbf{b}_p\|},
\end{aligned}
\quad (2)
$$

where the projection operator is defined as

$$
\mathrm{proj}_{\mathbf{b}}(\mathbf{a}) = \frac{<\mathbf{a}, \mathbf{b}>}{<\mathbf{b}, \mathbf{b}>}\mathbf{b}, \quad (3)
$$

and $<\mathbf{a}, \mathbf{b}>$ denotes the inner product of vectors **a** and **b**.

The sequence $\mathbf{b}_1, \ldots, \mathbf{b}_p$ in Equation (2) represents the set of orthogonal vectors generated by the Gram-Schmidt method, and thus, the normalized vectors $\mathbf{e}_1, \ldots, \mathbf{e}_p$ in (2) form an orthonormal set. As far as **B** spans the same p-dimensional subspace of $\mathbf{R}^n$ as **A**, an additional vector $\mathbf{b}_{p+1}$ computed by following the procedure stated at (2) is also orthogonal to all the vectors included in **A** and **B**. This algebraic assertion constitutes the cornerstone of the OSP method with Gram-Schmidt orthogonalization.

From the computational point of view, this algorithm has to be augmented with some sort of column pivoting that, at each step of the orthogonalization, detects the pixel with maximum projection value among those of the image (see [63] for details). Unfortunately, this requires that each projector is applied to all pixels of the scene, not only to p, yielding a significant increase in the arithmetic cost of the algorithm. Given the $3n$ flops required to apply the projector (3) to one pixel, and the p endmembers that have to be identified, the result is a total cost for the algorithm of *3mnp* flops.

N-FINDR

The *N*-FINDR algorithm [23] is one of the most widely used and successfully applied methods for automatically determining endmembers in hyperspectral image data without using *a priori* information. This algorithm looks for the set of pixels with the largest possible volume by *inflating* a simplex inside the data. The procedure begins with a random initial selection of pixels (see Figure 3a). Every pixel in the image must be evaluated in order to refine the estimate of endmembers, looking for the set of pixels that maximizes the volume of the simplex defined by the selected endmembers. The mathematical definition of the volume of a simplex formed by a set of endmember candidates is proportional to the determinant of the set augmented by a row of ones. The determinant is only defined in the case where the number of features is $p-1$, $p$being the number of desired endmembers [9]. Since in hyperspectral data typically $n \gg p$, a transformation that reduces the dimensionality of the input data is required. In this work, we use the principal component transform (PCT) [64] for this purpose. The corresponding volume is calculated for every pixel in each endmember position by replacing that endmember and finding the resulting volume. If the replacement results in an increase of volume, the pixel replaces the endmember. This procedure is repeated in iterative fashion until there are no more endmember replacements (see Figure 3b). The method can be summarized by a step-by-step algorithmic description which is given below for clarity:

- *Feature reduction.* Apply a dimensionality reduction transformation such as PCT to reduce the dimensionality of the data from n to $d=p-1$, where p is an input parameter to the algorithm (number of endmembers to be extracted). The basic idea of PCT is to orthogonally project the data into a new coordinate system, defined by the variance of the original data, i.e. the direction that accounts for the greatest variance of the original data will be the first coordinate (the principal component) of the transformed system, the second dimension will be the direction with the second largest variance, and so on. PCT requires the computation of the singular values and right singular vectors of $\tilde{\mathbf{Y}} = (\mathbf{Y} - \overline{\mathbf{Y}})^T(\mathbf{Y} - \overline{\mathbf{Y}})$. In particular, consider the singular value decomposition (SVD) $\tilde{\mathbf{Y}} = \mathbf{U}\Sigma\mathbf{V}^T$ [64, 65]. Then, the PCT performs the dimension reduction $n \rightarrow (d=p-1)$ by replacing $\mathbf{Y}$ with the first $p-1$ columns of Y˜V.

- *Initialization.* Let $\{e^{(0)}_1, e^{(0)}_2, \ldots, e^{(0)}_p\}$ be a set of endmembers randomly extracted from the input data.
- *Volume calculation.* At iteration $k \geq 0$, calculate the volume defined by the current set of endmembers as follows:

$$V\left(\mathbf{e}_1^{(k)}, \mathbf{e}_2^{(k)}, \ldots, \mathbf{e}_p^{(k)}\right) = \frac{\left|\det \begin{bmatrix} 1 & 1 & \ldots & 1 \\ \mathbf{e}_1^{(k)} & \mathbf{e}_2^{(k)} & \ldots & \mathbf{e}_p^{(k)} \end{bmatrix}\right|}{(p-1)!}. \quad (4)$$

- *Replacement.* For each pixel vector **y** in the input hyperspectral data, recalculate the volume by testing the pixel in all p endmember positions, i.e., first calculate $V(y, e^{(k)}_2, \ldots, e^{(k)}_p)$, then calculate $V(e^{(k)}_1, y, \ldots, e^{(k)}_p)$, and so on until $V(e(k)_1, e(k)_2, \ldots, y)$. If none of the p recalculated volumes is greater than $V(e^{(k)}_1, e^{(k)}_2, \ldots, e^{(k)}_p)$, then no endmember is replaced. Otherwise, the combination with maximum volume is retained. Let us assume that the endmember absent in the combination resulting in the maximum volume is denoted by $e(k+1)_i$. In this case, a new set of endmembers is produced by letting $e^{(k+1)}_i = y$ and $e^{(k+1)}_l = e^{(k)}_l$ for $l \neq i$. The replacement step is repeated for all the pixel vectors in the input data until all the pixels have been exhausted.

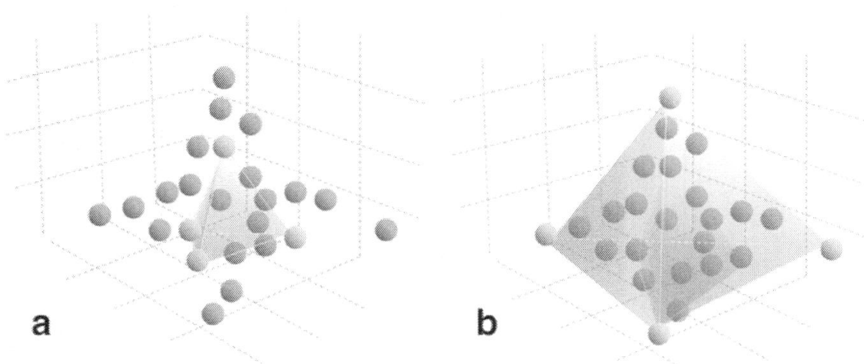

Figure 3: Graphical interpretation of the N-FINDR algorithm in a 3-dimensional space. (a) *N*-FINDR initialized randomly (p=4) and **(b)** final volume estimation by *N*-FINDR.

Computationally, this algorithm requires two major operations: feature reduction and volume calculation + replacement. Exploiting that $n \ll m$ and that only a few columns of Y˜V are required, determines that the PCT (first operation) can be computed in only $2m\, n^2$ flops, which are basically due to the calculation of the SVD of Y˜. The determination of the volumes are much more expensive. In particular, a straight-forward implementation of the computations of the determinants in steps (3)–(4), via e.g. the LU factorization (with partial pivoting), renders a total cost of $2m\ p^4/3$ flops, which results from having to compute mp factorizations of $p \times p$ matrices, with a cost of $2p^3/3$ flops per LU factorization. In [63], we describe a refined alternative that, by exploiting simple properties of the LU factorization, reduces this cost to $mp^3+2p^4/3$ flops.

As a final comment, it has been observed that different random initializations of *N*-FINDR may produce different final solutions. Thus, our *N*-FINDR algorithm was implemented in iterative fashion, so that each sequential run was initialized with the previous algorithm solution, until the algorithm converges to a simplex volume that cannot be further maximized. Our experiments show that, in practice, this approach allows the algorithm to converge in a few iterations only.

Methods for Abundance Estimation

This section introduces two different methods for estimating the abundance fractions: unconstrained least squares (ULS) [65] and non-negative constrained least squares (NCLS) [11].

Unconstrained Least Squares

Once a set of $E=\{e_j\}^p_{j=1}$ endmembers has been estimated using an endmember extraction algorithm, an unconstrained p-dimensional estimate of the endmember abundances in a given pixel in **y** can be simply obtained (in least squares sense) from the following expression [65]:

$$\hat{\mathbf{a}}^{\mathrm{UC}} = (\mathbf{E}^T\mathbf{E})^{-1}\mathbf{E}^T\mathbf{y}. \tag{5}$$

In the computation of (5), we can leverage that the term $M=(\mathbf{E}^T\mathbf{E})^{-1}\mathbf{E}^T$ remains fixed for all the pixels of the image. Thus, by explicitly obtaining M first, the cost of computing $\hat{a}^{UC}$ for all the scene pixels is basically reduced to $2m\ n\ p$ flops, since $n,\ p \ll m$ and, therefore, the number of arithmetic operations that are necessary to form M is negligible compared to that.

The main advantages of the unconstrained abundance estimation approach in Eq. (5) are the simplicity of its implementation and its fast execution. However, under this unconstrained model, the derivation of negative abundances is possible if the model endmembers are not pure or if they are affected by variability caused by spatial or temporal variations

[9]. To address this issue, two physical constrains can be introduced into the model described in Eq. (1), these are the abundance non-negativity constraint (ANC), i.e., $a_j \geq 0$, and the abundance sum-to-one constraint (ASC), i.e., $\sum_{j=1}^{p} a_j = 1$[12]. Imposing the ASC results in the following optimization problem:

$$\min_{\mathbf{a}\in\Delta}\left\{(\mathbf{y}-\mathbf{a}\cdot\mathbf{E})^T(\mathbf{y}-\mathbf{a}\cdot\mathbf{E})\right\},$$
$$\text{subject to: } \Delta=\left\{\mathbf{a}\,\middle|\,\sum_{j=1}^{p}a_j=1\right\}. \tag{6}$$

Similarly, imposing the ANC results in the optimization problem:

$$\min_{\mathbf{a}\in\Delta}\left\{(\mathbf{y}-\mathbf{a}\cdot\mathbf{E})^T(\mathbf{y}-\mathbf{a}\cdot\mathbf{E})\right\},$$
$$\text{subject to: } \Delta=\{\mathbf{a}\mid a_j\geq 0 \text{ for all } j\}. \tag{7}$$

As indicated in [12], a fully constrained (i.e. ASC-constrained and ANC-constrained) estimate can be obtained in least-squares sense by solving the optimization problems in Eq. (6) and Eq. (7) simultaneously. While partially constrained solutions imposing only the ANC have found success in the literature [11], the ASC is however prone to criticisms because, in a real image, there is a strong signature variability [66] that, at the very least, introduces positive scaling factors varying from pixel to pixel in the signatures present in the mixtures. As a result, the signatures are defined up to a scale factor, and thus, the ASC should be replaced with a generalized ASC of the form $\sum_{j=1}^{p} \xi_j \cdot a_j = 1$, in which the weights ξ_j denote the pixel-dependent scale factors [67]. What we conclude is that the non-negativity of the endmembers automatically imposes a generalized ASC. For this reason, in the following section we describe a solution that does not explicitly impose the ASC but only the ANC.

Non-negative Constrained Least Squares

A NCLS algorithm can be used to obtain a solution to the ANC-constrained problem described in Equation (7) in iterative fashion [11]. A successful approach for this purpose in different applications has been the image space reconstruction algorithm (ISRA) [68], a multiplicative algorithm for solving NCLS problems. The algorithm is based on the following iterative expression:

$$\hat{\mathbf{a}}^{k+1} = \hat{\mathbf{a}}^{k} \left(\frac{\mathbf{E}^T \cdot \mathbf{y}}{\mathbf{E}^T\mathbf{E} \cdot \hat{\mathbf{a}}^{k}} \right), \tag{8}$$

where the endmember abundances at pixel **y** are iteratively estimated, so that the abundances at the *k*+1-th iteration, $a^{\wedge k+1}$, depend on the abundances estimated at the *k*-th iteration, $a^{\wedge k}$. The procedure starts with an unconstrained abundance estimation $a^{\wedge UC}$ which is progressively refined in a given number of iterations. For illustrative purposes, Algorithm 3 shows the ISRA pseudocode for unmixing one hyperspectral pixel vector **y** using a set of **E** endmembers. For simplicity, in the pseudocode **y** is treated as an *n*-dimensional vector, and **E** is treated as a $n \times p$-dimensional matrix. The estimated abundance vector a^ is a *p*-dimensional vector, and variable *iters* denotes the number of iterations per pixel in the abundance estimation process (in this work, we set iters=200 as we have found good results empirically using this parameter setting). The pseudocode is subdivided into the numerator and denominator calculations in Equation (8). When these terms are obtained, they are divided and multiplied by the previous abundance vector. It is important to emphasize that the calculations of the fractional abundances for each pixel are independent, and therefore they can be calculated simultaneously without data dependencies, thus increasing the possibility of parallelization.

Algorithm 3 Pseudocode of ISRA algorithm for unmixing one hyperspectral pixel vector y using a set E of *p* endmembers

```
// For a certain number of iterations
for (k = 0; k < iters; k++) {
  // For all endmembers
  for (j = 0; j < p; j++)  {
    // For all bands
    for (i = 0; i < n; i++) {
      // Calculate the numerator of Equation (8)
      numerator = numerator + E[i][j]*y[i];
      // Calculate denominator of Equation (8) using â
```

```
from previous iteration (in first iteration, â = â^UC)
        for (s = 0; s < p; s++)  {
          dot += E[i][s]*â[s];
        } end for
        denominator += dot * E[i][j];
        dot = 0;
      } end for
      // Calculate the new â
      â[j] = â[j]*(numerator / denominator);
      numerator = 0;
      denominator = 0;
    } end for
  } for
```

The pseudocode for the ISRA algorithm reveals that this procedure is composed of very simple arithmetic operations, but also that the innermost loop, for variable s, dominates its arithmetic cost. In particular, as two arithmetic operations are performed at each iteration of this loop, this yields a total cost for the algorithm of $2np^2 \cdot$ iters flops.

MULTI-CORE IMPLEMENTATIONS

The six numerical methods introduced in the previous section for the different stages of hyperspectral unmixing can be decomposed into a collection of basic and advanced dense linear algebra operations. Among the basic ones, we can find, e.g, vector scalings, inner (dot) products, matrix-vector products, solution of triangular systems, matrix–matrix products, etc. The advanced ones comprise the solution of linear systems of equations, matrix inversion, eigenvalue problems, and singular values problems, among others. Fortunately, these operations are quite common to many other scientific and engineering applications, and nowadays there exist linear algebra libraries offering highly tuned and numerically reliable implementations of most of these operations for a variety of computer architectures, including multi-core processors.

In particular, Basic Linear Algebra Subprograms (BLAS) [69, 70, 71] defines the specification (the interface and functionality) of a collection of routines for basic linear algebra operations as those listed above. There is a legacy implementation of BLAS publicly available at http://www.netlib.org, but the aspect that makes BLAS really useful is the existence of implementations developed by most hardware vendors and highly tuned for their specific products. These developments include Intel MKL, AMD ACML, and IBM ESSL for their multi-core designs, but also more generic efforts like GotoBLAS2 and ATLAS. This approach has revealed so successful that NVIDIA, manufacturer of fancier hardware architectures such as GPUs, also offers their customers its own specialized implementation, CUBLAS. For the type of architectures considered in our work, i.e. multi-core processors, an appealing property of these libraries is that they can exploit the existence of hardware concurrency, in the form of several cores, by carefully using optimized multi-threaded codes. For example, the implementation of the matrix–matrix product kernel from MKL (routine_gemm), executed on a single core of an Intel Xeon core, attains more than 90% of the peak performance of the architecture when operating on matrices of moderate to large size. If several cores are used, and the problem dimension is scaled proportionally, the routine still achieves a similar performance rate.

The contents of BLAS are structured into three separate levels—BLAS-1, BLAS-2 and BLAS-3—according to the number of flops and memory operations (memops) carried out by the kernels. Thus, routines from BLAS-1 perform a linear number of flops on a linear number of data items and, therefore, memops; an example of a BLAS-1 routine is the inner product of two vectors. For BLAS-2, both flops and memops are quadratic on the amount of data items; the classical example for this level is the matrix-vector product. Finally, for BLAS-3 the flops are cubic while the memops are quadratic; e.g., the matrix–matrix product. The type of routine (level) has important implications on performance as current architectures feature a wide difference between the floating-point performance (flops/sec.) of the processor and the memory bandwidth (memops/sec.), and this gap continues growing. Concretely, only the routines from BLAS-3 exhibit enough data reuse so as to exploit the hierarchical structure of the memory subsystem of current computers, with several layers of cache, and thus hide the large latencies that requires the access to data that lie on the main memory. Developers leverage this property by designing so-called blocked algorithms for their implementations of BLAS-3 kernels that retrieve data from the main memory to the processor by blocks (square or rectangular submatrices), and operate with them as much as possible before returning the results back to memory. This is clearly not possible for BLAS-1 and BLAS-2 as the routines in these levels exhibit a flops/memops ratio that is $O(1)$. An additional advantage of BLAS-3 over the two other levels is that, in general, the use of multiple cores in a concurrent execution, is only justified if the arithmetic cost of the operation is cubic. In consequence, when implementing the spectral unmixing methods, it will be very important to identify numerical operations that can be casted in terms of the most convenient routine from BLAS, preferably BLAS-3.

Linear Algebra PACKage (LAPACK) [72] provides advanced methods for dense linear algebra operations as those mentioned above. There is a legacy implementation at http://wwww.netlib.org, but some hardware vendors also include tuned versions of certain routines in their mathematical libraries (e.g., Intel MKL and AMD ACML). The routines in LAPACK make a heavy use of kernels from BLAS, thus inheriting the performance (and parallelism) intrinsic to the latter. LAPACK provides specialized implementations that can leverage special matrix properties like symmetry, band structure, positive definiteness, etc., when solving linear systems or linear least squares problems as well as calculating the eigenvalues/singular values of a matrix. To improve numerical accuracy and increase performance, it is very important to select the appropriate routine from LAPACK as, in many cases, this library offers different solvers to tackle one particular problem.

Our task of developing high performance, possibly parallel, codes for the spectral unmixing methods started by (i) carefully selecting the appropriate data structures to hold the data (image and intermediate results); and (ii) developing an initial sequential implementation of the method, while simultaneously identifying basic and advanced linear algebra operations that could be performed by invoking the appropriate routines from BLAS and LAPACK. For those parts of the method that could be performed using kernels from these libraries, the implementation/optimization task was over.

However, during the implementation of the methods, we detected certain parts of the methods that had to be manually encoded, usually in the form of (nested) loops. For many of these code fragments, special care was taken to apply basic optimizations such as avoiding expensive operations, eliminating common subexpressions, avoiding branches, selecting the appropriate order in nested loops, using the appropriate compiler optimizations, etc. After this stage, the execution of the code was profiled to identify possible performance bottlenecks. For those fragments of code, in particular loops, that exhibited a significant cost (execution time), we analyzed the possibility of reducing their impact by leveraging loop concurrency via OpenMP [73]. This is a standard parallelization tool that is available in most current compilers (e.g., Intel icc, GNU gcc) and allows an easy and, in most cases, efficient parallelization of C/Fortran codes on multi-core processors.

Each one of these implementation and optimization stages was carefully monitored from the point of view of correctness, experimental accuracy, and performance. The result of this process was the spectral unmixing routines that we evaluate in the next section.

EXPERIMENTAL RESULTS

This section is organized as follows. Section 4.1 describes the hyperspectral data set used in experiments. In Section 4.2 we describe the multi-core processing platforms. Finally, Section 4.3 performs a detailed assessment of the performance versus energy consumption of the considered multi-core architectures when executing the different unmixing chains that can be formed with the processing modules described in Section 2.

Hyperspectral Data

The hyperspectral data set used in experiments was collected by the AVIRIS sensor over the Cuprite mining district in Nevada in the summer of 1997 (see Figure 4). It is available online (in reflectance units) after atmospheric correction [74]. The portion used in experiments corresponds to a 350×350-pixel subset of the sector labeled as f970619t01p02_r02_sc03.a.rfl in the online data, which comprise 188 spectral bands in the range from 400 to 2,500 nm and a total size of around 50 MB. Water absorption bands as well as bands with low signal-to-noise ratio (SNR) were removed prior to the analysis. The site is well understood mineralogically, and has several exposed minerals of interest, including *alunite*, *buddingtonite*, *calcite*, *kaolinite*, and *muscovite*. Reference ground signatures of the above minerals, available in the form of a USGS library [75] have been used in the literature for evaluation purposes [16].

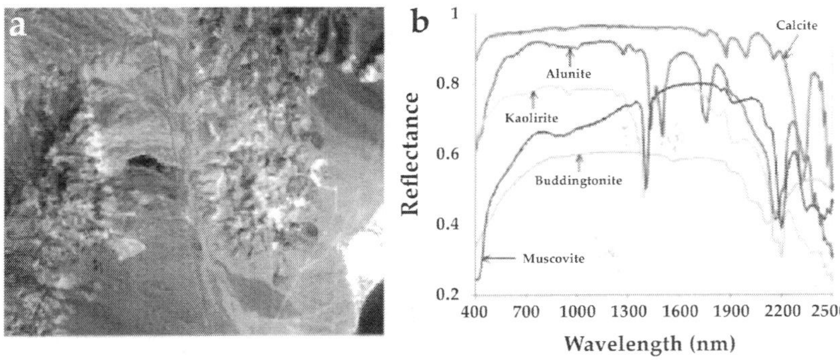

Figure 4: AVIRIS hyperspectral over the Cuprite mining district in Nevada. (a) False color composition and (b) U.S. Geological Survey mineral spectral signatures used for validation purposes.

For illustrative purposes, Table 2 provides the values of p (number of endmembers) estimated by the VD method for the considered hyperspectral scene, using different values of the false alarm probability (P_F). The number of endmembers estimated by HySime was $p=19$ for the Cuprite scene. As shown by Table 2, a consensus between VD and HySime was observed for $P_F=10^{-5}$ and $P_F=10^{-6}$. On the other hand, Table 3 shows the spectral angles (in degrees) between the most similar endmembers extracted by the OSP-GS and the reference USGS spectral signatures available for this scene. The range of values for the spectral angle is [0°, 90°], with values close to 0° indicating higher spectral similarity. As shown by Table 3, the endmembers extracted by both the OSP-GS and N-FINDR algorithms are very similar, spectrally, to the USGS reference signatures, despite the potential variations (due to posible interferers still remaining after the atmospheric correction process) between the ground signatures and the airbone data. For illustrative purposes, Figure 5 plots the estimated endmembers against the ground-truth spectra for the considered endmember extraction algorithms.

Table 2: VD estimates of p for various false alarm probabilities (P_F)

P_F	**Number of endmembers** p
10^{-1}	37
10^{-2}	28
10^{-3}	25
10^{-4}	20
10^{-5}	19

10^{-6}	19
10^{-7}	17
10^{-8}	16

HySime provided an estimate of p=19 for the Cuprite scene.

Table 3: Spectral angle values (in degrees) between the target pixels extracted by the OSP-GS algorithm and the reference USGS mineral signatures for the AVIRIS Cuprite scene

Algorithm	**Alunite (°)**	**Buddingtonite (°)**	**Calcite (°)**	**Kaolinite (°)**	**Muscovite (°)**	**Average (°)**
OSP-GS	5.48	4.08	5.87	11.14	5.68	6.45
N-FINDR	4.81	4.29	7.60	9.92	5.05	6.33

a

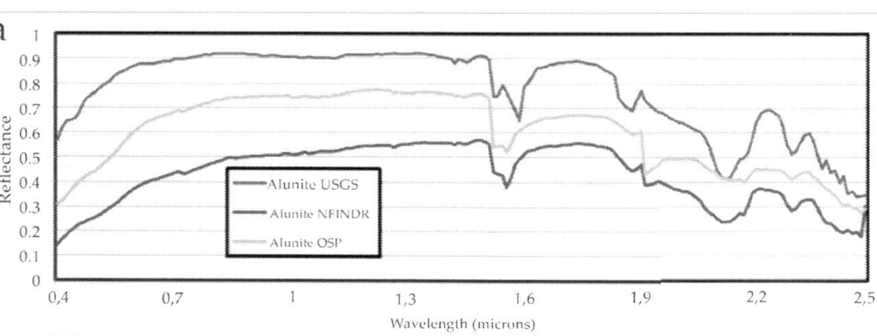

Alunite

b

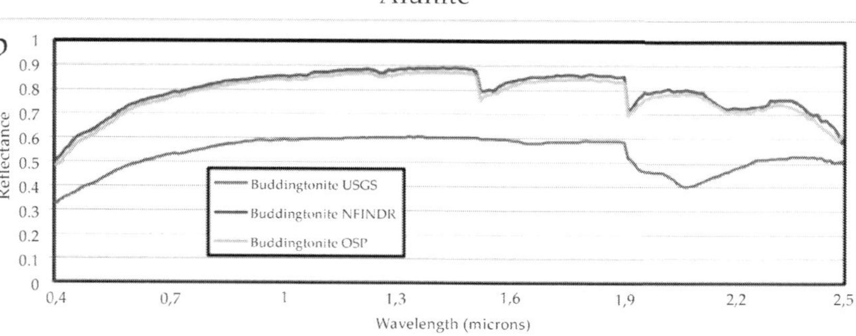

Muscovite

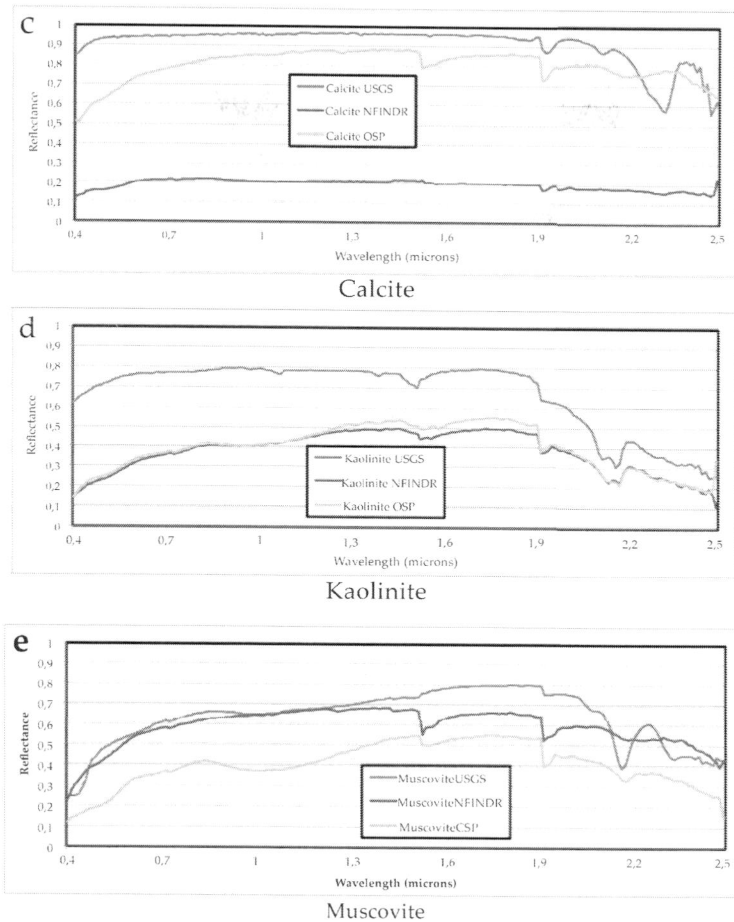

Figure 5: Estimated versus ground-truth endmembers for minerals: (a) alunite, (b) buddingtonite, (c) calcite, (d) kaolinite and (e) muscovite.

Multi-core Platforms

In 2004, the evolution of processor architecture shifted from a progressive increment of clock frequency to a growth in the number of cores. Thus, although current processors still feature a moderate number of cores (between 4 and 16), the trend indicates that next generations will include a larger number of cores. On the other hand, as-of-today it is possible to build a commodity shared-memory multiprocessor with four sockets that can accommodate 16-core processors each, for a total of 64 cores in a single desktop platform.

Following this trend towards high levels of hardware concurrency at the core-level, all the experiments were conducted on a platform equipped with 4 AMD Opteron 6172 processors, with 12 cores per processor, and a total of 48 cores in the platform. The software employed in the experiments included Intel MKL v10.3 implementation of the BLAS and LAPACK libraries, and Intel icc v12.1.3 compiler. The codes were compiled with the optimization flag -O3, and single-precision arithmetic was employed in all experiments. The explosion of hardware concurrency in multi-core processors requires the development of efficient parallel software that attains a significant fraction of the platform peak performance. However, in some cases the target method does not exhibit enough concurrency to efficiently exploit all the computational units in the platform. Therefore, when executing this kind of methods, the use of all the cores results in an increment of the execution time (due to the overhead introduced by the synchronization, communication and management of the cores) and the corresponding waste of energy. In this context, it is preferable to limit the number of cores employed, idling the rest of them so that the OS can move these cores into an energy-saving state (*C*-state) that yields a significant reduction of dynamic power.

In our study of the spectral unmixing methods, each code was evaluated separately to determine the optimal number of cores from the point of view of execution time. Given that the energy consumption equals the product of power dissipation times execution time, in general, reducing the execution time is an important step towards increasing energy efficiency.

In our experiments, the power consumption was measured using an internal DC powermeter. This device obtains 25 samples per second and is attached to the 12 V lines connecting the power supply with the motherboard (chipset plus processors) of the platform (see Figure 6). With this configuration, the results are not affected by inefficiencies of the power supply unit, or the "noise" due to the operation of other hardware components like fans, disks, etc. Also, samples from the powermeter are collected in a separate system, to prevent the measurements from impairing the accuracy of the tests. The error of the device is less than 5%. The powermeter is controlled via our library pmlib, which requires minimum changes to the code.

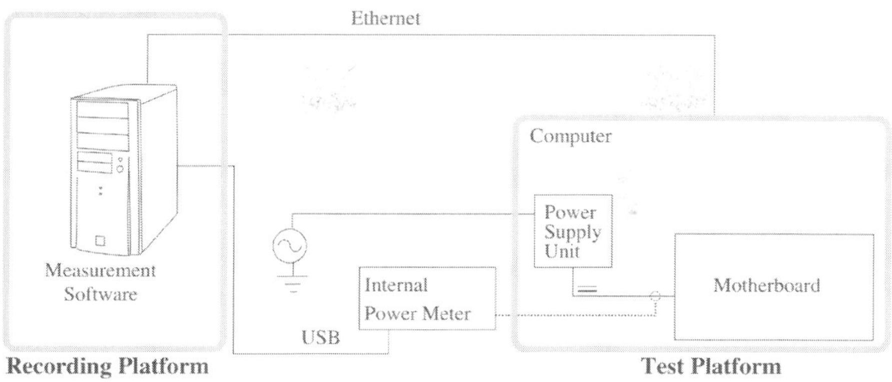

Figure 6: Experimental setup for power evaluation.

Assessment of Performance versus Energy Consumption

In this section, we analyze the processing times, energy consumption and maximum power obtained by different combinations of the modules reported in Section 2 for estimating the number of endmembers (VD and HySime), identifying the endmember signatures (OSP-GS and N-FINDR), and estimating the abundances (ULS and ISRA). In all cases, the number of endmembers to be extracted was set to $p=19$. Table 4 shows the processing times, energy consuption and maximum power for different full unmixing chains formed using combinations of the aforementioned modules. In Table 4, we also indicate (in the parentheses) the number of cores from the AMD Opteron 6172 system that were used in the multi-core implementation of each method. It is important to emphasize that, in order to satisfy the real-time processing constraint for the AVIRIS Cuprite scene, we should be able to process it in less than 1.98 s which is the time needed by the instrument to collect the data. As shown in Table 4, only the chain VD+OSP-GS+LSU achieved real-time performance, with two other chains (VD + OSP -GS + ISRA and VD + N-FINDR + ISRA) providing near real-time performance in the target multi-core architecture. It is remarkable that the inclusion of HySime for the identification of the number of endmembers significantly increments the processing times and also increases the energy consumption.

Table 4: Processing time (seconds), energy consumption (Watts-hour), and maximum power dissipation (Watts) of different full spectral unmixing chains applied to the AVIRIS Cuprite scene

Number of endmembers	**Endmember identification**	**Abundance estimation**	**Time**	**Energy**	**Maximum power**
VD (20)	OSP-GS (34)	ULS (20)	0.413	0.0468	492
VD (20)	OSP-GS (34)	ISRA (48)	2.957	0.4297	543
VD (20)	N-FINDR (20)	ULS (20)	2.332	0.2187	489
VD (20)	N-FINDR (20)	ISRA (48)	4.899	0.6093	563
HySime (48)	OSP-GS (34)	ULS (20)	15.508	2.1093	540
HySime (48)	OSP-GS (34)	ISRA (48)	16.762	2.2812	597
HySime (48)	N-FINDR (20)	ULS (20)	16.971	2.2343	540
HySime (48)	N-FINDR (20)	ISRA (48)	18.077	2.4140	571

The values inside the parentheses are the number of cores used in the implementation of each method.

In order to investigate the individual contributions of the methods in different parts of the unmixing chain to the total processing time, Table 5 reports the processing times measured for all the individual methods when processing the AVIRIS Cuprite scene. It can be observed that, by far, HySime is the most computationally expensive method while VD provides an alternative for this part of the chain which is about 100 × faster in comparison. Table 5 also reveals that the OSP-GS method used for the endmember identification is about 9 × faster than N-FINDR in the multi-core. Finally, ISRA is more than 100 × slower than ULS as a consequence of the fact that it imposes the non negativity constraint in the abundance estimation. The results in Table 5suggest that the combination VD+OSP-GS+ULS provides a good basis for fast spectral unmixing in the considered multi-core platform (in fact, this combination is the only one that results in real-time performance in our experiments).

Table 5: Processing time (seconds) by each different processing module for the AVIRIS Cuprite scene in a multi-core system

Number of endmembers		**Endmember identification**		**Abundance estimation**	
VD	HySime	OSP-GS	N-FINDR	ULS	ISRA
0.15	15.2	0.25	2.17	0.02	2.55

For comparative purposes, Table 6 shows the processing times measured for the same methods reported in Table 5 implemented in the NVidia ™

GeForce GTX 580 GPU [76], which features 512 processor cores operating at 1.544 GHz, with a total dedicated memory of 1,536 MB, at 2.004 MHz (with 384-bit GDDR5 interface) and memory bandwidth of 192.4 GB/s. The GPU is connected to an Intel core i7 920 CPU at 2.67 GHz with eight cores, which uses a motherboard Asus P6T7 WS SuperComputer. As shown in Table 6, the processing times in the GPU are comparatively similar to those reported for the multi-core for the OSP-GS and ULS algorithms. On the other hand, the times for HySime, *N*-FINDR and ISRA are sensibly lower in the GPU, with only the VD being slightly faster in the multi-core than in the GPU. However, the energy consumption is much higher in the GPU, which still makes the multi-core platform a more interesting platform from an operational point of view.

Table 6: Processing time (seconds) by each different processing module for the AVIRIS Cuprite scene in a GPU system

Number of endmembers		**Endmember identification**		**Abundance estimation**	
VD	HySime	OSP-GS	N-FINDR	ULS	ISRA
0.405	1.655	0.024	0.069	0.038	0.858

CONCLUSIONS

In this article, we have addressed hyperspectral imaging via spectral unmixing on multi-core processors, exposing a detailed evaluation of the performance and energy requirements of efficient parallel codes for all the stages in the spectral unmixing chain on multi-core processors. Specifically, we have implemented modules for (i) the estimation of the number of endmembers, (ii) the identification of a collection of these, and (iii) the estimation of the fractional abundances, using kernels from highly tuned linear algebra libraries and OpenMP directives on a platform equipped with 48 AMD cores.

Our study offers two major conclusions:

- Three of the unmixing chains attain real-time performance (VD + OSP - GS + ULS) or close to it (VD + OSP - GS + ISRA and VD + *N*-FINDR + ULS) as the underlying modules VD, OSP-GS, *N*-FINDR, LSU and ISRA exhibit a high degree algorithmic concurrency that can be leveraged to yield efficient parallel implementations on current multi-core processors. On the other hand, HySime presents a performance bottleneck that turns those chains that utilize this module inappropriate for real-time image processing.

- With the expected increase in the number of cores in future architectures, shared-memory platforms equipped with a few multi-core processors are a competitive approach to efficiently tackle computationally expensive hyperspectral imaging applications on cheap commodity hardware. Compared with FPGAs, conventional multi-core processors offer the plain advantage of being much easier to program, considerably improving the software development cycle. Furthermore, general-purpose cores offer an appealing performance-energy ratio and they clearly outperform GPUs in their tolerance to incorporate radiation-avoidance mechanisms.

As future work, we plan to analyze in more detail the energy consumption of other types of architectures, including FPGAs or GPUs, which are currently considered as candidate specialized hardware platforms for onboard hyperspectral image processing.

ACKNOWLEDGEMENTS

Enrique S Quintana-Ortí and Alfredo Remón were supported by the CICYT project TIN2011-23283 of the *Ministerio de Economía y Competitividad* and FEDER. Funding from the Spanish Ministry of Science and Innovation (CEOS-SPAIN project, reference AYA2011-29334-C02-02) is also gratefully acknowledged.

REFERENCES

1. AFH Goetz, G Vane, JE Solomon, BN Rock, Imaging spectrometry for earth remote sensing. Science. 228, 1147–1153 (1985)
2. RO Green, ML Eastwood, CM Sarture, TG Chrien, M Aronsson, BJ Chippendale, JA Faust, BE Pavri, CJ Chovit, M Solis, et al, Imaging spectroscopy and the airborne visible/infrared imaging spectrometer (AVIRIS). Remote Sens. Envir. 65(3), 227–248 (1998)
3. A Plaza, J Plaza, A Paz, S Sanchez, Parallel hyperspectral image and signal processing. IEEE Signal Process. Mag. 28(3), 119–126 (2011)
4. A Plaza, JA Benediktsson, J Boardman, J Brazile, L Bruzzone, G Camps-Valls, J Chanussot, M Fauvel, P Gamba, J Gualtieri, M Marconcini, TC Tilton, G Trianni, Recent advances in techniques for hyperspectral image processing. Remote Sens. Envir. 113, 110–122 (2009)
5. A Plaza, Q Du, JM Bioucas-Dias, X Jia, F Kruse, Foreword to the special issue on spectral unmixing of remotely sensed data. IEEE Trans. Geosci. Remote Sens. 49(11), 4103–4110 (2011)
6. PE Johnson, MO Smith, S Taylor-George, JB Adams, A semiempirical method for analysis of the reflectance spectra for binary mineral mixtures. J. Geophys. Res. 88, 3557–3561 (1983)
7. JB Adams, MO Smith, PE Johnson, Spectral mixture modeling: a new analysis of rock and soil types at the Viking Lander 1 site. J. Geophys. Res. 91, 8098–8112 (1986)
8. N Keshava, JF Mustard, Spectral unmixing. IEEE Signal Process. Mag. 19, 44–57 (2002)
9. A Plaza, P Martinez, R Perez, J Plaza, A quantitative and comparative analysis of endmember extraction algorithms from hyperspectral data. IEEE Trans. Geosci. Remote Sens. 42(3), 650–663 (2004)
10. Q Du, N Raksuntorn, NH Younan, RL King, End-member extraction for hyperspectral image analysis. Appl. Opt. 47, 77–84 (2008)
11. CI Chang, D Heinz, Constrained subpixel target detection for remotely sensed imagery. IEEE Trans Geosci. Remote Sens. 38, 1144–1159 (2000)
12. D Heinz, CI Chang, Fully constrained least squares linear mixture analysis for material quantification in hyperspectral imagery. IEEE Trans. Geosci. Remote Sens. 39, 529–545 (2001)

13. CC Borel, SAW Gerstl, Nonlinear spectral mixing model for vegetative and soil surfaces. Remote Sens. Envir. 47(3), 403–416 (1994)
14. W Liu, EY Wu, Comparison of non-linear mixture models. Remote Sens. Envir. 18, 1976–2003 (2004)
15. N Raksuntorn, Q Du, Nonlinear spectral mixture analysis for hyperspectral imagery in an unknown environment. IEEE Geosci. Remote Sens. Lett. 7(4), 836–840 (2010)
16. A Plaza, G Martin, J Plaza, M Zortea, S Sanchez, in Optical Remote Sensing, ed. by S Prasad, LM Bruce, and J Chanussot. Recent developments in spectral unmixing and endmember extraction (Springer, Berlin, Germany, 2011), pp. 235–267
17. JJ Settle, NA Drake, Linear mixing and the estimation of ground cover proportions. Int. J. Remote Sens. 14, 1159–1177 (1993)
18. JC Harsanyi, CI Chang, Hyperspectral image classification and dimensionality reduction: An orthogonal subspace projection. IEEE Trans. Geosci. Remote Sens. 32(4), 779–785 (1994)
19. JW Boardman, FA Kruse, RO Green, in Proc. JPL Airborne Earth Science Workshop. Mapping target signatures via partial unmixing of AVIRIS data (Pasadena, USA, 1995), pp. 23–26
20. JH Bowles, PJ Palmadesso, JA Antoniades, MM Baumback, LJ Rickard, Use of filter vectors in hyperspectral data analysis. Proc. SPIE Infrared Spaceborne Remote Sens. III. 2553, 148–157 (1995)
21. RA Neville, K Staenz, T Szeredi, J Lefebvre, P Hauff, in Proc. 21st Canadian Symp. Remote Sens. Automatic endmember extraction from hyperspectral data for mineral exploration (Ottawa, Canada, 1999), pp. 21–24
22. A Ifarraguerri, CI Chang, Multispectral and hyperspectral image analysis with convex cones. IEEE Trans. Geosci. Remote Sens. 37(2), 756–770 (1999)
23. ME Winter, N-FINDR: An algorithm for fast autonomous spectral endmember determination in hyperspectral data. Proc. SPIE. 3753, 266–277 (1999)
24. Q Du, H Ren, CI Chang, A comparative study for orthogonal subspace projection and constrained energy minimization. IEEE Trans. Geosci. Remote Sens. 41(6), 1525–1529 (2003)

25. M Berman, H Kiiveri, R Lagerstrom, A Ernst, R Dunne, JF Huntington, ICE: a statistical approach to identifying endmembers in hyperspectral images. IEEE Trans. Geosci. Remote Sens. 42(10), 2085–2095 (2004)
26. JMP Nascimento, JM Bioucas-Dias, Vertex component analysis: a fast algorithm to unmix hyperspectral data. IEEE Trans. Geosci. Remote Sens. 43(4), 898–910 (2005)
27. A Plaza, CI Chang, Impact of initialization on design of endmember extraction algorithms. IEEE Trans. Geosci. Remote Sens. 44(11), 3397–3407 (2006)
28. CI Chang, A Plaza, A fast iterative algorithm for implementation of pixel purity index. IEEE Geoscience. Remote Sens. Lett. 3, 63–67 (2006)
29. DM Rogge, B Rivard, J Zhang, J Feng, Iterative spectral unmixing for optimizing per-pixel endmember sets. IEEE Trans. Geosci. Remote Sens. 44(12), 3725–3736 (2006)
30. J Wang, CI Chang, Applications of independent component analysis in endmember extraction and abundance quantification for hyperspectral imagery. IEEE Trans. Geosci. Remote Sens. 44(9), 2601–2616 (2006)
31. CI Chang, CC Wu, W Liu, YC Ouyang, A new growing method for simplex-based endmember extraction algorithm. IEEE Trans. Geosci. Remote Sens. 44(10), 2804–2819 (2006)
32. A Zare, P Gader, Sparsity promoting iterated constrained endmember detection for hyperspectral imagery. IEEE Geosci. Remote Sens. Lett. 4(3), 446–450 (2007)
33. A Zare, P Gader, Hyperspectral band selection and endmember detection using sparsity promoting priors. IEEE. Remote Sens. Lett. 5(2), 256–260 (2008)
34. M Zortea, A Plaza, A quantitative and comparative analysis of different implementations of N-FINDR: a fast endmember extraction algorithm. IEEE Geosci. Remote Sens. Lett. 6, 787–791 (2009)
35. X Tao, B Wang, L Zhang, Orthogonal bases approach for the decomposition of mixed pixels in hyperspectral imagery. IEEE Geosci. Remote Sens. Lette. 6, 219–223 (2009)
36. A Zare, P Gader, PCE: piecewise convex endmember detection. IEEE Trans. Geosci. Remote Sens. 48(6), 2620–2632 (2010)

37. CI Chang, CC Wu, CS Lo, ML Chang, Real-time simplex growing algorithms for hyperspectral endmember extraction. IEEE Trans. Geosci. Remote Sens. 48(4), 1834–1850 (2010)
38. F Schmidt, A Schmidt, E Treandguier, M Guiheneuf, S Moussaoui, N Dobigeon, Implementation strategies for hyperspectral unmixing using Bayesian source separation. IEEE Trans. Geosci. Remote Sens. 48(11), 4003–4013 (2010)
39. O Duran, M Petrou, Robust endmember extraction in the presence of anomalies. IEEE Trans. Geosci. Remote Sens. 49(6), 1986–1996 (2011)
40. B Zhang, X Sun, L Gao, L Yang, Endmember extraction of hyperspectral remote sensing images based on the ant colony optimization (ACO) algorithm. IEEE Trans. Geosci. Remote Sens. 49(7), 2635–2646 (2011)
41. M Shoshany, F Kizel, N Netanyahu, N Goldshlager, T Jarmer, G Even-Tzur, An iterative search in end-member fraction space for spectral unmixing. IEEE Geosci. Remote Sens. Lett. 8(4), 706–709 (2011)
42. CI Chang, CC Wu, HM Chen, Random pixel purity index. IEEE Geosci. Remote Sens. Lett. 7(2), 324–328 (2010)
43. S Dowler, M Andrews, On the convergence of N-FINDR and related algorithms: to iterate or not to iterate. IEEE Geosci. Remote Sens. Lett. 8, 4–8 (2011)
44. A Plaza, D Valencia, J Plaza, P Martinez, Commodity cluster-based parallel processing of hyperspectral Imagery. J. Parall. Distr. Comput. 66(3), 345–358 (2006)
45. A Plaza, CI Chang, High Performance Computing in Remote Sensing. (Taylor & Francis, Boca Raton, FL, 2007)
46. A Plaza, Q Du, Y-L Chang, RL King, High performance computing for hyperspectral remote sensing. IEEE J. Sel. Top. Appl. Earth Obser. Remote Sens. 4(3), 528–544 (2011)
47. CA Lee, SD Gasster, A Plaza, CI Chang, B Huang, Recent developments in high performance computing for remote sensing: a review. IEEE J. Sel. Top. Appl. Earth Obser. Remote Sens. 4(3), 508–527 (2011)
48. A Plaza, CI Chang, Clusters versus FPGA for parallel processing of hyperspectral imagery. Int. J. High Perfor. Comput. Appl. 22(4), 366–385 (2008)
49. C Gonzalez, J Resano, D Mozos, A Plaza, D Valencia, FPGA implementation of the pixel purity index algorithm for remotely

sensed hyperspectral image analysis. EURASIP J. Adv. Signal Process. 969806, 1–13 (2010)

50. C Gonzalez, D Mozos, J Resano, A Plaza, FPGA implementation of the N-FINDR algorithm for remotely sensed hyperspectral image analysis. IEEE Trans. Geosci. Remote Sens. 50(2), 374–388 (2012)
51. J Setoain, M Prieto, C Tenllado, F Tirado, GPU for parallel on-board hyperspectral image processing. Int. J. High Perfor. Comput. Appl. 22(4), 424–437 (2008)
52. M Hsueh, CI Chang, Field programmable gate arrays (FPGA) for pixel purity index using blocks of skewers for endmember extraction in hyperspectral imagery. Int. J. High Perfor. Comput. Appl. 22, 408–423 (2008)
53. Y Tarabalka, TV Haavardsholm, I Kasen, T Skauli, Real-time anomaly detection in hyperspectral images using multivariate normal mixture models and GPU processing. J. Real-Time Image Process. 4, 1–14 (2009)
54. H Yang, Q Du, G Chen, Unsupervised hyperspectral band selection using graphics processing units. IEEE J. Sel. Top. Appl. Earth Obser. Remote Sens. 4(3), 660–668 (2011)
55. JA Goodman, D Kaeli, D Schaa, Accelerating an imaging spectroscopy algorithm for submerged marine environments using graphics processing units. IEEE J. Sel. Top. Appl. Earth Obser. Remote Sens. 4(3), 669–676 (2011)
56. J Mielikainen, B Huang, A Huang, GPU-accelerated multi-profile radiative transfer model for the infrared atmospheric sounding interferometer. IEEE J. Sel. Top. Appl. Earth Obser. Remote Sens. 4(3), 691–700 (2011)
57. E Christophe, J Michel, J Inglada, Remote sensing processing: from multicore to GPU. IEEE J. Sel. Top. Appl. Earth Obser. Remote Sens. 4(3), 643–652 (2011)
58. S Sanchez, A Paz, G Martin, A Plaza, Parallel unmixing of remotely sensed hyperspectral images on commodity graphics processing units. Concur. Comput. Pract. Exp. 23(13), 1538–1557 (2011)
59. CI Chang, Q Du, Estimation of number of spectrally distinct signal sources in hyperspectral imagery. IEEE Trans. Geosci. Remote Sens. 42(3), 608–619 (2004)

60. JM Bioucas-Dias, JMP Nascimento, Hyperspectral subspace identification. IEEE Trans. Geosci. Remote Sens. 46(8), 2435–2445 (2008)
61. S Lopez, P Horstrand, GM Callico, JF Lopez, R Sarmiento, A low-computational-complexity algorithm for hyperspectral endmember extraction: modified vertex component analysis. IEEE Geosci. Remote Sens. Lett. 9(3), 502–506 (2012)
62. S Bernabe, S L ´ opez, A Plaza, R Sarmiento, GPU implementation of an ´ automatic target detection and classification algorithm for hyperspectral image analysis. IEEE Geosci. Remote Sens. Lett. 10(2), 221–225 (2013). doi:10.1109/LGRS.2012.2198790
63. A Remon, S Sanchez, A Paz, ES Quintana-Ort ´ ´ı, A Plaza, Real-time endmember extraction on multi-core processors. IEEE Geosci. Remote Sens. Lett. 8(5), 924–928 (2011)
64. JA Richards, X Jia, Remote Sensing Digital Image Analysis: An Introduction. (Springer-Verlag, New York, 2006)
65. CI Chang, Hyperspectral Imaging: Techniques for Spectral Detection and Classification. (Kluwer Academic/Plenum Publishers, New York, 2003)
66. CA Bateson, GP Asner, CA Wessman, Endmember bundles: a new approach to incorporating endmember variability into spectral mixture analysis. IEEE Trans. Geosci. Remote Sens. 38(2), 1083–1094 (2000)
67. MD Iordache, J Bioucas-Dias, A Plaza, Sparse unmixing of hyperspectral data. IEEE Trans. Geosci. Remote Sens. 49(6), 2014–2039 (2011)
68. ME Daube-Witherspoon, G Muehllehner, An iterative image space reconstruction algorithm suitable for volume ECT. IEEE Trans. Med. Imag. 5, 61–66 (1986)
69. CL Lawson, RJ Hanson, DR Kincaid, FT Krogh, Basic linear algebra subprograms for Fortran usage. ACM Trans. Math. Soft. 5(3), 308–323 (1979)
70. JJ Dongarra, J Du Croz, S Hammarling, RJ Hanson, An extended set of FORTRAN basic linear algebra subprograms. ACM Trans. Math. Soft. 14, 1–17 (1988)
71. JJ Dongarra, J Du Croz, S Hammarling, I Duff, A set of level 3 basic linear algebra subprograms. ACM Trans. Math. Soft. 16, 1–17 (1990)

72. E Anderson, Z Bai, C Bischof, LS Blackford, J Demmel, JJ Dongarra, JD Croz, S Hammarling, A Greenbaum, A McKenney, D Sorensen, LAPACK Users' guide, 3rd edn. (SIAM, Philadelphia, 1999)
73. The OpenMP API specification for parallel programming. [http://www.openmp.org/]
74. NASA, Aviris (Airbone Visible infrared Imaging Spectrometer) - Data. aviris.jpl.nasa.gov/data/free data.html
75. U.S. Geological Survey (U.S. Dep. of the Interior), USGS Spectroscopy Lab - Spectral Library. speclab.cr.usgs.gov/spectral-lib.html 76. NVIDIA Corporation, NVIDIA GTX580 specifications. [www.nvidia.com/ object/product-geforce-gtx-580-us.html]

SECTION V

CHAPTER 10

A UNIFIED VIEW OF ADAPTIVE VARIABLE-METRIC PROJECTION ALGORITHMS

Masahiro Yukawa[1] **and Isao Yamada**[2]

[1]Mathematical Neuroscience Laboratory, BSI, RIKEN, 2-1 Hirosawa, Wako, Saitama 351-0198, Japan

[2]Department of Communications and Integrated Systems, Tokyo Institute of Technology, Meguro-ku, Tokyo 152-8552, Japan

ABSTRACT

We present a unified analytic tool named *variable-metric adaptive projected subgradient method (V-APSM)* that encompasses the important family of adaptive variable-metric projection algorithms. The family includes the transform-domain adaptive filter, the Newton-method-based adaptive filters such as quasi-Newton, the proportionate adaptive filter, and the

Citation (APA): Yukawa, M., & Yamada, I. (2009). A unified view of adaptive variable-metric projection algorithms. EURASIP Journal on Advances in Signal Processing, 2009(1), 589260. (13 pages).

Krylov-proportionate adaptive filter. We provide a rigorous analysis of V-APSM regarding several invaluable properties including *monotone approximation*, which indicates stable tracking capability, and convergence to an asymptotically optimal point. Small metric-fluctuations are the key assumption for the analysis. Numerical examples show (i) the robustness of V-APSM against violation of the assumption and (ii) the remarkable advantages over its constant-metric counterpart for colored and nonstationary inputs under noisy situations.

Keywords: Information Technology, Analytic Tool, Quantum Information, Optimal Point

INTRODUCTION

The adaptive projected subgradient method (APSM) [1– 3] serves as a unified guiding principle of many existing projection algorithms including the normalized least mean square (NLMS) algorithm [4, 5], the affine projection algorithm (APA) [6, 7], the projected NLMS algorithm [8], the constrained NLMS algorithm [9], and the adaptive parallel subgradient projection algorithm [10, 11]. Also, APSM has been proven a promising tool for a wide range of engineering applications: interference suppression in the code-division multiple access (CDMA) and multi-input multioutput (MIMO) wireless communication systems [12, 13], multichannel acoustic echo cancellation [14], online kernel-based classification [15], nonlinear adaptive beamforming [16], peak-to-average power ratio reduction in the orthogonal frequency division multiplexing (OFDM) systems [17], and online learning in diffusion networks [18]. However, APSM does not cover the important family of algorithms that are based on iterative projections with its metric controlled adaptively for better performance. Such a family of variable-metric projection algorithms includes the transform-domain adaptive filter (TDAF) [19–21], the LMSNewton adaptive filter (LNAF) [22–24] (or quasi-Newton adaptive filter (QNAF) [25, 26]), the proportionate adaptive filter (PAF) [27–33], and Krylov-proportionate adaptive filter (KPAF) [34–36]; it has been shown, respectively, in [34, 37] that TDAF and PAF perform iterative projections onto hyperplanes (the same as used by NLMS) with variable metric. The variable-metric projection algorithms enjoy significantly faster convergence compared to their constant-metric counterparts with reasonable computational complexity. At the same time, however, the variability of metric causes major difficulty in analyzing this family of algorithms. It is of great interests and importance to reveal the convergence mechanism.

The goal of this paper is to build a unified analytic tool that encompasses the family of adaptive variablemetric projection algorithms. The key to achieve this goal is the assumption of small metric-fluctuations. We extend APSM into the variable-metric adaptive projected subgradient method (V-APSM) that allows the metric to change in time. V-APSM includes TDAF, LNAF/QNAF, PAF, and KPAF as its particular examples. We present a rigorous analysis of V-APSM regarding several properties. First, we show that V-APSM enjoys monotone approximation, which indicates stable tracking capability. Second, we prove that the vector sequence generated by V-APSM converges to a point in a certain desirable set. Third, we prove that both the vector sequence and its limit point minimize a sequence of cost functions to be designed by the user asymptotically; each cost function determines each iteration procedure of the algorithm. The analysis gives us an interesting view that TDAF, LNAF/QNAF, PAF, or KPAF asymptotically minimizes the metric distance to the data-dependent hyperplane which makes the instantaneous output-error be zero. The impacts of metric-fluctuations on the performance of adaptive filter are investigated by simulations.

The remainder of the paper is organized as follows. Preliminary to the major contributions, we present a brief review of APSM starting with a connection to the widely used NLMS algorithm in Section 2. We present V-APSM and its examples in Section 3, the analysis in Section 4, the numerical examples in Section 5, and the conclusion in Section 6.

ADAPTIVE PROJECTED SUBGRADIENT METHOD: ASYMPTOTIC MINIMIZATION OF A SEQUENCE OF COST FUNCTIONS

Throughout the paper, R and N denote the sets of all real numbers and nonnegative integers, respectively, and vectors (matrices) are represented by bold-faced lower-case (uppercase) letters. Let $<\cdot,\cdot>$ be an inner product defined on the N-dimensional Euclidean space R^N and $||\cdot||$ its induced norm. The projected gradient method [38, 39] is a simple extension of the popular gradient method (also known as the steepest descent method) to convexly constrained optimization problems. Precisely, it solves the minimization problem of a differentiable convex function $\phi : R^N \rightarrow R$ over a given closed convex set $C \subset R^N$, based on the metric projection:

$$P_C : \mathbb{R}^N \longrightarrow C, \quad \mathbf{x} \longmapsto P_C(\mathbf{x}) \in \arg\min_{\mathbf{a}\in C} \|\mathbf{a} - \mathbf{x}\|. \tag{1}$$

To deal with a (possibly nondifferentiable) continuous convex function, a generalized method named the projected subgradient method has been developed in [40]. For convenience, a brief review of the projected gradient and projected subgradient methods is given in Appendix A.

In 2003, Yamada has started to investigate the generalized problem in which ϕ is replaced by a sequence of continuous convex functions$(\phi k)_{k\in N}$ [1]. We begin by explaining how this formulation is linked to the adaptive filtering.

NLMS from a Viewpoint of Asymptotic Minimization

Let $<\cdot,\cdot>\cdot_2$ and $\|\cdot\|_2$ be the standard inner product and the Euclidean norm, respectively. We consider the following linear system [41, 42]:

$$d_k := \mathbf{u}_k^T \mathbf{h}^* + n_k, \quad k \in \mathbb{N}. \tag{2}$$

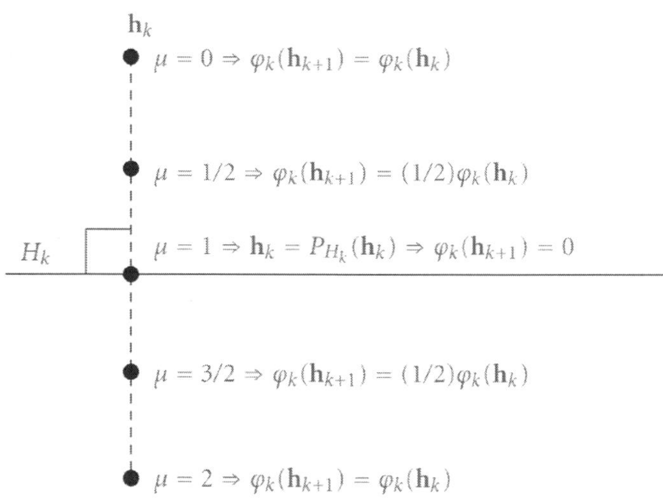

Figure 1: Reduction of the metric distance function $\phi_k(x) := d(x, H_k)$ by the relaxed projection.

Here, $u_k := [u_k, u_{k-1}, \ldots, u_{k-N+1}]^T \in R^N$ is the input vector at time k with $(u_k)_{k\in N}$ being the observable input process, $h^* \in R^N$ the unknown system, $(n_k)_{k\in N}$ the noise process, and $(d_k)_{k\in N}$ the observable output process. In the parameter estimation problem, for instance, the goal is to estimate h^*. Given an initial $h_0 \in R^N$, the NLMS algorithm [4, 5] generates the vector sequence $(h_k)_{k\in N}$ recursively as follows:

$$\mathbf{h}_{k+1} := \mathbf{h}_k - \mu \frac{e_k(\mathbf{h}_k)}{\|\mathbf{u}_k\|_2^2}\mathbf{u}_k \quad (3)$$

$$= \mathbf{h}_k + \mu(P_{H_k}(\mathbf{h}_k) - \mathbf{h}_k), \quad k \in \mathbb{N}, \quad (4)$$

where $\mu \in [0, 2]$ is the step size (In the presence of noise, $\mu > 1$ would never be used in practice due to its unacceptable misadjustment without increasing the speed of convergence.) and

$$e_k(\mathbf{h}) := \langle \mathbf{u}_k, \mathbf{h}\rangle_2 - d_k, \quad \mathbf{h} \in \mathbb{R}^N,\ k \in \mathbb{N}, \quad (5)$$

$$H_k := \left\{\mathbf{h} \in \mathbb{R}^N : e_k(\mathbf{h}) = 0\right\}, \quad k \in \mathbb{N}. \quad (6)$$

The right side of (4) is called the relaxed projection due to the presence of μ, and it is illustrated in Figure 1. We see that for any $\mu \in (0, 2)$ the update of NLMS decreases the value of the metric distance function:

$$\varphi_k(\mathbf{x}) := d(\mathbf{x}, H_k) := \min_{\mathbf{a}\in H_k}\|\mathbf{x} - \mathbf{a}\|_2, \quad \mathbf{x} \in \mathbb{R}^N,\ k \in \mathbb{N}. \quad (7)$$

Figure 2 illustrates several steps of NLMS for $\mu = 1$. In noiseless case, it is readily verified that $\phi_k(h^*) = d(h^*, H_k) = 0$, for all $k \in N$, implying that (i) $h^* \in {}_{k\in N} H_k$ and (ii) $\|h_{k+1} - h^*\|_2 \le \|h_k - h^*\|_2$, for all $k \in N$, due to the Pythagorean theorem. The figure suggests that $(h_k)_{k\in N}$ would converge to h^*; namely, it would minimize $(\phi_k)_{k\in N}$ asymptotically. In noisy case, the properties (i) and (ii) shown above are not guaranteed, and NLMS can only compute an approximate solution. APA [6, 7] can be viewed in a similar way [10]. The APSM presented below is an extension of NLMS and APA.

A Brief Review of Adaptive Projected Subgradient Method

We have seen above that asymptotic minimization of a sequence of functions is a natural formulation in the adaptive filtering. The task we consider now is asymptotic minimization of a sequence of (general) continuous convex functions $(\phi_k)_{k\in N}$, $\phi_k : R^N \to [0, \infty)$, over a possible constraint set $(\emptyset \neq)C \subset R^N$, which is assumed to be closed and convex. In [2], it has been proven that APSM achieves this task under certain mild conditions by generating a sequence $(h_k)_{k\in N} \subset R^N$ (for an initial vector $h_0 \in R^N$) recursively by

$$\mathbf{h}_{k+1} := P_C\left[\mathbf{h}_k + \lambda_k\left(T_{\mathrm{sp}(\varphi_k)}(\mathbf{h}_k) - \mathbf{h}_k\right)\right], \quad k \in \mathbb{N}, \quad (8)$$

where $\lambda_k \in [0, 2]$, $k \in N$, and $T_{sp(\phi k)}$ denotes the subgradient projection relative to ϕ_k (see Appendix A). APSM reproduces NLMS by letting $C := R^N$ and

$\phi_k(x) := d(x, H_k)$, $x \in R^N$, $k \in N$, with the standard inner product. A useful generalization has been presented in [3]; this makes it possible to take into account multiple convex constraints in the parameter space [3] and also such constraints in multiple domains [43, 44].

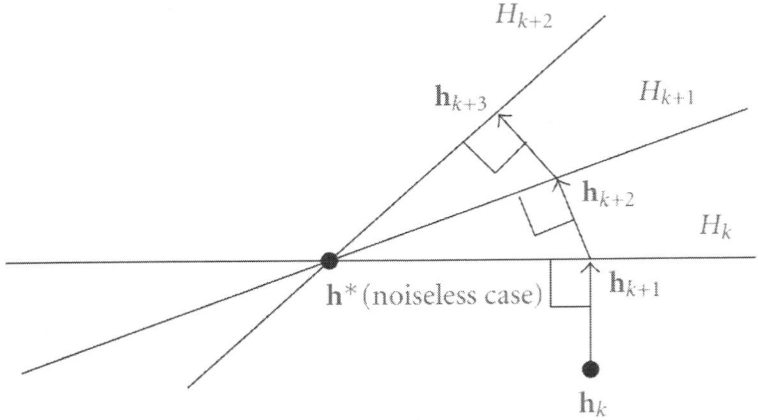

Figure 2: NLMS minimizes the sequence of the metric distance functions $\phi_k(x) := d(x, H_k)$ asymptotically under certain conditions.

VARIABLE-METRIC EXTENSION OF APSM

We extend APSM such that it encompasses the family of adaptive variable-metric projection algorithms, which have remarkable advantages in performance over their constantmetric counterparts. We start with a simplified version of the variable-metric APSM (V-APSM) and show that it includes TDAF, LNAF/QNAF, PAF, and KPAF as its particular examples. We then present the V-APSM that can deal with a convex constraint (the reader who has no need to consider any constraint may skip Section 3.3).

Variable-Metric Adaptive Projected Subgradient Method without Constraint

We present the simplified V-APSM which does not take into account any constraint (The full version will be presented in Section 3.3). Let $(R^{N\times N} \ni) G_k \succ 0$, $k \in N$; we express by $A \succ 0$ that a matrix A is symmetric and positive definite. Define the inner product and its induced norm, respectively, as <x>, $yG_k := x^T G_k y$, for all $(x, y) \in R^N \times R^N$, and $\|x\|_{G_k} := \sqrt{\langle x, x\rangle_{G_k}}$, for all $x \in R^N$. For convenience, we regard G_k as a metric. Recalling the definition, the subgradient projection depends on the inner product (and the norm), thus depending on the metric G_k (see (A.3) and (A.4) in Appendix A). We therefore specify the metric G_k employed in the subgradient projection by $T^{(G_k)}_{sp(\varphi_k)}$. The simplified variable-metric APSM is given as follows.

Scheme 1 (Variable-metric APSM without constraint). Let $\phi_k : R^N \rightarrow [0, \infty)$, $k \in N$, be continuous convex functions. Given an initial vector $h_0 \in R^N$, generate $(h_k)_{k\in N} \subset R^N$ by

$$\mathbf{h}_{k+1} := \mathbf{h}_k + \lambda_k \left(T^{(\mathbf{G}_k)}_{\mathrm{sp}(\varphi_k)}(\mathbf{h}_k) - \mathbf{h}_k \right), \quad k \in \mathbb{N}, \tag{9}$$

where $\lambda_k \in [0, 2]$, for all $k \in N$.

Recalling the linear system model presented in Section 2.1, a simple example of Scheme 1 is given as follows.

Example 1 (Adaptive variable-metric projection algorithms). An application of Scheme 1 to

$$\varphi_k(\mathbf{x}) := d_{\mathbf{G}_k}(\mathbf{x}, H_k) := \min_{\mathbf{a}\in H_k} \|\mathbf{x} - \mathbf{a}\|_{\mathbf{G}_k}, \quad \mathbf{x} \in \mathbb{R}^N, k \in \mathbb{N} \tag{10}$$

yields

$$\begin{aligned}\mathbf{h}_{k+1} &:= \mathbf{h}_k + \lambda_k \left(P^{(\mathbf{G}_k)}_{H_k}(\mathbf{h}_k) - \mathbf{h}_k \right) \\ &= \mathbf{h}_k - \lambda_k \frac{e_k(\mathbf{h}_k)}{\mathbf{u}_k^T \mathbf{G}_k^{-1} \mathbf{u}_k} \mathbf{G}_k^{-1} \mathbf{u}_k, \quad k \in \mathbb{N}.\end{aligned} \tag{11}$$

Equation (11) is obtained by noting that the normal vector of H_k with respect to the G_k-metric is $G_k^{-1} u_k$ because $H_k = \{h \in R^N : \langle G_k^{-1} u_k, h \rangle_{Gk} = d_k\}$. More sophisticated algorithms than Example 1 can be derived by following the way in [2, 37]. To keep this work as simple as possible for better accessibility, such sophisticated algorithms will be investigated elsewhere.

Examples of the Metric Design

The TDAF, LNAF/QNAF, PAF, and KPAF algorithms have the common form of (11) with individual design of G_k; interesting relations among TDAF, PAF, and KPAF are given in [34] based on the socalled error surface analysis. The G_k-design in each of the algorithms is given as follows.

(1) Let $V \in R^{N\times N}$ be a prespecified transformation matrix such as the discrete cosine transform (DCT) and discrete Fourier transform

(DFT). Given $s_0^{(i)} > 0, i=1,2,\ldots,N$, define $s_{k+1}^{(i)} := \gamma s_k^{(i)} + (\widetilde{u}_k^{(i)})^2$, where $\gamma \in (0, 1)$ and $[\widetilde{u}_k^{(1)}, \widetilde{u}_k^{(2)}, \ldots, \widetilde{u}_k^{(N)}]^T$ is the transform-domain input vector. Then, G_k for TDAF [19, 20] is given as follows:

$$\mathbf{G}_k := \mathbf{V}^T \operatorname{diag}\left(s_k^{(1)}, s_k^{(2)}, \ldots, s_k^{(N)}\right)\mathbf{V}. \tag{12}$$

Here, diag(a) denotes the diagonal matrix whose diagonal entries are given by the components of a vector $a \in R^N$. This metric is useful for colored input signals.

(2) G_ks for LNAF in [23] and QNAF in [26] are given by $G_k := \hat{R}_{k,LN}$ and $G_k := \hat{R}_{k,QN}$, respectively, where for some initial matrices $\hat{R}_{0,LN}$ and $\hat{R}_{0,QN}$ their inverses are updated as follows:

$$\hat{\mathbf{R}}_{k+1,\mathrm{LN}}^{-1} := \frac{1}{1-\alpha}\left(\hat{\mathbf{R}}_{k,\mathrm{LN}}^{-1} - \frac{\hat{\mathbf{R}}_{k,\mathrm{LN}}^{-1}\mathbf{u}_k\mathbf{u}_k^T\hat{\mathbf{R}}_{k,\mathrm{LN}}^{-1}}{(1-\alpha)/\alpha + \mathbf{u}_k^T\hat{\mathbf{R}}_{k,\mathrm{LN}}^{-1}\mathbf{u}_k}\right),$$

$$\alpha \in (0,1),$$

$$\hat{\mathbf{R}}_{k+1,\mathrm{QN}}^{-1} := \hat{\mathbf{R}}_{k,\mathrm{QN}}^{-1} + \left(\frac{1}{2\mathbf{u}_k^T\hat{\mathbf{R}}_{k,\mathrm{QN}}^{-1}\mathbf{u}_k} - 1\right)\frac{\hat{\mathbf{R}}_{k,\mathrm{QN}}^{-1}\mathbf{u}_k\mathbf{u}_k^T\hat{\mathbf{R}}_{k,\mathrm{QN}}^{-1}}{\mathbf{u}_k^T\hat{\mathbf{R}}_{k,\mathrm{QN}}^{-1}\mathbf{u}_k}. \tag{13}$$

The matrices $\hat{R}_{k,LN}$ and $\hat{R}_{k,QN}$ well approximate the autocorrelation matrix of the input vector u_k, which coincides with the Hessian of the mean squared error (MSE) cost function. Therefore, LNAF/QNAF is a stochastic approximation of the Newton method, yielding faster convergence than the LMS-type algorithms based on the steepest descent method.

(3) Let $h_k =: [h_k^{(1)}, h_k^{(2)}, \ldots, h_k^{(N)}]^T$, $k \in N$. Given small constants $\sigma > 0$ and $\delta > 0$, define $L^{max}_k := \max\{\delta, |h^{(1)}_k|, |h^{(2)}_k|, \ldots, |h^{(N)}_k|\} > 0$, $\gamma^{(n)}_k := \max\{\sigma L^{max}_k, |h^{(n)}_k|\} > 0$, $n = 1, 2, \ldots, N$, and $\alpha^{(n)}_k := \gamma_k^{(n)} / \sum_{i=1}^{N} \gamma_k^{(i)}$ $\gamma^{(i)}_k$, $n = 1, 2, \ldots, N$. Then, G_k for the PNLMS algorithm [27, 28] is as follows:

$$\mathbf{G}_k := \operatorname{diag}^{-1}\left(\alpha_k^{(1)}, \alpha_k^{(2)}, \ldots, \alpha_k^{(N)}\right). \tag{14}$$

This metric is useful for sparse unknown systems h∗. The improved proportionate NLMS (IPNLMS) algorithm [31] employs $\gamma^{(n)}_{\mathrm{ip},k} := 2[(1 - \omega)\|h_k\|_1/N + \omega|h^{(n)}_k|]$, $\omega \in [0, 1)$, for n = 1, 2, ... , N in place of $\gamma^{(n)}_k$; $\|\cdot\|_1$ denotes the 1 norm. IPNLMS is reduced to the standard NLMS algorithm when ω := 0. Another modification has been proposed in, for example, [32].

(4) Let $\hat{R}$ and $\hat{p}$ be the estimates of $R := E\{u_k u^T_k\}$ and $p := E\{u_k d_k\}$. Also let $Q \in \mathbb{R}^{N\times N}$ be a matrix obtained by orthonormalizing (from left to right) the Krylov matrix $[\hat{\mathbf{p}}, \hat{\mathbf{R}}\hat{\mathbf{p}}, \ldots, \hat{\mathbf{R}}^{N-1}\hat{\mathbf{p}}]$. Define $[\widetilde{h}^{(1)}_k, \widetilde{h}^{(2)}_k, \ldots, \widetilde{h}^{(N)}_k]^T := Q^T h_k$, $k \in \mathbb{N}$. Given a proportionality factor $\omega \in [0, 1)$ and a small constant $\varepsilon > 0$, define

$$\beta^{(n)}_k := \frac{1-\overline{\omega}}{N} + \overline{\omega}\frac{\left|\widetilde{h}^{(n)}_k\right|}{\sum_{i=1}^{N}\left|\widetilde{h}^{(i)}_k\right| + \varepsilon} > 0,$$

$$n = 1, 2, \ldots, N, \quad k \in \mathbb{N}. \tag{15}$$

Then, G_k for KPNLMS [34] is given as follows:

$$\mathbf{G}_k := \mathbf{Q}\mathrm{diag}^{-1}\left(\beta^{(1)}_k, \beta^{(2)}_k, \ldots, \beta^{(N)}_k\right)\mathbf{Q}^T. \tag{16}$$

This metric is useful even for dispersive unknown systems h^*, as Q^T sparsifies it. If the input signal is highly colored and the eigenvalues of its autocorrelation matrix are not clustered, then this metric is used in combination with the metric of TDAF (see [34]). We mention that this is not exactly the one proposed in [34]. The transformation Q^T makes the optimal filter into a special sparse system of which only a few first components would have large magnitude and the rest is nearly zero. This information (which is much more than only that the system is sparse) is exploited to reduce the computational complexity

Finally, we present below the full version of V-APSM, which is an extension of Scheme 1 for dealing with a convex constraint.

The Variable-Metric Adaptive Projected Subgradient Method—A Treatment of Convex Constraint

We generalize Scheme 1 slightly so as to deal with a constraint set $K \subset \mathbb{R}^N$, which is assumed to be closed and convex. Given a mapping $T : \mathbb{R}^N \to \mathbb{R}^N$, $\mathrm{Fix}(T) := \{x \in \mathbb{R}^N : T(x) = x\}$ is called the fixed point set of T. The operator

$P^{(Gk)}_K$, $k \in N$, which denotes the metric projection onto K with respect to the G_k metric, is 1-attracting nonexpansive (with respect to the G_k metric) with $\mathrm{Fix}(P^{(Gk)}_K) = K$, for all $k \in N$ (see Appendix B). It holds moreover that $P^{(Gk)}_K(x) \in K$ for any $x \in R^N$. For generality, we let $T_k : R^N \to R^N$, $k \in N$, be an η-attracting nonexpansive mapping (η > 0) with respect to the G_k-metric satisfying

$$T_k(\mathbf{x}) \in \mathfrak{K} = \mathrm{Fix}(T_k), \quad \forall k \in \mathbb{N}, \forall \mathbf{x} \in \mathbb{R}^N. \tag{17}$$

The full version of V-APSM is then given as follows.

Scheme 2 (The Variable-metric APSM). Let $\phi_k : R^N \to [0, \infty)$, $k \in N$, be continuous convex functions. Given an initial vector $h_0 \in R^N$, generate $(h_k)_{k \in N} \subset R^N$ by

$$\mathbf{h}_{k+1} := T_k\left[\mathbf{h}_k + \lambda_k\left(T^{(\mathbf{G}_k)}_{\mathrm{sp}(\varphi_k)}(\mathbf{h}_k) - \mathbf{h}_k\right)\right], \quad k \in \mathbb{N}, \tag{18}$$

where $\lambda_k \in [0, 2]$, for all $k \in N$.

Scheme 2 is reduced to Scheme 1 by letting $T_k := I$ ($K = R^N$), for all $k \in N$, where I denotes the identity mapping. The form given in (18) was originally presented in [37] without any consideration of the convergence issue. Moreover, a partial convergence analysis for $T_k := I$ was presented in [45] with no proof. In the following section, we present a more advanced analysis for Scheme 2 with a rigorous proof.

A DETERMINISTIC ANALYSIS

We present a deterministic analysis of Scheme 2. In the analysis, small metric-fluctuations is the key assumption to be employed. The reader not intending to consider any constraint may simply let $K := R^N$.

Monotone Approximation in the Variable-Metric Sense

We start with the following assumption.

Assumption 1. (a) (Assumption in [2]). There exists $K_0 \in N$ s.t.

$$\varphi_k^* := \min_{\mathbf{x} \in \mathfrak{K}} \varphi_k(\mathbf{x}) = 0, \quad \forall k \geq K_0,$$

$$\Omega := \bigcap_{k \geq K_0} \Omega_k \neq \emptyset, \tag{19}$$

where

$$\Omega_k := \left\{\mathbf{x} \in \mathfrak{K} : \varphi_k(\mathbf{x}) = \varphi_k^*\right\}, \quad k \in \mathbb{N}. \tag{20}$$

(b) There exist $\varepsilon_1, \varepsilon_2 > 0$ s.t. $\lambda_k \in [\varepsilon_1, 2 - \varepsilon_2] \subset (0, 2)$, $k \geq K_0$.

The following fact is readily verified.

Fact 1. Under Assumption 1(a), the following statements are equivalent (for $k \geq K_0$):

- $h_k \in \Omega_k$,
- $h_{k+1} = h_k$,
- $\phi_k(h_k) = 0$,
- $0 \in \partial_{G_k}\phi_k(h_k)$.

V-APSM enjoys a sort of monotone approximation in the G_k-metric sense as follows.

Proposition 1. Let $(h_k)_{k\in N}$ be the vectors generated by Scheme 2. Under Assumption 1, for any $z^*_k \in \Omega_k$,

$$\left\|\mathbf{h}_k - \mathbf{z}_k^*\right\|_{\mathbf{G}_k}^2 - \left\|\mathbf{h}_{k+1} - \mathbf{z}_k^*\right\|_{\mathbf{G}_k}^2 \geq \varepsilon_1\varepsilon_2 \frac{\varphi_k^2(\mathbf{h}_k)}{\left\|\varphi_k'(\mathbf{h}_k)\right\|_{\mathbf{G}_k}^2}$$

$$(\forall k \geq K_0 \text{ s.t. } \mathbf{h}_k \notin \Omega_k), \tag{21}$$

$$\left\|\mathbf{h}_k - \mathbf{z}_k^*\right\|_{\mathbf{G}_k}^2 - \left\|\mathbf{h}_{k+1} - \mathbf{z}_k^*\right\|_{\mathbf{G}_k}^2$$

$$\geq \frac{\eta\varepsilon_2}{\varepsilon_2 + (2 - \varepsilon_2)\eta} \|\mathbf{h}_k - \mathbf{h}_{k+1}\|_{\mathbf{G}_k}^2, \quad \forall k \geq K_0. \tag{22}$$

Proof. See Appendix C.

Proposition 1 will be used to prove the theorem in the following.

Analysis under Small Metric-Fluctuations

To prove the deterministic convergence, we need the property of monotone approximation in a certain "constant-metric" sense [2]. Unfortunately, this property is not ensured automatically for the adaptive variable-metric projection algorithm unlike the constant-metric one. Indeed, as described in Proposition 1, the monotone approximation is only ensured in the G_k-metric sense at each iteration; this is because the strongly attracting nonexpansivity of T_k and the subgradient projection $T^{(\mathbf{G}_k)}_{sp(\varphi_k)}$ are both dependent on G_k. Therefore, considerably different metrics may result in totally different directions of update, suggesting that under large metric-fluctuations it would be impossible to ensure the monotone approximation in the "constant-metric" sense. Small metric-fluctuations is thus the key assumption to be made for the analysis.

Given any matrix $A \in R^{N \times N}$, its spectral norm is defined by $||A||_2 := \sup_{x \in RN} ||Ax||_2/||x||_2$ [46]. Given $A \succ 0$, let $\sigma^{min}{}_A > 0$ and $\sigma^{max}{}_A > 0$ denote its minimum and maximum eigenvalues, respectively; in this case $||A||_2 = \sigma^{max}{}_A$. We introduce the following assumptions.

Assumption 2. (a) Boundedness of the eigenvalues of Gk. There exist $\delta_{min}, \delta_{max} \in (0, \infty)$ s.t. $\delta_{min} < \sigma^{min}{}_{Gk} \leq \sigma^{max}{}_{Gk} < \delta_{max}$, for all $k \in N$.

(b) Small metric-fluctuations. There exist $(R^{N \times N} \ni) G \succ 0$, $K_1 \geq K_0$, $\tau > 0$, and a closed convex set $\Gamma \subseteq \Omega$ s.t. $E_k := G_k - G$ satisfies

$$\frac{||\mathbf{h}_{k+1} + \mathbf{h}_k - 2\mathbf{z}^*||_2 ||\mathbf{E}_k||_2}{||\mathbf{h}_{k+1} - \mathbf{h}_k||_2} < \frac{\varepsilon_1 \varepsilon_2 \sigma_{\mathbf{G}}^{\min} \delta_{\min}^2}{(2-\varepsilon_2)^2 \sigma_{\mathbf{G}}^{\max} \delta_{\max}} - \tau \tag{23}$$

We now reach the convergence theorem.

Theorem 1. Let $(h_k)_{k \in N}$ be generated by Scheme 2. Under Assumptions 1 and 2, the following holds.

(a) Monotone approximation in the constant-metric sense. For any $z^* \in \Gamma$,

$$||\mathbf{h}_k - \mathbf{z}^*||_{\mathbf{G}}^2 - ||\mathbf{h}_{k+1} - \mathbf{z}^*||_{\mathbf{G}}^2 \geq \frac{(2-\varepsilon_2)^2 \sigma_{\mathbf{G}}^{\max}}{\delta_{\min}^2} \tau \frac{\varphi_k^2(\mathbf{h}_k)}{\left\|\varphi_k'(\mathbf{h}_k)\right\|_{\mathbf{G}}^2} \quad (\forall k \geq K_1 \text{ s.t. } \mathbf{h}_k \notin \Omega_k) \tag{24}$$

$$||\mathbf{h}_k - \mathbf{z}^*||_{\mathbf{G}}^2 - ||\mathbf{h}_{k+1} - \mathbf{z}^*||_{\mathbf{G}}^2 \geq \frac{\tau}{\sigma_{\mathbf{G}}^{\max}} ||\mathbf{h}_k - \mathbf{h}_{k+1}||_{\mathbf{G}}^2, \quad \forall k \geq K_1. \tag{25}$$

(b) Asymptotic minimization. Assume that $(\phi'_k(h_k))_{k \in N}$ is bounded. Then,

$$\lim_{k \to \infty} \varphi_k(\mathbf{h}_k) = 0. \tag{26}$$

(c) Convergence to an asymptotically optimal point. Assume that Γ has a relative interior with respect to a hyperplane $\Pi \subset R^N$; that is, there exists $\tilde{h} \in \Pi \cap \Gamma$ s.t. $\{x \in \Pi : \|x - \tilde{h}\| < \varepsilon_{r.i.}\} \subset \Gamma$ for some $\varepsilon_{r.i.} > 0$. (The norm $\|\cdot\|$ can be arbitrary due to the norm equivalency for finitedimensional vector spaces.) Then, $(h_k)_{k\in N}$ converges to a point $\hat{h} \in K$. In addition, under the assumption in Theorem 1(b),

$$\lim_{k \to \infty} \varphi_k\left(\hat{\mathbf{h}}\right) = 0 \tag{27}$$

provided that there exists bounded $(\phi'_k(\hat{h}))_{k\in N}$ where $\phi'_k(\hat{h}) \in \partial_{G_k}\phi_k(\hat{h})$, for all $k \in N$.

(d) Characterization of the limit point. Assume the existence of some interior point $\tilde{h}$ of Ω. In this case, under the assumptions in (c), if for all $\varepsilon > 0$, for all $r > 0$, $\exists \delta > 0$ s.t.

$$\inf_{\substack{d(\mathbf{h}_k, \mathrm{lev}_{\leq 0}\varphi_k) \geq \varepsilon, \\ \|\tilde{\mathbf{h}} - \mathbf{h}_k\| \leq r, \\ k \geq K_1}} \varphi_k(\mathbf{h}_k) \geq \delta, \tag{28}$$

then $\hat{h} \in \overline{\liminf_{k \to \infty} \Omega_k}$, where $\liminf_{k \to \infty} \Omega_k := \bigcup_{k=0}^{\infty} \bigcap_{n \geq k} \Omega_n$ and the overline denotes the closure (see Appendix A for the definition of $\mathrm{lev}_{\leq 0}\phi_k$). Note that the metric for $\|\cdot\|$ and $d(\cdot, \cdot)$ is arbitrary.

Proof. See Appendix D.

We conclude this section by giving some remarks on the assumptions and the theorem.

Remark 1 (On Assumption 1). (a) Assumption 1(a) is required even for the simple NLMS algorithm [2].

(b) Assumption 1(b) is natural because the step size is usually controlled so as not to become too large nor small for obtaining reasonable performance.

Remark 2 (On Assumption 2). (a) In the existing algorithms mentioned in Example 1, the eigenvalues of G_k are controllable directly and usually bounded. Therefore, Assumption 2(a) is natural.

(b) Assumption 2(b) implies that the metric-fluctuations $\|E_k\|_2$ should be sufficiently small to satisfy (23). We mention that the constant metric (i.e., $G_k := G \succ 0$, for all $k \in N$, thus $\|E_k\|_2 = 0$) surely satisfies (23): note that $\|h_{k+1} - h_k\|_2 \neq 0$ by Fact 1. In the algorithms presented in Example 1, the fluctuations of G_k tend to become small as the filter adaptation proceeds. If in particular a constant step size $\lambda_k := \lambda \in (0, 2)$, for all $k \in N$, is used, we have $\varepsilon_1 = \lambda$ and $\varepsilon_2 = 2 - \lambda$ and thus (23) becomes

$$\frac{\|\mathbf{h}_{k+1} + \mathbf{h}_k - 2\mathbf{z}^*\|_2 \|\mathbf{E}_k\|_2}{\|\mathbf{h}_{k+1} - \mathbf{h}_k\|_2} < \left(\frac{2}{\lambda} - 1\right) \frac{\sigma_{\mathbf{G}}^{\min} \delta_{\min}^2}{\sigma_{\mathbf{G}}^{\max} \delta_{\max}} - \tau. \quad (29)$$

This implies that the lower the value of λ is, the larger amount of metric-fluctuations would be acceptable in the adaptation. In Section 5, it will be shown that the use of small λ makes the algorithm relatively insensitive to large metric-fluctuations. Finally, we mention that multiplication of G_k by any scalar $\xi > 0$ does not affect the assumption, because (i) $\sigma^{\min}{}_G$, $\sigma^{\max}{}_G$, $\delta_{\min}$, $\delta_{\max}$, and $\|E_k\|_2$ in (23) are equally scaled, and (ii) the update equation (23) is unchanged (as $\phi'_k(x)$ is scaled by $1/\xi$ by the definition of subgradient).

Remark 3 (On Theorem 1). (a) Theorem 1(a) ensures the monotone approximation in the "constant" G-metric sense; that is, $\|h_{k+1} - z^*\|_G \leq \|h_k - z^*\|_G$ for any $z^* \in \Gamma$. This remarkable property is important for stability of the algorithm.

(b) Theorem 1(b) tells us that the variable-metric adaptive filtering algorithm in (11) asymptotically minimizes the sequence of the metric distance functions $\phi_k(x) = d_{Gk}(x, H_k)$, $k \in N$. This intuitively means that the output error $e_k(h_k)$ diminishes, since H_k is the zero output-error hyperplane. Note however that this does not imply the convergence of the sequence $(h_k)_{k\in N}$ (see Remark 3(c)). The condition of boundedness is automatically satisfied for the metric distance functions [2].

(c) Theorem 1(c) ensures the convergence of the sequence $(h_k)_{k\in N}$ to a point $\hat{h} \in K$. An example that the NLMS algorithm does not converge without the assumption in Theorem 1(c) is given in [2]. Theorem 1(c) also tells us that the limit point $\hat{h}$ minimizes the function sequence ϕ_k asymptotically; that is, the limit point is asymptotically optimal. In the special case where $n_k = 0$ (for all $k \in N$) and the autocorrelation matrix of u_k is nonsingular, h^* is the unique point that makes $\phi_k(h^*) = 0$ for all $k \in N$. The condition of boundedness is automatically satisfied for the metric distance functions [2].

(d) From Theorem 1(c), we can expect that the limit point h should be characterized by means of the intersection of Ω_ks, because Ω_k is the set of minimizers of ϕ_k on K. This intuition is verified by

Theorem 1(d), which provides an explicit characterization of $\hat{h}$. The condition in (28) is automatically satisfied for the metric distance functions [2].

NUMERICAL EXAMPLES

We first show that V-APSM outperforms its constant-metric (or Euclidean-metric) counterpart with the design of G_k presented in Section 3.2. We then examine the impacts of metric-fluctuations on the performance of adaptive filter by taking PAF as an analogy; recall here that metric fluctuations were the key in the analysis. We finally consider the case of nonstationary inputs and present numerical studies on the properties of the monotone approximation and the convergence to an asymptotically optimal point (see Theorem 1).

Variable Metric versus Constant Euclidean Metric

First, we compare TDAF [19, 20] and PAF (specifically, IPNLMS) [31] with their constant-metric counterpart, that is, NLMS. We consider a sparse unknown system $h* \in RN$ depicted in Figure 3(a) with N = 256. The input is the colored signal called USASI and the noise is white Gaussian with the signal-to-noise ratio (SNR) 30 dB, where SNR := 10 log10$(E\{z_k^2\}/E\{n_k^2\})$ with $z_k := u_k, h^*$ (The USASI signal is a wide sense stationary process and is modeled on the autoregressive moving average (ARMA) process characterized by $H(z) := (1 - z^{-2})/(1 - 1.70223z^{-1} + 0.71902z^{-2})$, $z \in C$, where C denotes the set of all complex numbers. In the experiments, the average eigenvalue-spread of the input autocorrelation-matrix was 1.20×10^6.). We set $\lambda_k = 0.2$, for all $k \in N$, for all algorithms. For TDAF, we set $\gamma = 1 - 10^{-3}$ and employ the DCT matrix for V. For PAF (IPNLMS), we set $\omega = 0.5$. We use the performance measure of MSE 10 log10$(E\{e_k^2\}/E\{z_k^2\})$. The expectation operator is approximated by an arithmetic average over 300 independent trials. The results are depicted in Figure 3(b).

Next, we compare QNAF [26] and KPAF [34] with NLMS. We consider the noisy situation of SNR 10 dB and nonsparse unknown systems h* drawn from a normal distribution N (0, 1) randomly at each trial. The other conditions are the same as the first experiment. We set $\lambda_k = 0.02$, for all $k \in N$, for KPAF and NLMS, and use the same parameters for KPAF as in [34]. Although the use of $\lambda_k = 1.0$ for QNAF is implicitly suggested in [26], we instead use $\lambda_k = 0.04$ with $\widehat{\mathbf{R}}^{-1}_{0,QN} = I$ to attain the same steady-state error as the other algorithms (I denotes the identity matrix). The results are depicted in Figure 4.

Figures 3 and 4 clearly show remarkable advantages of the V-APSM-based algorithms (TDAF, PAF, QNAF, and KPAF) over the constant-metric NLMS. In both experiments, NLMS suffers from slow convergence because of the high correlation of the input signals. The metric designs of TDAF and QNAF accelerate the convergence by reducing the correlation. On the other hand, the metric design of PAF accomplishes it by exploiting the sparse structure of h∗, and that of KPAF does it by sparsifying the nonsparse h∗.

Impacts of Metric-Fluctuations on the MSE Performance

We examine the impacts of metric-fluctuations on the MSE performance under the same simulation conditions as the first experiment in Section 5.1. We take IPNLMS because of its convenience in studying the metric-fluctuations as seen below. The metric employed in IPNLMS can be obtained by replacing h∗ in

$$\mathbf{G}_{\text{ideal}} := 2\left(\frac{1}{N}\mathbf{I} + \frac{\operatorname{diag}(|\mathbf{h}^*|)}{\|\mathbf{h}^*\|_1}\right)^{-1} \tag{30}$$

by its instantaneous estimate h_k, where $|\cdot|$ denotes the element wise absolute-value operator. We can thus interpret that IPNLMS employs an approximation of G_{ideal}. For ease of evaluating the metric-fluctuations $\|E_k\|_2$, we employ a test algorithm which employs the metric G_{ideal} with cyclic fluctuations as follows:

$$\mathbf{G}_k^{-1} := \mathbf{G}_{\text{ideal}}^{-1} + \frac{\varrho}{N}\operatorname{diag}(\hat{\mathbf{e}}_{\iota(k)}), \quad k \in \mathbb{N}. \tag{31}$$

Here, $\iota(k) := (k \bmod N)+1 \in \{1, 2, \ldots, N\}$, $k \in N$, $\rho \geq 0$ determines the amount of metric-fluctuations, and $\hat{e}_j \in R^N$ is a unit vector with only one nonzero component at the jth position. Letting $G := G_{ideal}$, we have

$$\|\mathbf{E}_k\|_2 = \frac{\varrho\left(g_{\text{ideal}}^{\iota(k)}\right)^2}{N + \varrho g_{\text{ideal}}^{\iota(k)}} \in \left[0, g_{\text{ideal}}^{\iota(k)}\right), \quad \forall k \in \mathbb{N}, \tag{32}$$

where g^n_{ideal}, $n \in \{1, 2, \ldots, N\}$, denotes the nth diagonal element of G_{ideal}. It is seen that (i) for a given $\iota(k)$, $\|E_k\|_2$ is monotonically increasing in terms of $\rho \geq 0$, and (ii) for a given ρ, $\|E_k\|_2$ is maximized by $g_{\text{ideal}}^{\iota(k)} = \min_{j=1}^{N} g_{\text{ideal}}^{j}$ideal.

First, we set $\lambda_k = 0.2$, for all $k \in N$, and examine the performance of the algorithm for $\rho = 0, 10, 40$. Figure 5(a) depicts the learning curves. Since the test algorithm has the knowledge about G_{ideal} (subject to the fluctuations depending on the ρ value) from the beginning of adaptation, it achieves faster convergence than PAF (and of course than NLMS). There is a fractional difference between $\rho = 0$ and $\rho = 10$, indicating robustness of the algorithm against a moderate amount of metric-fluctuations. The use of $\rho = 40$, on the other hand, causes the increase of steady-state error and the instability at the end. Meanwhile, the good steady state performance of IPNLMS suggests that the amount of its metric-fluctuations is sufficiently small.

Next, we set $\lambda_k = 0.1, 0.2, 0.4$, for all $k \in N$, and examine the MSE performance in the steady-state for each value of $\rho \in [0, 50]$. For each trial, the MSE values are averaged over 5000 iterations after convergence. The results are depicted in Figure 5(b). We observe the tendency that the use of smaller λ_k makes the algorithm less sensitive to metric-fluctuations. This should not be confused with the well-known relations between the step size and steady-state performance in the standard algorithms such as NLMS. Focusing on $\rho = 25$ in Figure 5(b), the steady-state MSE of $\lambda_k = 0.2$ is slightly higher than that of $\lambda_k = 0.1$, while the steady-state MSE of $\lambda_k = 0.4$ is unacceptably high compared to that of $\lambda_k = 0.2$. This does not usually happen in the standard algorithms. The analysis presented in the previous section offers a rigorous theoretical explanation for the phenomena observed in Figure 5. Namely, the larger the metric-fluctuations or the step size, the more easily Assumption 2(b) is violated, resulting in worse performance. Also, the analysis clearly explains that the use of smaller λ_k allows a larger amount of metric-fluctuations $\|E_k\|_2$ [see (29)].

Performance for Nonstationary Input

In the previous subsection, we changed the amount of metric-fluctuations in a cyclic fashion and studied its impacts on the performance. We finalize our numerical studies by considering more practical situations in which Assumption 2(b) is easily violated. Specifically, we examine the performance of TDAF and NLMS for nonstationary inputs of female speech sampled at 8 kHz (see Figure 6(a)). Indeed, TDAF controls its metric to reduce the correlation of inputs, whose statistical properties change dynamically due to the nonstationarity. The metric therefore would tend to fluctuate dynamically by reflecting the change of statistics. For better controllability of the metric-fluctuations, we slightly modify the update of $s^{(i)}_k$ in (12) into

$\widehat{s}_{k+1}^{(i)} := \widehat{\gamma}\widehat{s}_k^{(i)} + (1 - \widehat{\gamma})(\widetilde{u}_k^{(i)})^2$ for $\widehat{\gamma} \in (0, 1)$, i = 1, 2, ... , N. The amount of metric-fluctuations can be reduced by increasing γ up to one. Considering the acoustic echo cancellation problem (e.g., [33]), we assume SNR 20 dB and use the impulse response $h* \in R^N$ (N = 1024) described in Figure 6(b), which was recorded in a small room.

For all algorithms, we set $\lambda_k = 0.02$. For TDAF, we set (A) $\widehat{\gamma} = 1 - 10^{-4}$, (B) $\widehat{\gamma} = 1 - 10^{-4.5}$, and (C) $\widehat{\gamma} = 1 - 10^{-5}$, and were employ the DCT matrix for V. In noiseless situations, V-APSM enjoys the monotone approximation of h∗ and the convergence to the asymptotically optimal point h∗ under Assumptions 1 and 2 (see Remark 3). To illustrate how these properties are affected by the violation of the assumptions due mainly to the noise and the input nonstationarity, Figure 6(c) plots the system mismatch $10 \log_{10} (\|\mathbf{h}_k - \mathbf{h}^*\|_2^2 / \|\mathbf{h}^*\|_2^2)$ for one trial. We mention that, although Theorem 1(a) indicates the monotone approximation in the G-metric sense, G is unavailable and thus we employ the standard Euclidean metric (note that the convergence does not depend on the choice of metric). For (B) $\widehat{\gamma} = 1 - 10^{-4.5}$ and (C) $\widehat{\gamma} = 1 - 10^{-5}$, it is seen that hk is approaching h∗ monotonically. This implies that the monotone approximation and the convergence to h∗ are not seriously affected from a practical point of view. For (A) $\widehat{\gamma} = 1 - 10^{-4}$, on the other hand, h_k is approaching h∗but not monotonically. This is because the use of $\widehat{\gamma} = 1 - 10^{-4}$ makes Assumption 2(b) violated easily due to the relatively large metric-fluctuations. Nevertheless, the observed nonmonotone approximation of (A) $\widehat{\gamma} = 1 - 10^{-4}$ would be acceptable in practice; on its positive side, it yields the great benefit of faster convergence because it reflects the statistics of latest data more than the others.

CONCLUSION

This paper has presented a unified analytic tool named variable-metric adaptive projected subgradient method (VAPSM). The small metric-fluctuations has been the key for the analysis. It has been proven that V-APSM enjoys the invaluable properties of monotone approximation and convergence to an asymptotically optimal point. Numerical examples have demonstrated the remarkable advantages of V-APSM and its robustness against a moderate amount of metric-fluctuations. Also the examples have shown that the use of small step size robustifies the algorithm against a large amount of metric-fluctuations. This phenomenon should be distinguished from the well-known relations between the step size and steady-state performance, and our analysis has offered a rigorous theoretical explanation for the phenomenon. The results give us a useful insight that, in case an adaptive variable-metric projection algorithm suffers from poor steady-state performance, one could either reduce the step size or control the variable-metric such that its fluctuations become smaller. We believe—and it is our future task to prove—that V-APSM serves as a guiding principle to derive effective adaptive filtering algorithms for a wide range of applications.

APPENDICES

A. Projected Gradient and Projected Subgradient Methods

Let us start with the definitions of a convex set and a convex function. A set $C \subset R^N$ is said to be convex if $\nu x + (1-\nu)y \in C$, for all $(x, y) \in C \times C$, for all $\nu \in (0, 1)$. A function $\phi : R^N \to R$ is said to be convex if $\phi(\nu x+ (1-\nu)y) \leq \nu\phi(x) + (1-\nu)\phi(y)$, for all $(x, y) \in R^N \times R^N$, for all $\nu \in (0, 1)$.

Projected Gradient Method

The projected gradient method [38, 39] is an algorithmic solution to the following convexly constrained optimization:

$$\min_{\mathbf{h}\in C} \varphi(\mathbf{h}), \tag{A.1}$$

where $C \subset R^N$ is a closed convex set and $\phi : R^N \to R$ a differentiable convex function with its derivative $\phi' : R^N \to R^N$ being κ-Lipschitzian: that is, there exists $\kappa > 0$ s.t. $\|\phi'(x) - \phi'(y)\| \leq \kappa\|x - y\|$, for all $x, y \in R^N$. For an initial vector $h_0 \in R^N$ and the step size $\lambda \in (0, 2/\kappa)$, the projected gradient method generates a sequence $(h_k)_{k\in N} \subset R^N$ by

$$\mathbf{h}_{k+1} := P_C[\mathbf{h}_k - \lambda\varphi'(\mathbf{h}_k)], \quad k \in \mathbb{N}. \tag{A.2}$$

It is known that the sequence $(h_k)_{k\in N}$ converges to an arbitrary solution to the problem (A.1). If, however, ϕ is nondifferentiable, how should we do? An answer to this question has been given by Polyak in 1969 [40], which is described below.

Projected Subgradient Method

For a continuous (but not necessarily differentiable) convex function $\phi : R^N \to R$, it has been proven that the so-called projected subgradient method solves the problem (A.1) iteratively under certain conditions. The interested reader is referred to, for example, [3] for its detailed results. We only explain the method itself, as it is helpful to understand APSM.

What is subgradient, and does it always exist? The subgradient is a generalization of gradient, and it always exists for any continuous (possibly nondifferentiable) convex function (To be precise, the subgradient is a generalization of Gateaux di ˆ fferential.). In a differentiable case, the gradient $\phi'(y)$ at an arbitrary point $y \in R^N$ is characterized as the unique vector satisfying $x - y, \phi'(y) + \phi(y) \leq \phi(x)$, for all $x \in R^N$. In a nondifferentiable case, however, such a vector is nonunique in general, and the set of such vectors

$$\partial\varphi(\mathbf{y}) := \left\{\mathbf{a} \in \mathbb{R}^N : \langle \mathbf{x} - \mathbf{y}, \mathbf{a} \rangle + \varphi(\mathbf{y}) \leq \varphi(\mathbf{x}), \forall \mathbf{x} \in \mathbb{R}^N \right\} \neq \varnothing \quad \text{(A.3)}$$

is called subdifferential of ϕ at $y \in R^N$. Elements of the subdifferential $\partial\phi(y)$ are called subgradients of ϕ at y.

The projected subgradient method is based on subgradient projection, which is defined formally as follows (see Figure 7 for its geometric interpretation). Suppose that $\text{lev}_{\leq 0}\phi := \{x \in R^N : \phi(x) \leq 0\} \neq \emptyset$. Then, the mapping $T_{sp(\phi)} : R^N \to R^N$ defined as

$$T_{\text{sp}(\varphi)} : \mathbf{x} \longmapsto \begin{cases} \mathbf{x} - \dfrac{\varphi(\mathbf{x})}{\|\varphi'(\mathbf{x})\|^2}\varphi'(\mathbf{x}) & \text{if } \varphi(\mathbf{x}) > 0, \\ \mathbf{x} & \text{otherwise} \end{cases} \quad \text{(A.4)}$$

is called subgradient projection relative to ϕ, where $\phi'(x) \in \partial\phi(x)$, for all $x \in R^N$. For an initial vector $h_0 \in R^N$, the projected subgradient method generates a sequence $(h_k)_{k\in N} \subset R^N$ by

$$\mathbf{h}_{k+1} := P_C\left[\mathbf{h}_k + \lambda_k\left(T_{\text{sp}(\varphi)}(\mathbf{h}_k) - \mathbf{h}_k\right)\right], \quad k \in \mathbb{N}, \quad \text{(A.5)}$$

where $\lambda_k \in [0, 2]$, $k \in N$. Comparing (A.2) with (A.4) and (A.5), one can see similarity between the two methods. However, it should be emphasized that $\phi'(h_k)$ is (not the gradient but) a subgradient.

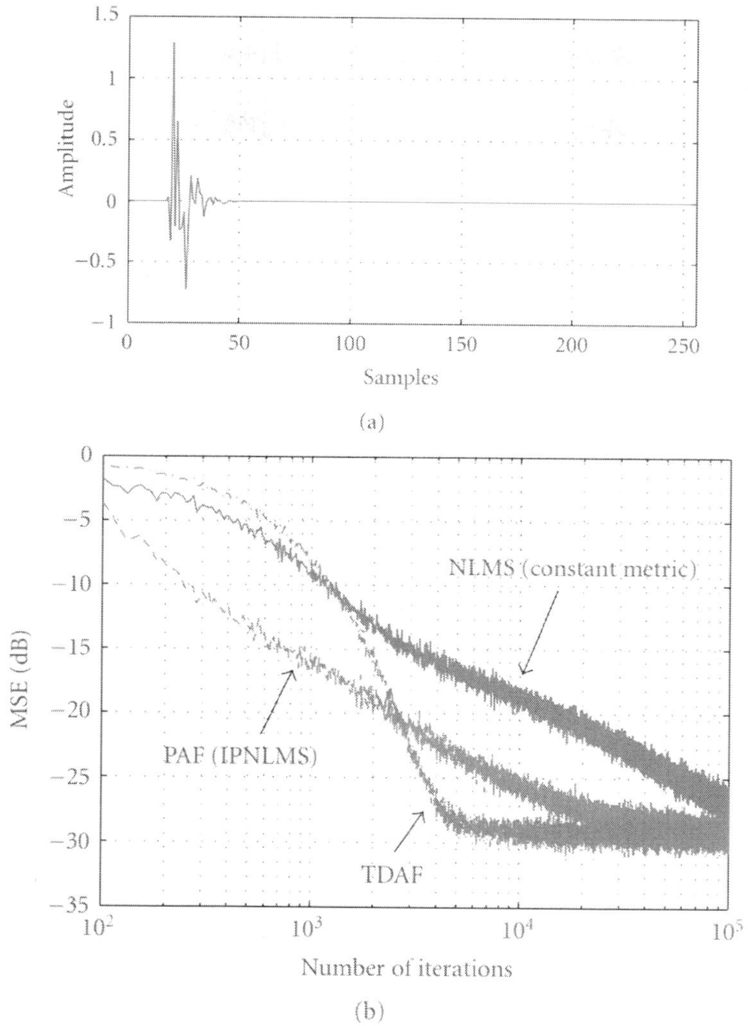

Figure 3: (a) Sparse impulse response and (b) MSE performance of NLMS, TDAF, and IPNLMS for $\lambda_k = 0.2$. SNR = 30 dB, N = 256, and colored inputs (USASI).

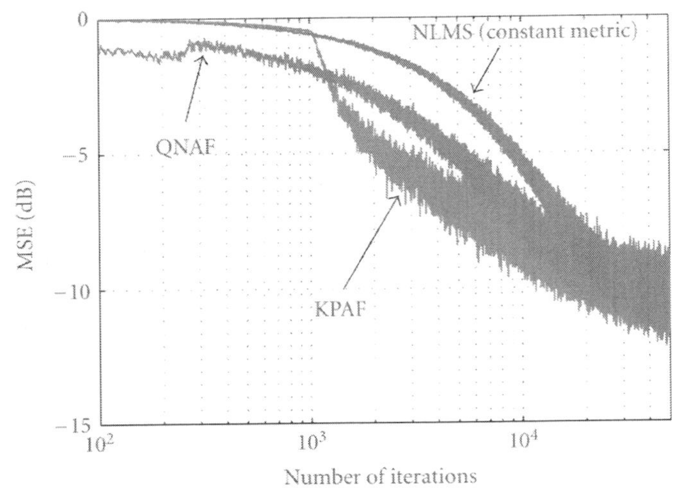

Figure 4: MSE performance of NLMS (λ_k = 0.02), QNAF (λ_k = 0.04), and KPAF (λ_k = 0.02) for nonsparse impulse responses and colored inputs (USASI). SNR = 10 dB, N = 256.

B. Definitions of Nonexpansive Mappings

(a) A mapping T is said to be nonexpansive if $\|T(x) - T(y)\| \leq \|x - y\|$, for all $(x, y) \in R^N \times R^N$; intuitively, T does not expand the distance between any two points x and y.

(b) A mapping T is said to be attracting nonexpansiveif T is nonexpansive with $Fix(T) \neq \emptyset$ and $\|T(x) - f\|^2 < \|x - f\|^2$, for all $(x,f) \in [RN \setminus Fix(T)] \times Fix(T)$; intuitively, T attracts any exterior point x to Fix(T).

(c) A mapping T is said to be strongly attracting nonexpansive or η- attracting nonexpansive if T is nonexpansive with $Fix(T) \neq \emptyset$ and there exists $\eta > 0$ s.t. $\eta\|x - T(x)\|^2 \leq \|x - f\|^2 - \|T(x) - f\|^2$, for all $(x,f) \in RN \times Fix(T)$. This condition is stronger than that of attracting nonexpansivity, because, for all $(x,f) \in [R^N \setminus Fix(T)] \times Fix(T)$, the difference $\|x - f\|^2 - \|T(x) - f\|^2$ is bounded by $\eta\|x - T(x)\|^2 > 0$.

A mapping $T : RN \to RN$ with $Fix(T) \neq \emptyset$ is called quasi-nonexpansive if $\|T(x)-T(f)\| \leq \|x-f\|$ for all $(x,f) \in R^N \times Fix(T)$.

C. Proof of Proposition 1

Due to the nonexpansivity of T_k with respect to the G_k-metric, (21) is verified by following the proof of [2, Theorem 2]. Noticing the property of the subgradient projection $\mathrm{Fix}(T^{(\mathbf{G}_k)}_{\mathrm{sp}(\varphi_k)}) = \mathrm{lev}_{\leq 0}\phi_k$, we can verify that the mapping $\hat{T}_k := T_k[I + \lambda_k(T^{(\mathbf{G}_k)}_{\mathrm{sp}(\varphi_k)}) - I)]$ is $(2 - \lambda_k)\eta/(2 - \lambda_k(1 - \eta))$-attracting quasi-nonexpansive with respect to G_k with $\mathrm{Fix}(\hat{T}_k) = K \cap \mathrm{lev}\leq 0\phi k = \Omega_k$ (cf. [3]). Because $((2 - \lambda_k)\eta)/(2 - \lambda_k (1 - \eta)) = [1/\eta + (\lambda_k/(2 - \lambda_k))]{-}^{1} = [1/\eta + (2/\lambda k - 1){-}1\ ]^{-1} \geq (\eta\varepsilon_2)/(\varepsilon 2 + (2 - \varepsilon_2)\eta)$, (22) is verified.

D. Proof of Theorem 1

Proof of (a). In the case of $h_k \in \Omega_k$, Fact 1 suggests $h_{k+1} = h_k$; thus (25) holds with equality. In the following, we assume $h_k \notin \Omega_k (\Leftrightarrow h_{k+1} \neq h_k)$. For any x $\in R^N$, we have

$$\mathbf{x}^T\mathbf{G}_k\mathbf{x} = \left(\frac{\mathbf{y}^T\mathbf{H}_k\mathbf{y}}{\mathbf{y}^T\mathbf{y}}\right)\mathbf{x}^T\mathbf{G}\mathbf{x}, \tag{D.1}$$

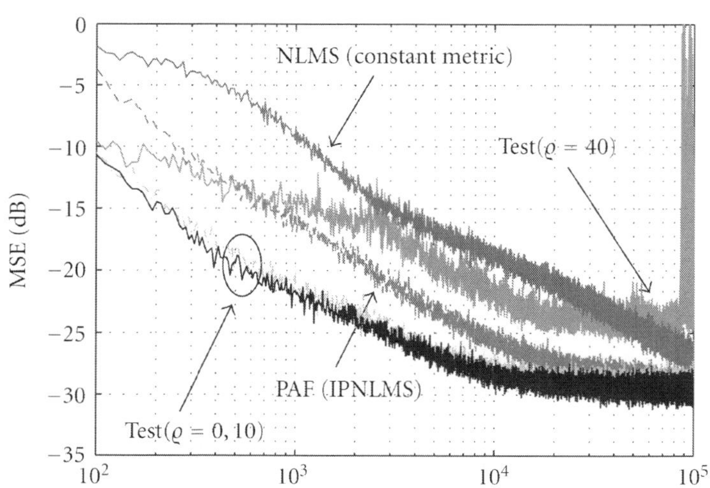

(a)

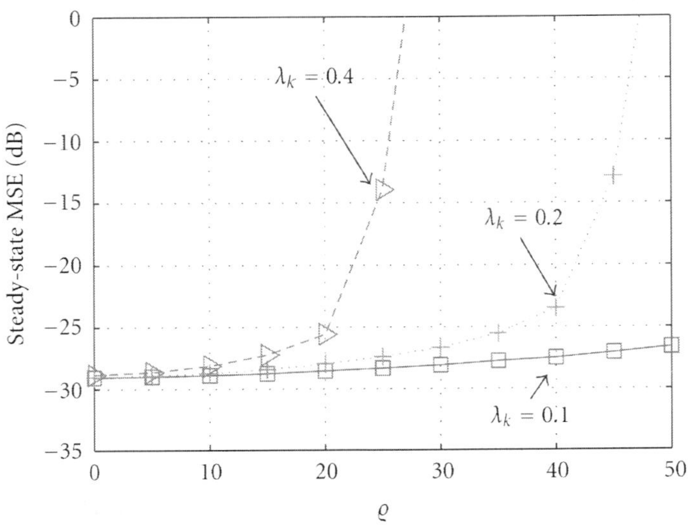

(b)

Figure 5: (a) MSE learning curves for $\lambda_k = 0.2$ and (b) steady-state MSE values for $\lambda_k = 0.1, 0.2, 0.4$. SNR = 30 dB, N = 256, and colored inputs (USASI).

where y := $G^{1/2}$x and Hk := $G^{-1/2}$Gk$G^{-1/2} \succ 0$. By Assumption 2(a), we obtain

$$\sigma_{\mathbf{H}_k}^{\max} = \|\mathbf{H}_k\|_2 \leq \left\|\mathbf{G}^{-1/2}\right\|_2 \|\mathbf{G}_k\|_2 \left\|\mathbf{G}^{-1/2}\right\|_2 = \frac{\sigma_{\mathbf{G}_k}^{\max}}{\sigma_{\mathbf{G}}^{\min}} < \frac{\delta_{\max}}{\sigma_{\mathbf{G}}^{\min}}$$

$$\left(\sigma_{\mathbf{H}_k}^{\min}\right)^{-1}$$

$$= \left\|\mathbf{H}_k^{-1}\right\|_2 \leq \left\|\mathbf{G}^{1/2}\right\|_2 \left\|\mathbf{G}_k^{-1}\right\|_2 \left\|\mathbf{G}^{1/2}\right\|_2 = \frac{\sigma_{\mathbf{G}}^{\max}}{\sigma_{\mathbf{G}_k}^{\min}} < \frac{\sigma_{\mathbf{G}}^{\max}}{\delta_{\min}}. \quad \text{(D.2)}$$

By (D.1) and (D.2), it follows that

$$\frac{\delta_{\min}}{\sigma_{\mathbf{G}}^{\max}} \|\mathbf{x}\|_{\mathbf{G}}^2 < \|\mathbf{x}\|_{\mathbf{G}_k}^2 < \frac{\delta_{\max}}{\sigma_{\mathbf{G}}^{\min}} \|\mathbf{x}\|_{\mathbf{G}}^2, \quad \forall k \geq K_1, \ \forall \mathbf{x} \in \mathbb{R}^N. \quad \text{(D.3)}$$

Noting $E^T{}_k = E_k$, for all $k \geq K_1$ (because $G^T{}_k = G_k$ and $G^T = G$), we have, for all $z* \in \Gamma \subseteq \Omega \subset \Omega_k$ and (for all $k \geq K_1$ s.t. $h_k \notin \Omega_k$),

$$\begin{aligned}
&\|\mathbf{h}_k - \mathbf{z}^*\|_{\mathbf{G}}^2 - \|\mathbf{h}_{k+1} - \mathbf{z}^*\|_{\mathbf{G}}^2 \\
&= \|\mathbf{h}_k - \mathbf{z}^*\|_{\mathbf{G}_k}^2 - \|\mathbf{h}_{k+1} - \mathbf{z}^*\|_{\mathbf{G}_k}^2 \\
&\quad - (\mathbf{h}_k - \mathbf{z}^*)^T \mathbf{E}_k (\mathbf{h}_k - \mathbf{z}^*) + (\mathbf{h}_{k+1} - \mathbf{z}^*)^T \mathbf{E}_k (\mathbf{h}_{k+1} - \mathbf{z}^*) \\
&\geq \varepsilon_1 \varepsilon_2 \frac{\varphi_k^2(\mathbf{h}_k)}{\left\|\varphi_k'(\mathbf{h}_k)\right\|_{\mathbf{G}_k}^2} + (\mathbf{h}_{k+1} + \mathbf{h}_k - 2\mathbf{z}^*)^T \mathbf{E}_k (\mathbf{h}_{k+1} - \mathbf{h}_k) \\
&\geq \frac{\varepsilon_1 \varepsilon_2 \sigma_{\mathbf{G}}^{\min}}{\delta_{\max}} \frac{\varphi_k^2(\mathbf{h}_k)}{\left\|\varphi_k'(\mathbf{h}_k)\right\|_{\mathbf{G}}^2} - \|\mathbf{h}_{k+1} + \mathbf{h}_k - 2\mathbf{z}^*\|_2 \|\mathbf{E}_k\|_2 \\
&\quad \times \|\mathbf{h}_{k+1} - \mathbf{h}_k\|_2.
\end{aligned} \tag{D.4}$$

The first inequality is verified by Proposition 1 and the second one is verified by (D.3), the Cauchy-Schwarz inequality, and the basic property of induced norms. Here, $\delta_{\min} < \sigma_{\mathbf{G}_k}^{\min} \leq (\mathbf{x}^T \mathbf{G}_k \mathbf{x})/(\mathbf{x}^T \mathbf{x})$ implies

$$\begin{aligned}
\|\mathbf{h}_{k+1} - \mathbf{h}_k\|_2^2 &< (\delta_{\min})^{-1} \|\mathbf{h}_{k+1} - \mathbf{h}_k\|_{\mathbf{G}_k}^2 \\
&\leq (\delta_{\min})^{-1} \lambda_k^2 \frac{\varphi_k^2(\mathbf{h}_k)}{\left\|\varphi_k'(\mathbf{h}_k)\right\|_{\mathbf{G}_k}^2} \\
&< \frac{(2 - \varepsilon_2)^2 \sigma_{\mathbf{G}}^{\max}}{\delta_{\min}^2} \frac{\varphi_k^2(\mathbf{h}_k)}{\left\|\varphi_k'(\mathbf{h}_k)\right\|_{\mathbf{G}}^2},
\end{aligned} \tag{D.5}$$

where the second inequality is verified by substituting $h_{k+1} = T_k[\mathbf{h}_k - \lambda_k(\varphi_k(\mathbf{h}_k)/\|\varphi_k'(\mathbf{h}_k)\|_{\mathbf{G}_k}^2)\varphi_k'(\mathbf{h}_k)]$ and $h_k = T_k(h_k)$ ($\Leftarrow h_k \in K = \mathrm{Fix}(T_k)$; see (17)) and noticing the nonexpansivity of T_k with respect to the G_k-metric. By (D.4), (D.5), and Assumption 2(b), it follows that, for all $z^* \in \Gamma$, for all $k \geq K_1$ s.t. $h_k \notin \Omega_k$,

$$\|\mathbf{h}_k - \mathbf{z}^*\|_{\mathbf{G}}^2 - \|\mathbf{h}_{k+1} - \mathbf{z}^*\|_{\mathbf{G}}^2$$
$$\geq \left(\frac{\varepsilon_1 \varepsilon_2 \sigma_{\mathbf{G}}^{\min}}{\delta_{\max}} - \frac{\|\mathbf{h}_{k+1} + \mathbf{h}_k - 2\mathbf{z}^*\|_2 \|\mathbf{E}_k\|_2}{\|\mathbf{h}_{k+1} - \mathbf{h}_k\|_2} \frac{(2-\varepsilon_2)^2 \sigma_{\mathbf{G}}^{\max}}{\delta_{\min}^2} \right)$$
$$\times \frac{\varphi_k^2(\mathbf{h}_k)}{\left\|\varphi_k'(\mathbf{h}_k)\right\|_{\mathbf{G}}^2} > \frac{(2-\varepsilon_2)^2 \sigma_{\mathbf{G}}^{\max}}{\delta_{\min}^2} \tau \frac{\varphi_k^2(\mathbf{h}_k)}{\left\|\varphi_k'(\mathbf{h}_k)\right\|_{\mathbf{G}}^2} \tag{D.6}$$

which verifies (24). Moreover, from (D.3) and (D.5), it is verified that

$$\frac{\varphi_k^2(\mathbf{h}_k)}{\left\|\varphi_k'(\mathbf{h}_k)\right\|_{\mathbf{G}}^2} > \frac{\delta_{\min}}{(2-\varepsilon_2)^2 \sigma_{\mathbf{G}}^{\max}} \|\mathbf{h}_{k+1} - \mathbf{h}_k\|_{\mathbf{G}_k}^2$$
$$> \frac{1}{(2-\varepsilon_2)^2} \left(\frac{\delta_{\min}}{\sigma_{\mathbf{G}}^{\max}} \right)^2 \|\mathbf{h}_{k+1} - \mathbf{h}_k\|_{\mathbf{G}}^2. \tag{D.7}$$

By (D.6) and (D.7), we can verify (25).

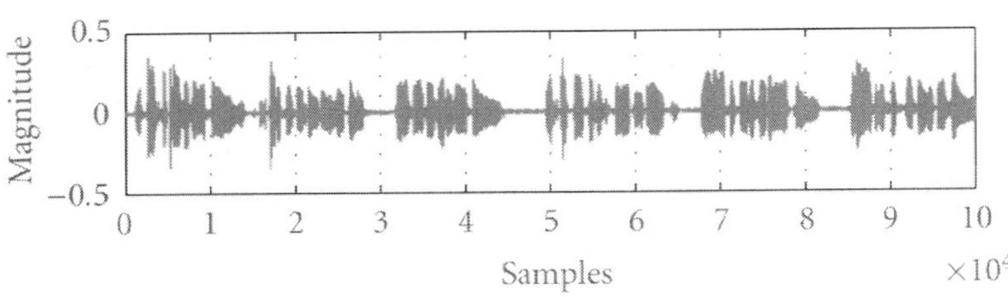

(a)

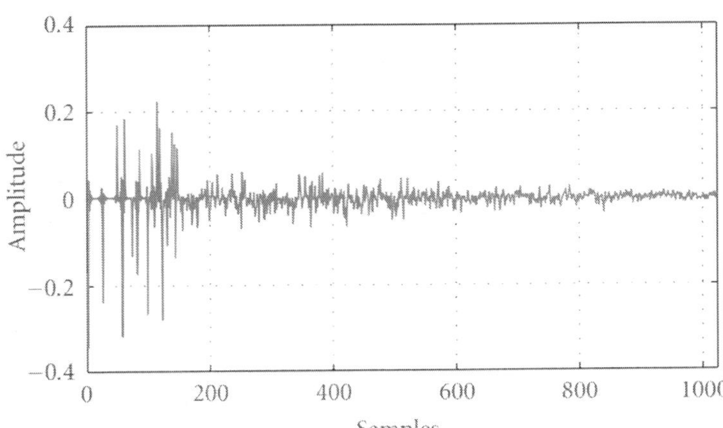

(b)

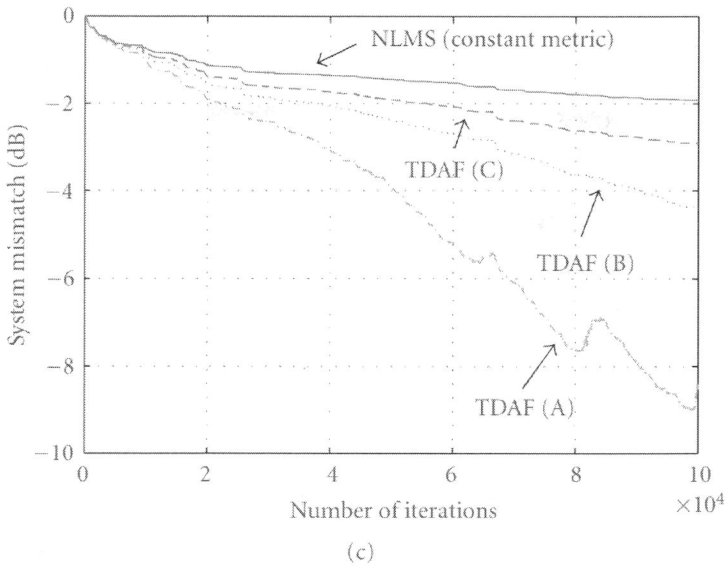

Figure 6: (a) Speech input signal, (b) recorded room impulse response, and (c) system mismatch performance of NLMS and TDAF for $\lambda_k = 0.02$, SNR = 20 dB, and N = 1024. For TDAF, (A) $\widehat{\gamma} = 1 - 10^{-4}$, (B) $\widehat{\gamma} = 1 - 10^{-4.5}$, and (C) $\widehat{\gamma} = 1 - 10^{-5}$.

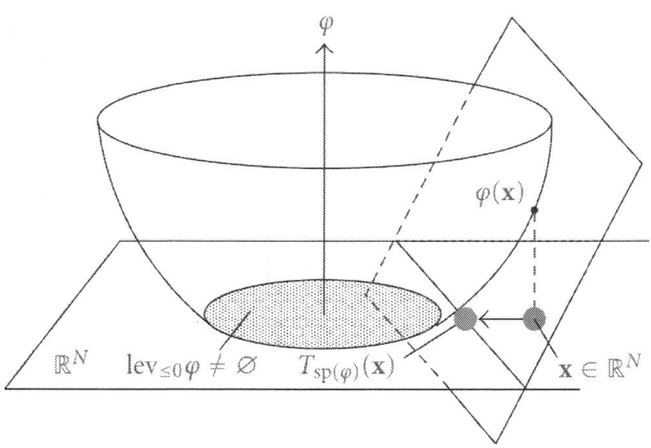

Figure 7: Subgradient projection $T_{sp(\phi)}(x) \in R^N$ is the projection of x onto the separating hyperplane (the thick line), which is the intersection of R^N and the tangent plane at $(x, \phi(x)) \in R^N \times R$.

Proof of (b). From Fact 1, for proving $\lim_{k\to\infty}\phi_k(h_k) = 0$, it is sufficient to check the case $h_k \notin \Omega_k (\Rightarrow \phi'_k(h_k) \neq 0)$. In this case, by Theorem 1(a),

$$\|\mathbf{h}_k - \mathbf{z}^*\|_G^2 - \|\mathbf{h}_{k+1} - \mathbf{z}^*\|_G^2$$

$$\geq \frac{(2-\varepsilon_2)^2 \sigma_G^{\max}}{\delta_{\min}^2} \tau \frac{\varphi_k^2(\mathbf{h}_k)}{\left\|\varphi_k'(\mathbf{h}_k)\right\|_G^2} \geq 0. \tag{D.8}$$

For any $z* \in \Gamma$, the nonnegative sequence $(\|h_k - z*\|_G)_{k\geq K1}$ is monotonically nonincreasing, thus convergent. This implies that

$$\lim_{\substack{k\to\infty \\ \varphi_k'(\mathbf{h}_k)\neq \mathbf{0}}} \frac{\varphi_k^2(\mathbf{h}_k)}{\left\|\varphi_k'(\mathbf{h}_k)\right\|_G^2} = 0; \tag{D.9}$$

hence the boundedness of $(\phi'_k(h_k))_{k\in N}$ ensures $\lim_{k\to\infty}\phi_k(h_k) = 0$.

Proof of (c). By Theorem 1(a) and [2, Theorem 1], the sequence $(h_k)_{k\geq K1}$ converges to a point $\hat{h} \in R^N$. The closedness of K(h_k, for all $k \in N \setminus \{0\}$) ensures $\hat{h} \in K$.

By the definition of subgradients and Assumption 2(a), we obtain

$$0 \leq \varphi_k(\hat{\mathbf{h}}) \leq \varphi_k(\mathbf{h}_k) - \left\langle \mathbf{h}_k - \hat{\mathbf{h}}, \varphi_k'(\hat{\mathbf{h}})\right\rangle_{G_k}$$

$$\leq \varphi_k(\mathbf{h}_k) + \left\|\mathbf{h}_k - \hat{\mathbf{h}}\right\|_2 \|\mathbf{G}_k\|_2 \left\|\varphi_k'(\hat{\mathbf{h}})\right\|_2$$

$$< \varphi_k(\mathbf{h}_k) + \delta_{\max}\left\|\mathbf{h}_k - \hat{\mathbf{h}}\right\|_2 \left\|\varphi_k'(\hat{\mathbf{h}})\right\|_2. \tag{D.10}$$

Hence, noticing (i) Theorem 1(b) under the assumption, (ii) the convergence $h_k \to \hat{h}$, and (iii) the boundedness of $(\phi' k(\hat{h}))_{k\in N}$, it follows that $\lim_{k\to\infty}\phi_k(\hat{h}) = 0$.

Proof of (d). The claim can be verified in the same way as in [2, Theorem 2(d)].

ACKNOWLEDGMENT

The authors would like to thank the anonymous reviewers for their invaluable suggestions which improved particularly the simulation part.

REFERENCES

1. I. Yamada, "Adaptive projected subgradient method: a unified view for projection based adaptive algorithms," The Journal of IEICE, vol. 86, no. 8, pp. 654–658, 2003 (Japanese).
2. I. Yamada and N. Ogura, "Adaptive projected subgradient method for asymptotic minimization of sequence of nonnegative convex functions," Numerical Functional Analysis and Optimization, vol. 25, no. 7-8, pp. 593–617, 2004.
3. K. Slavakis, I. Yamada, and N. Ogura, "The adaptive projected subgradient method over the fixed point set of strongly attracting nonexpansive mappings," Numerical Functional Analysis and Optimization, vol. 27, no. 7-8, pp. 905–930, 2006.
4. J. Nagumo and J. Noda, "A learning method for system identification," IEEE Transactions on Automatic Control, vol. 12, no. 3, pp. 282–287, 1967.
5. A. E. Albert and L. S. Gardner Jr., Stochastic Approximation and Nonlinear Regression, MIT Press, Cambridge, Mass, USA, 1967.
6. T. Hinamoto and S. Maekawa, "Extended theory of learning identification," Transactions of IEE of Japan, vol. 95, no. 10, pp. 227–234, 1975 (Japanese).
7. K. Ozeki and T. Umeda, "An adaptive filtering algorithm using an orthogonal projection to an affine subspace and its properties," Electronics & Communications in Japan A, vol. 67, no. 5, pp. 19–27, 1984.
8. S. C. Park and J. F. Doherty, "Generalized projection algorithm for blind interference suppression in DS/CDMA communications," IEEE Transactions on Circuits and Systems II, vol. 44, no. 6, pp. 453–460, 1997.
9. J. A. Apolinario Jr., S. Werner, P. S. R. Diniz, and T. I. ´ Laakso, "Constrained normalized adaptive filters for CDMA mobile communications," in Proceedings of the European Signal Processing Conference (EUSIPCO '98), vol. 4, pp. 2053– 2056, Island of Rhodes, Greece, September 1998.
10. I. Yamada, K. Slavakis, and K. Yamada, "An efficient robust adaptive filtering algorithm based on parallel subgradient projection techniques,"

IEEE Transactions on Signal Processing, vol. 50, no. 5, pp. 1091–1101, 2002.

11. M. Yukawa and I. Yamada, "Pairwise optimal weight realization—acceleration technique for set-theoretic adaptive parallel subgradient projection algorithm," IEEE Transactions on Signal Processing, vol. 54, no. 12, pp. 4557–4571, 2006.
12. M. Yukawa, R. L. G. Cavalcante, and I. Yamada, "Efficient blind MAI suppression in DS/CDMA systems by embedded constraint parallel projection techniques," IEICE Transactions on Fundamentals of Electronics, Communications and Computer Sciences, vol. E88-A, no. 8, pp. 2062–2071, 2005.
13. R. L. G. Cavalcante and I. Yamada, "Multiaccess interference suppression in orthogonal space-time block coded MIMO systems by adaptive projected subgradient method," IEEE Transactions on Signal Processing, vol. 56, no. 3, pp. 1028–1042, 2008.
14. M. Yukawa, N. Murakoshi, and I. Yamada, "Efficient fast stereo acoustic echo cancellation based on pairwise optimal weight realization technique," EURASIP Journal on Applied Signal Processing, vol. 2006, Article ID 84797, 15 pages, 2006.
15. K. Slavakis, S. Theodoridis, and I. Yamada, "Online kernelbased classification using adaptive projection algorithms," IEEE Transactions on Signal Processing, vol. 56, no. 7, part 1, pp. 2781–2796, 2008.
16. K. Slavakis, S. Theodoridis, and I. Yamada, "Adaptive constrained learning in reproducing kernel Hilbert spaces: the robust beamforming case," IEEE Transactions on Signal Processing, vol. 57, no. 12, pp. 4744–4764, 2009.
17. R. L. G. Cavalcante and I. Yamada, "A flexible peak-to-average power ratio reduction scheme for OFDM systems by the adaptive projected subgradient method," IEEE Transactions on Signal Processing, vol. 57, no. 4, pp. 1456–1468, 2009.
18. R. L. G. Cavalcante, I. Yamada, and B. Mulgrew, "An adaptive projected subgradient approach to learning in diffusion networks," IEEE Transactions on Signal Processing, vol. 57, no. 7, pp. 2762–2774, 2009.
19. S. S. Narayan, A. M. Peterson, and M. J. Narasimha, "Transform domain LMS algorithm," IEEE Transactions on Acoustics, Speech, and Signal Processing, vol. 31, no. 3, pp. 609–615, 1983.

20. D. F. Marshall, W. K. Jenkins, and J. J. Murphy, "The use of orthogonal transforms for improving performance of adaptive filters," IEEE Transactions on Circuits and Systems, vol. 36, no. 4, pp. 474–484, 1989.
21. F. Beaufays, "Transform-domain adaptive filters: an analytical approach," IEEE Transactions on Signal Processing, vol. 43, no. 2, pp. 422–431, 1995.
22. B. Widrow and S. D. Stearns, Adaptive Signal Processing, Prentice Hall, Englewood Cliffs, NJ, USA, 1985.
23. P. S. R. Diniz, M. L. R. de Campos, and A. Antoniou, "Analysis of LMS-Newton adaptive filtering algorithms with variable convergence factor," IEEE Transactions on Signal Processing, vol. 43, no. 3, pp. 617–627, 1995.
24. B. Farhang-Boroujeny, Adaptive Filters: Theory and Applications, John Wiley & Sons, Chichester, UK, 1998.
25. D. F. Marshall and W. K. Jenkins, "A fast quasi-Newton adaptive filtering algorithm," IEEE Transactions on Signal Processing, vol. 40, no. 7, pp. 1652–1662, 1992.
26. M. L. R. de Campos and A. Antoniou, "A new quasi-Newton adaptive filtering algorithm," IEEE Transactions on Circuits and Systems II, vol. 44, no. 11, pp. 924–934, 1997.
27. D. L. Duttweiler, "Proportionate normalized least-meansquares adaptation in echo cancelers," IEEE Transactions on Speech and Audio Processing, vol. 8, no. 5, pp. 508–517, 2000.
28. S. L. Gay, "An efficient fast converging adaptive filter for network echo cancellation," in Proceedings of the 32nd Asilomar Conference on Signals, Systems and Computers, pp. 394–398, Pacific Grove, Calif, USA, November 1998.
29. T. Gansler, S. L. Gay, M. M. Sondhi, and J. Benesty, "Double- ¨ talk robust fast converging algorithms for network echo cancellation," IEEE Transactions on Speech and Audio Processing, vol. 8, no. 6, pp. 656–663, 2000.
30. J. Benesty, T. Gansler, D. R. Morgan, M. M. Sondhi, and S. ¨ L. Gay, Advances in Network and Acoustic Echo Cancellation, Springer, Berlin, Germany, 2001.
31. J. Benesty and S. L. Gay, "An improved PNLMS algorithm," in Proceedings of the IEEE International Conference on Acoustics,

Speech and Signal Processing (ICASSP ’02), pp. 1881–1884, Orlando, Fla, USA, May 2002.

32. H. Deng and M. Doroslovacki, “Proportionate adaptive ˇ algorithms for network echo cancellation,” IEEE Transactions on Signal Processing, vol. 54, no. 5, pp. 1794–1803, 2006.
33. Y. Huang, J. Benesty, and J. Chen, Acoustic MIMO Signal Processing—Signals and Communication Technology, Springer, Berlin, Germany, 2006.
34. M. Yukawa, “Krylov-proportionate adaptive filtering techniques not limited to sparse systems,” IEEE Transactions on Signal Processing, vol. 57, no. 3, pp. 927–943, 2009.
35. M. Yukawa and W. Utschick, “Proportionate adaptive algorithm for nonsparse systems based on Krylov subspace and constrained optimization,” in Proceedings of the IEEE International Conference on Acoustics, Speech and Signal Processing (ICASSP ’09), pp. 3121–3124, Taipei, Taiwan, April 2009.
36. M. Yukawa and W. Utschick, “A fast stochastic gradient algorithm: maximal use of sparsification benefits under computational constraints,” to appear in IEICE Transactions on Fundamentals of Electronics, Communications and Computer Sciences, vol. E93-A, no. 2, 2010.
37. M. Yukawa, K. Slavakis, and I. Yamada, “Adaptive parallel quadratic-metric projection algorithms,” IEEE Transactions on Audio, Speech and Language Processing, vol. 15, no. 5, pp. 1665– 1680, 2007.
38. A. A. Goldstein, “Convex programming in Hilbert space,” Bulletin of the American Mathematical Society, vol. 70, pp. 709– 710, 1964.
39. E. S. Levitin and B. T. Polyak, “Constrained minimization methods,” USSR Computational Mathematics and Mathematical Physics, vol. 6, no. 5, pp. 1–50, 1966.
40. B. T. Polyak, “Minimization of unsmooth functionals,” USSR Computational Mathematics and Mathematical Physics, vol. 9, no. 3, pp. 14–29, 1969.
41. S. Haykin, Adaptive Filter Theory, Prentice Hall, Upper Saddle River, NJ, USA, 4th edition, 2002.
42. A. H. Sayed, Fundamentals of Adaptive Filtering, John Wiley & Sons, Hoboken, NJ, USA, 2003.

43. M. Yukawa, K. Slavakis, and I. Yamada, "Signal processing in dual domain by adaptive projected subgradient method," in Proceedings of the 16th International Conference on Digital Signal Processing (DSP '09), pp. 1–6, Santorini-Hellas, Greece, July 2009.
44. M. Yukawa, K. Slavakis, and I. Yamada, "Multi-domain adaptive learning based on feasibility splitting and adaptive projected subgradient method," to appear in IEICE Transactions on Fundamentals of Electronics, Communications and Computer Sciences, vol. E93-A, no. 2, 2010.
45. M. Yukawa and I. Yamada, "Adaptive parallel variable-metric projection algorithm—an application to acoustic ECHO cancellation," in Proceedings of the IEEE International Conference on Acoustics, Speech and Signal Processing (ICASSP '07), vol. 3, pp. 1353–1356, Honolulu, Hawaii, USA, May 2007.
46. R. A. Horn and C. R. Johnson, Matrix Analysis, Cambridge University Press, New York, NY, USA, 1985.

CHAPTER 11

NEW TECHNIQUES FOR LINEAR ARITHMETIC: CUBES AND EQUALITIES

Martin Bromberger[1,2], Christoph Weidenbach[1]

[1] Max-Planck-Institut für Informatik, Saarland Informatics Campus E1 4, 66123 Saarbrücken, Germany

[2] Graduate School of Computer Science, Saarland Informatics Campus E1 3, 66123 Saarbrücken, Germany

ABSTRACT

We present several new techniques for linear arithmetic constraint solving. They are all based on the linear cube transformation, a method presented here, which allows us to efficiently determine whether a system of linear arithmetic constraints contains a hypercube of a given edge length. Our first

Citation (APA): Bromberger, M., & Weidenbach, C. (2017). New techniques for linear arithmetic: cubes and equalities. Formal Methods in System Design, 51(3), 433-461. (29 pages).

findings based on this transformation are two sound tests that find integer solutions for linear arithmetic constraints. While many complete methods search along the problem surface for a solution, these tests use cubes to explore the interior of the problems. The tests are especially efficient for constraints with a large number of integer solutions, e.g., those with infinite lattice width. Inside the SMT-LIB benchmarks, we have found almost one thousand problem instances with infinite lattice width. Experimental results confirm that our tests are superior on these instances compared to several state-of-the-art SMT solvers. We also discovered that the linear cube transformation can be used to investigate the equalities implied by a system of linear arithmetic constraints. For this purpose, we developed a method that computes a basis for all implied equalities, i.e., a finite representation of all equalities implied by the linear arithmetic constraints. The equality basis has several applications. For instance, it allows us to verify whether a system of linear arithmetic constraints implies a given equality. This is valuable in the context of Nelson–Oppen style combinations of theories.

Keywords: Linear arithmetic, SMT, Integer arithmetic, Constraint solving, Equalities, Combination of theories

INTRODUCTION

Polyhedra and the systems of linear arithmetic constraints $Ax \leq b$ defining them have a vast number of theoretical and real-world applications [5, 19]. It is, therefore, no surprise that the theory of linear arithmetic is one of the most popular and best investigated theories for *satisfiability modulo theories* (SMT) solving [14, 15, 16].

This paper serves as a collection of our results based on the linear cube transformation. On its own, the linear cube transformation allows us to efficiently determine whether a system of linear arithmetic constraints contains a hypercube of a given edge length. We were able to develop several techniques based on this transformation that allow us to investigate linear arithmetic constraints in various ways. Here, we present our previous results [7, 8] on the linear cube transformation in more detail as well as some new applications (e.g., quantifier elimination).

Finding an integer solution for a polyhedron that is defined by a system of linear inequalities $Ax \leq b$ is a well-known NP-complete problem [25]. This problem has been investigated in different research areas, e.g., in optimization via *(mixed) integer linear programming*(MILP) [19] and in constraint solving via *satisfiability modulo theories* (SMT) [4, 6, 11, 16]. For commercial MILP implementations, it is standard to integrate preprocessing techniques, heuristics, and specialized tests [19]. Although these techniques are not complete, they are much more efficient on their designated target systems of linear inequalities than a complete algorithm alone.

The SMT community is still in the process of developing a variety of specialized tests. A big challenge is to adopt the tests from the MILP community so that they still fit the requirements of SMT solving. SMT theory solvers have to solve a large number of incrementally connected, small systems of linear inequalities. Exploiting this incremental connection is key for making SMT theory solvers efficient [15]. In contrast, MILP solvers typically target one large system. The same holds for their specialized tests, which are not well suited to exploit incremental connections.

Based on the linear cube transformation, we present two tests tailored for SMT solvers: the *largest cube test* and the *unit cube test* [8]. The largest cube test finds a hypercube with maximum edge length contained in the input polyhedron, determines its rational valued center, and rounds it to a potential integer solution. The unit cube test determines if a polyhedron contains a hypercube with edge length one, which is the minimal edge length that guarantees an integer solution. Due to computational complexity, we restrict ourselves to those hypercubes that are parallel to the coordinate axes.

Most SMT linear integer arithmetic theory solvers are based on a branch-and-bound algorithm on top of the simplex algorithm. They search for a solution at the surface of a polyhedron. In contrast, our tests search in the interior of the polyhedron. This gives them an advantage on polyhedra with a large number of integer solutions, e.g., polyhedra with infinite lattice width [20].

SMT theory solvers are designed to efficiently exchange bounds [14]. This efficient exchange is the main reason why SMT theory solvers exploit the incremental connection between the different polyhedra so well. Our unit cube test also requires only an exchange of bounds. After applying the test, we can easily recover the original polyhedron by reverting to the original bounds. In doing so, the unit cube test conserves the incremental connection between the different original polyhedra. We make a similar observation about the largest cube test.

Equalities are a special instance of linear arithmetic constraints. They are useful in simplifying systems of arithmetic constraints [16], and they are essential for the Nelson–Oppen style combinations of theories [9]. However, they are also an obstacle for our fast cube tests. If a system of linear arithmetic constraints implies an equality, then it has only a surface and no interior; so our cube tests cannot explore an interior and will certainly fail. In order to expand the applicability of our cube tests, we have to develop methods that find, isolate, and eliminate implied equalities from systems of linear arithmetic constraints [7].

We can detect the existence of an implied equality by searching for a hypercube in our polyhedron. If the maximal edge length of such a hypercube is zero, there exists an implied equality. This test can be further simplified. By turning all inequalities into strict ones, the interior of the original polyhedron remains while the surface disappears. If the strict system is unsatisfiable, the original system has no interior and implies an equality. Based on an explanation of unsatisfiability for the strict system, the method generates an implied equality as an explanation.

We are also able to extend the above method into an algorithm that computes an equality basis, i.e., a finite representation of all equalities implied by a satisfiable system of linear arithmetic constraints. For this purpose, the algorithm repeatedly applies the above method to find, collect, and eliminate equalities from our system of constraints. When the system contains no more equalities, then the collected equalities represent an equality basis, i.e., any implied equality can be obtained by a linear combination of the equalities in the basis. The equality basis has many applications. If transformed into a substitution, it eliminates all equalities implied by our system of constraints, which results in a system of constraints with an interior and, therefore, improves the applicability of our cube tests. The equality basis also allows us to test whether a system of linear arithmetic constraints implies a given equality. We even extend this test into an efficient method that computes all pairs of equivalent variables inside a system of constraints. These pairs are necessary for the Nelson–Oppen style combination of theories.

While Hillier [17] was aware of the unit cube test, he applied it only to cones (a special class of polyhedra) as a subroutine in a new heuristic. His work never mentioned applications beyond cones, nor did he prove any structural properties connected to hypercubes. Hillier's heuristic tailored for MILP optimization lost popularity as soon as interior point methods [21] became efficient in practice. Nonetheless, our cube tests remain relevant for SMT theory solvers because there are no competitive incremental interior point methods known.

Also, Bobot et al. [4] discuss relations between hypercubes and polyhedra including infinite lattice width and positive linear combinations between inequalities. Our largest cube test can also detect these relations because it is, with some minor changes, the dual of the linear optimization problem of Bobot et al. In contrast to the linear optimization problem of Bobot et al., our tests are closer to the original polyhedron and, therefore, easier to construct. Our cube tests also produce sample points and find solutions for polyhedra with finite lattice width.

Another method that provides a sufficient condition for the existence of an integer solution is the dark shadow of the omega test [26]. The dark shadow is based on Fourier–Motzkin elimination and its worst case runtime is double exponential. Although not practically advantageous, formulating the unit cube test through Fourier–Motzkin elimination allows us to put the sufficient conditions of the two methods in relation. Fourier–Motzkin elimination eliminates the variable x from a problem by combining each pair of inequalities $ax \leq p$ and $q \leq bx$ (with $a,b>0$) into a new inequality $aq-bp \leq 0$. The dark shadow creates a stronger version ($aq-bp \leq a+b-ab$) of the combined inequality to guarantee the existence of an integer solution for x. Formulating the unit cube test through Fourier–Motzkin elimination makes the combined inequality even stronger ($aq-bp \leq -ab$). This means that the sufficient condition of the dark shadow subsumes the condition of the unit cube test. Still, our unit cube test is definable as a linear program and it is, therefore, computable in polynomial time. So the better condition of the dark shadow comes at the cost of being much harder to compute.

There also already exist several methods that find, isolate, and eliminate implied equalities [3, 27, 31, 32]. Hentenryck and Graf [32] define unique normal forms for systems of linear constraints with non-negative variables. To compute a normal form, they first eliminate all implied equalities from the system. To this end, they determine the lower bound for each inequality by solving one linear optimization problem. Similarly, Refalo [27] describes several incremental methods that use optimization to turn a satisfiable system of linear constraints in "revised solved form" into a system without any implied equalities. Rueß and Shankar also use this optimization scheme to determine a basis of implied equalities [28]. Additionally, they present a necessary but not sufficient condition for an inequality to be part of an equality explanation. During preprocessing, all inequalities not fulfilling this condition are eliminated, thus, reducing the number of optimization problems their method has to compute. However, this preprocessing step might be in itself expensive because it relies on a non-trivial fixed-point scheme. The method presented by Telgen [31] does not require optimization. He presents criteria to detect implied equalities based on the tableau used in the simplex algorithm, but he was not able to formulate an algorithm that efficiently computes these criteria. In the worst case, he has to pivot the simplex tableau until he has computed all possible tableaux for the given system of constraints. Another method that detects implied equalities was presented by Bjørner [3]. He uses Fourier Motzkin variable elimination to compute linear combinations that result in implied equalities.

Our methods that detect implied equalities do not require optimization, which is advantageous because SMT solvers are usually not fine-tuned for optimization. Moreover, we defined our methods for a rather general formulation of linear constraints, which allows us to convert our results into other representations, e.g., the tableau-and-bound representation used in Dutertre and de Moura's version of the simplex algorithm (see Sect. 7), while preserving efficiency. Finally, our method efficiently searches for implied equalities. We neither have to check each inequality independently nor do we have to blindly pivot the simplex tableau. This also makes potentially expensive preprocessing techniques obsolete.

The paper is organized as follows: we define in Sect. 3 the linear cube transformation (Proposition 3) that allows us to efficiently compute whether a polyhedron contains a hypercube of a given edge length by solely changing the bounds of the inequalities. Based on this transformation, we develop in Sect. 4 two tests: the largest cube test and the unit cube test. Both tests find integer solutions for linear arithmetic constraints. For polyhedra with infinite lattice width, both tests always succeed (Lemma 4). Inside the SMT-LIB benchmarks, there are almost one thousand problem instances with infinite lattice width, and we show the advantage of our cube tests on these instances by comparing our implementation of the cube test with several state-of-the-art SMT solvers in Sect. 5. In Sect. 6, we show how to investigate equalities with the linear cube transformation. First, we introduce an efficient method for testing whether a system of linear arithmetic constraints implies a given equality (Sect. 6.1). Then, we extend the method so that it computes an equality basis for our system of constraints (Sect. 6.2). In Sect. 7 we start with an implementation of our methods as an extension of Dutertre and de Moura's version of the simplex algorithm [14], which is integrated in many SMT solvers. The implementation generates justifications and preserves incrementality. The efficient computation of an equality basis can then be used in identifying equivalent variables for the Nelson–Oppen combination of theories. Section 8 concludes the paper including a further application of the linear cube transformation to quantifier elimination.

PRELIMINARIES

While the difference between matrices, vectors, and their components is always clear in context, we generally use upper case letters for matrices (e.g., *A*), lower case letters for vectors (e.g., *x*), and lower case letters with an index *i* or *j* (e.g., b_i, x_j) as components of the associated vector at position *i* or *j*, respectively. The only exceptions are the row vectors $a^T_i=(a_{i1},\ldots,a_{in})$ of a matrix $A=(a_1,\ldots,a_m)^T$, which already contain an index *i* that indicates the row's position inside *A*. In order to save space, we write vectors only implicitly as columns via the transpose operator $()^T$, which turns all rows $(b_1,\ldots,b_m)$ into columns $(b_1,\ldots,b_m)^T$ and vice versa. We also abbreviate the *n*-dimensional origin $(0,\ldots,0)^T$ as 0^n. Likewise, we abbreviate $(1,\ldots,1)^T$ as 1^n.

In the context of SMT solvers, we have to deal with strict inequalities $a^T_i x<b_i$ and non-strict inequalities $a^T_i x\leq b_i$ as our constraints, where $a_i \in Q^n$ and $b_i \in Q$. A system of constraints is, therefore, just a set of inequalities and the rational solutions of this system are exactly those points $x\in Q^n$ that satisfy all inequalities in this set. The rational solutions of this system also define a polyhedron in the Q^n, where each rational solution is equivalent to a point in the polyhedron. For this reason, we treat *polyhedra* and their definitions through a *system of inequalities* as interchangeable. In the case that our system contains only non-strict inequalities, the polyhedron is even *closed convex*, which entails two very useful properties: firstly, the closed convex polyhedron has a surface if it is neither empty nor encompasses the whole Q^n; secondly, any supremum $h_{max}=\sup\{h^Tx : x\in Q^n \text{ satisfies } Ax\leq b\}$ is either $h_{max}=-\infty$ because there exists no point satisfying our constraints, $h_{max}=\infty$ because the supremum is unbounded, or there exists an actual maximum, i.e., there exists an $x\in Q^n$ that satisfies the constraints and its cost h^Tx is equivalent to our supremum h_{max}.

If we also consider strict inequalities, then our polyhedron is no longer necessarily a closed convex set. This means we lose the above properties, which poses a problem when we want to adapt algorithms that originally deal only with non-strict inequalities so they can also deal with strict inequalities. For instance, the classical dual simplex algorithm [29] returns only rational solutions on the surface of the polyhedron defined by $Ax\leq b$. It is, therefore, not trivial to adapt the classical dual simplex algorithm to also handle strict inequalities.

To avoid these problems, we instead model strict inequalities as non-strict inequalities by generalizing the field Q for our inequality bounds bi and variables xi to Qδ [14].

Lemma 1 ([14]) Let ai ∈ Q^n and bi ∈ Q. Then a set of linear arithmetic constraints S containing strict inequalities S' = $\{a^T{}_1\, x < b_1, \ldots, a^T{}_m\, x < b_m\}$ is satisfiable iff there exists a rational number δ > 0 such that $S_{\delta'} = (S \cup S'_{\delta'}) \setminus S'$ is satisfiable for all δ' with 0 < δ' ≤ δ, where $S'_{\delta'} = \{a^T{}_1\, x \le b_1 - \delta', \ldots, a^T{}_m\, x \le b_m - \delta'\}$.

As a result of this observation, δ is expressed symbolically as an infinitesimal parameter. This leads to the ordered vector space Qδ that has pairs of rationals as elements (p,q)∈Q×Q, representing p+qδ, with the following operations:

$$(p_1, q_1) + (p_2, q_2) \equiv (p_1 + p_2, q_1 + q_2)$$

$$a \cdot (p, q) \equiv (a \cdot p, a \cdot q)$$

$$(p_1, q_1) \le (p_2, q_2) \equiv (p_1 < p_2) \vee (p_1 = p_2 \wedge q_1 \le q_2)$$

$$(p_1, q_1) < (p_2, q_2) \equiv (p_1 < p_2) \vee (p_1 = p_2 \wedge q_1 < q_2),$$

where a∈Q [14]. Now we can represent $a^T{}_i\, x < bi$ by $a^T{}_i\, x \le bi - \delta$, where $a_i \in Q^n$ and $b_i \in Q$. Since we also let the assignments for our variables range over Qδ, the solutions of our inequalities describe a closed convex polyhedron in the $Q^n{}_\delta$ and methods like the classical simplex algorithm are again complete. It is also easy to extract a purely rational solution $v' \in Q^n$ from a δ-rational solution $v \in Q^n{}_\delta$ of Ax≤b. We just have to choose a small enough value δ′∈Q and replace the parameter δ with this value (Lemma 1). Therefore, we call a δ-rational solution just a rational solution. Representing our strict inequalities as non-strict inequalities also allows us to use the second property listed above for closed convex polyhedra: any supremum $h_{max} = \sup\{h^T x : x \in Q^n{}_\delta$ satisfies $Ax \le b\}$ is either $h_{max} = -\infty$, $h_{max} = \infty$, or there exists an actual maximum and not just a limit. This property is especially useful on techniques based on optimization like the largest cube test (see Sect. 4.1).

With the δ-rationals, we are now able to formally define a system of constraints as Ax≤b. Ax≤b is just an abbreviation for the set of inequalities $\{a^T{}_1 x \le b_1, \ldots, a^T{}_m x \le b_m\}$. The row coefficients are given by $A = (a_1, \ldots, a_m)^T \in Q^{m \times n}$, the variables are given by $x = (x_1, \ldots, x_n)^T$, and the inequality bounds are given by $b = (b_1, \ldots, b_m)^T \in Q^m{}_\delta$. Moreover, we assume that any constant rows $a_i = 0^n$ were eliminated from our system during an implicit preprocessing step. This is a trivial task and eliminates some unnecessarily complicated corner cases. The δ-coefficients q_i in the bounds $b_i=p_i+q^i_\delta$ can take on any value in Qδ. In case $q_i=0$, the inequality $a^T_i\ x\le b_i$ is equivalent to the non-strict inequality $a^T_i\ x\le p_i$. In case $q_i<0$, the inequality $a^T_i\ x\le b_i$ is equivalent to the strict inequality $a^T_i\ x<p_i$. In case qi>0, we have no clear interpretation over the actual rationals (compare also Lemma 1). For instance, the two inequalities $x\le\delta$ and $-x\le-\delta$ describe a satisfiable system of constraints in Qδ, but there is no clear way of interpreting $x\le\delta$ in Q. Beware also that our linear cube transformation can introduce positive δ-coefficients in the bounds. But since we derive the transformation via a semantically clear construction, the semantic interpretation over the rationals is still discernible if the original system has only non-positive δ-coefficients in its inequality bounds before the transformation.

For the remainder of the paper, we abbreviate with b^δ_i the strict version of a given bound $b_i\in Q\delta$. If the bound b_i is non-strict, i.e., $b_i=(p_i,0)$, then the strict version is $b^\delta_i:=(p_i,-1)$. Otherwise, the bound b_i is already strict, i.e., $b_i=(p_i,q_i)$ with $q_i<0$, and we just standardize the δ-coefficient to −1, i.e, $b^\delta_i:=(p_i,-1)$.

Since $Ax\le b$ and $A'x\le b'$ are just sets, we can write their combination as $(Ax\le b)\cup(A'x\le b')$. A special system of inequalities is a system of equations $Dx=c$, which is equivalent to the combined system of inequalities $(Dx\le c)\cup(-Dx\le-c)$. For such a system of equalities, the row coefficients are given by $D=(d_1,\ldots,d_m)^T\in Q^{m\times n}$, the variables are given by $x=(x_1,\ldots,x_n)^T$, and the equality bounds are given by $c=(c_1,\ldots,c_m)^T\in Q^m$.

We denote by $P^A_b=\{x\in Q^n_\delta : Ax\le b\}$ the *set of δ-rational solutions* to the system of inequalities $Ax\le b$ and, therefore, the points inside the polyhedron. Similarly, we denote by $Ce(z)=\{x\in Q^n_\delta : \forall j\in 1,\ldots,n. |xj-zj|\le e/2\}$ the set of points contained in the *n-dimensional hypercube* that is parallel to the coordinate axes, has *edge length* $e\in Q\delta$ (with $e\ge 0$), and has *center* $z\in Q^n_\delta$. For the remainder of this paper, we consider only hypercubes that are parallel to the coordinate axes. For simplicity, we call these restricted hypercubes *cubes*.

Besides cubes and polyhedra, we use two norms in this paper. The first norm we use is the *maximum norm*. It is defined by: $\|x\|_\infty=\max\{|x_1|,\ldots,|x_n|\}$, and we use it because it is closely related to our definition of cubes Ce(z), i.e., the condition in the definition of Ce(z) can also be expressed with the maximum norm: $(\|x-z\|_\infty\le e/2)$□$(\forall j\in 1,\ldots,n. |x_j-z_j|\le e/2)$. The second norm we use is the *1-norm*, which is defined by: $\|x\|_1=(|x_1|+\ldots+|x_n|)$. We use it in Sect. 3 to define the linear cube transformation.

Using the maximum norm, we define a *closest integer* to x as a point $x' \in Zn$ with minimal distance $\|x-x'\|\infty$. We also define the operators $\lceil x_j \rfloor$ and $\lceil x \rfloor$ such that they describe a *closest integer* to x_j and x, respectively. Formally, this means that $\lceil x \rfloor = (\lceil x_1 \rfloor, \ldots, \lceil x_n \rfloor)^T$ and

$$\lceil x_j \rfloor = \begin{cases} \lfloor x_j \rfloor & \text{if } x_j - \lfloor x_j \rfloor < 0.5\,, \\ \lceil x_j \rceil & \text{if } x_j - \lfloor x_j \rfloor \geq 0.5\,. \end{cases}$$

This definition of $\lceil x \rfloor \lceil x \rfloor$ is also known as *simple rounding.*

Proposition 1 For $x \in Q^n_\delta$, $\lceil x \rfloor$ is a closest integer to x, or formally:

$$\forall x' \in \mathbb{Z}^n.\ \|x - \lceil x \rfloor\|_\infty \leq \|x - x'\|_\infty\,.$$

We say that a polyhedron implies an inequality $h^T x \leq g$, where $h \in Qn$, $h \neq 0^n$, and $g \in Q\delta$, if $h^T x \leq g$ holds for all $x \in P A b$. In the same manner, a polyhedron implies an equality $h^T x = g$, where $h \in Q^n$, $h \neq 0^n$, and $g \in Q$, if $h^T x = g$ holds for all $x \in P A b$. An equality implied by $Ax \leq b$ is explicit if the inequalities $h^T x \leq g$ and $-h^T x \leq -g$ appear in $Ax \leq b$. Otherwise, the equality is implicit. Polyhedra implying equalities have only surface points and, therefore, neither an interior nor a center. Thus, all cubes that fit into a polyhedron implying an equality $d^T x = c$ with $d \neq 0^n$ have edge length zero.

In Sect. 6, we present a method that detects whether a polyhedron implies an equality at all and returns one such equality. To prove the correctness of this method, we use Farkas' Lemma [5]. But first we have to proof that Farkas' Lemma works with δ-rationals:

Lemma 2 $Ax \leq b$ is unsatisfiable iff there exists a $y \in Q^m$ with $y \geq 0^m$ and $y^T A = 0^n$ so that $y^T b < 0$, i.e., there exists a non-negative linear combination of inequalities in $Ax \leq b$ that results in an inequality $y^T Ax \leq y^T b$ that is constant and unsatisfiable.

Proof

Let us first consider the case where $Ax \leq b$ is unsatisfiable. Dutertre and de Moura's version of the dual simplex algorithm is a complete and correct algorithm for determining the satisfiability of a linear arithmetic problem over the δ-rationals [14]. In case the problem is unsatisfiable, the algorithm returns a conflict explanation, which can be turned, together with the final simplex tableau, into the non-negative linear combination $y \in Q^m$ we are looking for. Let us now consider the case where $x \in Q^n_\delta$ is a solution for $Ax \leq b$. If x is a solution to the inequalities in $Ax \leq b$, then it is also a solution to any non-negative linear combination of inequalities in $Ax \leq b$.

Our method that detects implied equalities transforms our original polyhedron $Ax \leq b$ into a second polyhedron $A'x \leq b'$ that is unsatisfiable if $Ax \leq b$ implies an equality. We also show how to extract an equality implied by $Ax \leq b$ from a *minimal* set C of unsatisfiable inequalities in $A'x \leq b'$. We call an unsatisfiable set C of inequalities *minimal* if every proper subset $C' \subset C$ is satisfiable. If a polyhedron $Ax \leq b$ is unsatisfiable, there exists a minimal set C of unsatisfiable inequalities so that every inequality in C appears also in $Ax \leq b$ [14]. We call such a minimal set C an *explanation* for $Ax \leq b$'s unsatisfiability. In case we are investigating a minimal set of unsatisfiable inequalities, we can refine Farkas' Lemma:

Lemma 3 ([7]) Let $C=\{ a^T_i x \leq b_i : 1 \leq i \leq m\}$ be a minimal set of unsatisfiable constraints. Let $A=(a_1,\ldots,a_m)^T$ and $b=(b_1,\ldots,b_m)^T$. Then it holds for every $y \in Q^m$ with $y \geq 0^n$, $y^T A=0^n$, and $y^T b<0$ that yi>0 for all $i \in \{1,\ldots,m\}$.

FITTING CUBES INTO POLYHEDRA

We say that a cube $Ce(z)$ *fits* into a polyhedron defined by $Ax \leq b$ if all points inside the cube $Ce(z)$ are solutions of $Ax \leq b$, or formally: $Ce(z) \subseteq P^A_b$. In order to compute this, we transform the polyhedron $Ax \leq b$ into another polyhedron $Ax \leq b'$. For this new polyhedron, we merely have to test whether the cube's center point z is a solution ($z \in P^A_{b'}$) in order to also determine whether the cube $Ce(z)$ fits into the original polyhedron. This is a simple test that requires only evaluation. We call this entire transformation the *linear cube transformation*.

We start explaining the linear cube transformation by looking at the case where the polyhedron is defined by a single inequality $a^T_i x \leq b_i$. A cube $Ce(z)$ fits into the inequality $a^T_i x \leq b_i$ if all points inside the cube $Ce(z)$ are solutions of $a^T_i x \leq b_i$, or formally: $\forall x \in Ce(z).\ a^T_i x \leq bi \forall x \in Ce(z).aiTx \leq bi$.

We can think of $a^T_i x$ as an objective function that we want to maximize and see bi as a guard for the maximum objective of any solution in the cube. Thus, we can express the universal quantifier in the above equation as an optimization problem (see Figure 1a): $\max\{ a^T_i x : x \in Ce(z)\} \leq b_i$. This also means that all points in $x \in Ce(z)$ satisfy the inequality $a^T_i x \leq b_i$ if a point $x* \in Ce(z)$ with maximum value $a^T_i x*=\max\{ a^T_i x : x \in Ce(z)\}$ for the objective function $a^T_i x$ satisfies the inequality $a^T_i x* \leq b_i$. Since every cube is a bounded polyhedron, one of the points with maximum objective value is a vertex $x^v \in Ce(z)$. A vertex xv of the cube $Ce(z)$ is one of the points with maximum distance to the center z (see Figure 1b), or formally: $x^v=(z_1 \pm e/2,\ldots,z_n \pm e/2)^T$. If we insert the above equation into the objective function $a^T_i x$, we get: a^T_i

$(z_1 \pm e/2,\ldots,z_n \pm e/2)^T = a_i^T z + \frac{e}{2}\sum_{j=1}^{n} \pm a_{ij}$, which in turn is maximal if we choose x^v such that $\pm a_{ij}$ is always positive:

$$a_i^T x^v = a_i^T z + \frac{e}{2}\sum_{j=1}^{n} |a_{ij}| = a_i^T z + \frac{e}{2}\|a_i\|_1 \ .$$

Hence, we transform the inequality $a^T_i x \le b_i$ into $a^T_i x \le b_i - e/2\|a_i\|_1$, and Ce(z) fits into $a^T_i x \le bi$ if $a^T_i z \le b_i - e/2\|a_i\|_1$.

Proposition 2

Let Ce(z) be a cube and $a^T_i x \le bi$ be an inequality. All $x \in Ce(z)$ fulfill $a^T_i x \le b_i$ if and only if $a^T_i z \le bi - e/2\|a_i\|_1$.

Next, we look at the case where multiple inequalities $a^T_i x \le b_i$ (for i=1,…,m) define the polyhedron $Ax \le b$. Since P^A_b is the intersection of all P^{ai}_{bi}, the cube fits into $Ax \le b$ if and only if it fits into all inequalities $a^T_i x \le b_i$:

$$\forall i \in \{1,\ldots,m\}.\ \forall x \in C_e(z).\ a_i^T x \le b_i \ .$$

We can express this by *m* optimization problems:

$$\forall i \in \{1,\ldots,m\}.\ \max\{a_i^T x : x \in C_e(z)\} \le b_i$$

and, after applying Proposition 2, by the following *m* inequalities:

$$\forall i \in \{1,\ldots,m\}.\ a_i^T z \le b_i - \frac{e}{2}\|a_i\|_1 \ .$$

Hence, the linear cube transformation transforms the polyhedron $Ax \le b$ into the polyhedron $Ax \le b'$, where $b'_i = b_i - \frac{e}{2}\|a_i\|_1$, and Ce(z) fits into $Ax \le b$ if $Az \le b'$.

Proposition 3 Let Ce(z) be a cube and $Ax \le b$ be a polyhedron. $Ce(z) \subseteq P^A_b$ if and only if $Az \le b'$, where $b'_i = b_i - \frac{e}{2}\|a_i\|_1$.

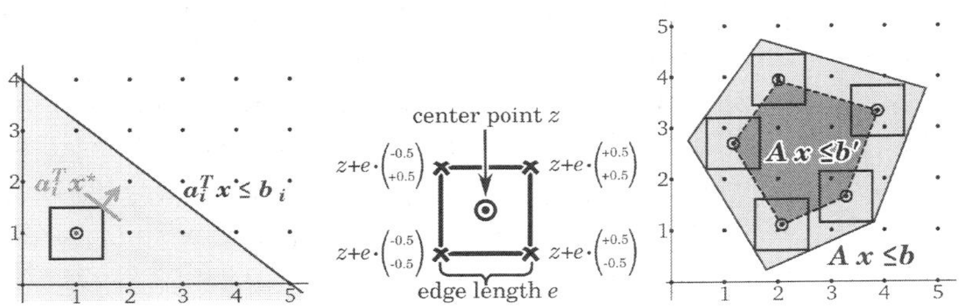

Figure 1: a A square (two-dimensional cube) fitting into an inequality $a^T_i x \le b_i$ and the cube's maximum $a^T_i x*$ for the objective $a^T_i x$. **b** The vertices of an ar-

bitrary square parallel to the coordinate axes (two-dimensional cube with edge length e and center z). **c** The transformed polyhedron $Ax \leq b'$ for edge length 1 together with the original polyhedron $Ax \leq b$.

Until now, we have discussed how to use the linear cube transformation to determine if one cube Ce(z) with fixed center point z fits into a polyhedron. A generalization of this problem determines whether a polyhedron $Ax \leq b$ contains a cube of edge length e at all. Actually, a closer look at the transformed polyhedron $Ax \leq b'$ reveals that the linear cube transformation $b'_i = b_i - \frac{e}{2} \|a_i\|_1$ is dependent only on the edge length e of the cube. Therefore, the solutions $P^A_{b'}$ of the transformed polyhedron $Ax \leq b'$ are exactly all center points of cubes with edge length e that fit into the original polyhedron $Ax \leq b$ (see Figure 1c). By determining the satisfiability of the transformed polyhedron $Ax \leq b'$, we can now also determine whether a polyhedron $Ax \leq b$ contains a cube of edge length e at all. If we choose a suitable algorithm, e.g., the simplex algorithm, then we even get the center point z of a cube Ce(z) that fits into $Ax \leq b$. This observation is the foundation for the cube tests that we present in Sect. 4.

FAST CUBE TESTS

A polyhedron $Ax \leq b$ has an integer solution if and only if $P^A_b \cap Z^n \neq \emptyset$, i.e., if the set of rational solutions contains an integer point. In this section, we show how to use the linear cube transformation to find such an integer solution. In contrast to arbitrary polyhedra, determining whether a cube Ce(z) contains an integer point is easy. Because of the cubes symmetry, it is enough to test whether it contains a closest integer point $\lceil z \rfloor$ to the center z.

Proposition 4

A cube Ce(z) contains an integer point if and only if it contains a closest integer point $\lceil z \rfloor$ to the center z.

Note that every point $z \in Q^n_\delta$ is also a cube $C_0(z)$ of edge length 0. In order to be efficient, our tests look only at cubes with special properties. In the case of the largest cube test, we check for an integer solution in one of the largest cubes fitting into the polyhedron $Ax \leq b$. In the case of the unit cube test, we look for a cube of edge length one, which always guarantees an integer solution. Due to these restrictions, both tests are not complete but very fast to compute.

Largest Cube Test

A well-known test, implemented in most ILP solvers, is *simple rounding*. For simple rounding, the ILP solver computes a rational solution x for a set of inequalities, *rounds* it to a closest integer ⌈x⌋⌈x⌋, and determines whether this point is an integer solution. Not all types of rational solutions are good candidates for this test to be successful. Especially *surface points*, such as *vertices*, the usual output of the simplex algorithm, are not good candidates for rounding. For many polyhedra, *center and interior points* z are a better choice because all integer points adjacent to z are solutions, including a closest integer point ⌈z⌋.

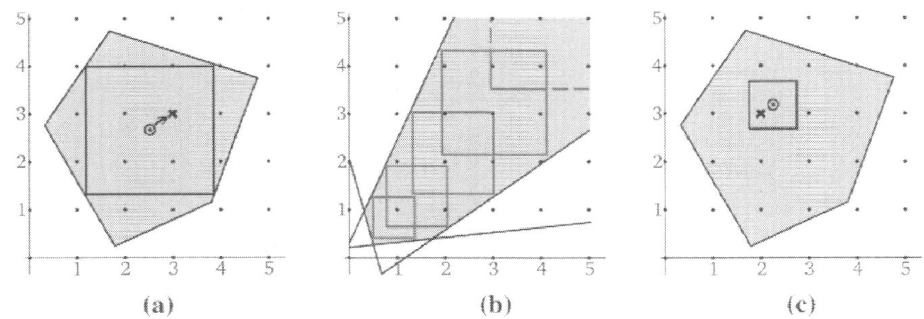

Figure 2: a The largest cube inside a polyhedron, its center point, and a closest integer point to the center. **b** An infinite lattice width polyhedron, containing cubes for every edge length e>0. **c** A unit cube inside a polyhedron, its center point, and a closest integer point to the center.

We now use the linear cube transformation (Sect. 3) to calculate a rational center point with the simplex algorithm. The center point we calculate is the center point of a largest cube that fits into the polyhedron Ax≤b (see Figure 2a). We determine the center z of this largest cube and the associated edge length e with the following linear program (LP):

$$\begin{aligned} &\text{maximize} && x_e \\ &\text{subject to} && Ax + a'\tfrac{x_e}{2} \le b, \ \text{where } a_i' = \|a_i\|_1 \\ & && x_e \ge 0\,. \end{aligned}$$

This linear program employs the linear cube transformation from Sect. 3. The only generalization is a variable xe for the edge length instead of a constant value e. Additionally, this linear program maximizes the edge length as an optimization goal. If the resulting maximum edge length is unbounded, the original polyhedron contains cubes of arbitrary edge length (see Figure 2b) and, thus, infinitely many integer solutions. Since the linear program contains all solutions of the original polyhedron (see xe=0), the original polyhedron is empty if and only if the above linear program is infeasible. If the maximum edge length is a finite value *e*, we use the resulting assignment *z* for the variables *x* as a center point and Ce(z) is a largest cube that fits into the polyhedron. From the center point, we round to a closest integer point ⌈z⌋ and determine if it fits into the original polyhedron. If this is the case, we are done because we have found an integer solution for $Ax \leq b$. Otherwise, the largest cube test does not know whether or not $Ax \leq b$ has an integer solution. An example for the latter case, are the following inequalities: $3x_1 - x_2 \leq 0$, $-2x_1 - x_2 \leq -2$, and $-2x_1 + x_2 \leq 1$. These inequalities have exactly one integer solution $(1,3)^T$, but the largest cube contained by the inequalities has edge length e=3/17 and center point $(3/17,3/2)^T$, which rounds to $(0,2)^T$.

The largest cube test also upholds the incremental advantages of Dutertre and de Moura's version of the dual simplex algorithm [14]. The only difference is the extra column $a'_{xe/2}$, which the theory solver can internally create while it is notified of all potential arithmetic literals. Adding this column from the start does not influence the correctness of the solution because $xe \geq 0$ guarantees that the largest cube test is satisfiable exactly when the original inequalities $Ax \leq b$ are satisfiable. Even for explanations of unsatisfiability, it suffices to remove the bound $xe \geq 0$ to obtain an explanation for the original inequalities $Ax \leq b$. The only disadvantage is the additional variable xe, which only shrinks the search space when it is increased. Therefore, increasing xe can never resolve any conflicts during the satisfiability search. The simplex solver recognizes this with at least one additional pivot that sets xe to 0. Hence, adding the extra column $a'_{xe/2}$ from the beginning has only constant influence on the theory solver's run-time, and is therefore negligible.

Unit Cube Test

Most SMT solvers implement a simplex algorithm that is specialized towards feasibility and not towards optimization [1, 14, 16, 24]. Therefore, a test based on optimization, such as the largest cube test, does not fit well with existing implementations. As an alternative, we have developed a second test based on cubes that does not need optimization.

We avoid optimization by fixing the edge length *e* to the value 1 for all the cubes Ce(z) we consider (see Figure 2c). We do so because cubes $C_1(z)$ of edge length 1 are the smallest cubes to always guarantee an integer solution, independent of the center point *z*. A cube with edge length 1 is also called a *unit cube*. To prove this guarantee, we first fix e=1 in the definition of cubes, $C_1(z)=\{x\in Q^n_\delta:\forall j\in 1,\dots,n.|xj-zj|\leq 1/2\}$, and look at the following property for the rounding operator $\lceil . \rfloor$: $\forall z_j \in Q\delta.|\lceil z_j \rfloor - z_j|\leq 1/2$. We see that any unit cube contains a closest integer $\lceil z \rfloor$ to its center point z. Furthermore, 1 is the smallest edge length that guarantees an integer solution for a cube with center point $z=(\dots,1/2,\dots)^T$. Thus, 1 is the smallest value that we can fix as an edge length to guarantee an integer solution for all cubes $C_1(z)$.

Our second test tries to find a unit cube that fits into the polyhedron $Ax\leq b$ and, thereby, a guarantee for an integer solution for $Ax\leq b$. Again, we employ the linear cube transformation from Sect. 3 and obtain the linear program:

$$Az \leq b', \text{ where } b'_i = b_i - \tfrac{1}{2}\|a_i\|_1 \ .$$

In addition to being a linear program without an optimization objective, we only have to change the row bounds b'_i of the original inequalities. In Dutertre and de Moura's version of the dual simplex algorithm [14], which is implemented in many SMT solvers [1, 14, 16, 24], such a change of bounds is already part of the framework so that integrating the unit cube test into theory solvers is possible with only minor adjustments to the existing implementation. Since our unit cube test requires only an exchange of bounds, we can easily return to the original polyhedron by reverting the bounds. In doing so, the unit cube test upholds the incremental connection between the different original polyhedra.

Mixed Linear Integer and Rational Arithmetic

We can also extend our cube tests to the theory of mixed linear integer and rational arithmetic. In this theory, we partition our variables $x = (x_1,\dots, x_n)^T$ into two vectors: the integer variables $x_Z = (x^Z{}_1,\dots, x^Z{}_k)T$ and the rational variables $xQ = (xQ_1,\dots, xQ_t)^T$. Based on this partitioning, we also split the coefficient matrix A into two matrices A = (S, R), where $S = (s_1,\dots,s_m)^T \in Q^{m\times k}$ defines the coefficients for the integer variables and $R = (r_1,\dots,r_m)^T \in Q^{m\times t}$ defines the coefficients for the rational variables. The system has a solution if there exists an integer assignment for the variables x^Z and a rational assignment for the variables x^Q that satisfies our system of inequalities $s_i^T x^{\mathbb{Z}} + r_i^T x^{\mathbb{Q}} \leq b_i$ (for i = 1,..., m).

Because only integer variables need to be assigned to integer values, tests like simple rounding should be restricted to integer variables. For instance, if z is a rational solution for the overall polyhedron, then simple rounding applies ⌈.⌋⌈.⌋ only to the components z^Z of z that correspond to integer variables. The same holds for our fast cube tests. Instead of looking for hypercubes of the same dimension n as the number of total variables, we are looking for hypercubes of dimension k that expand in the directions that correspond to integer variables, but are flat in the directions that correspond to rational variables. Such a hypercube of dimension n1 with center point z is defined as the set:

$$F_e(z) = \left\{x \in \mathbb{Q}_\delta^k : \forall j \in 1, \ldots, k.\ |x_j^{\mathbb{Z}} - z_j^{\mathbb{Z}}| \leq \frac{e}{2}\right\} \times \left\{z^{\mathbb{Q}}\right\}.$$

We can also modify the linear cube transformation so that we can compute whether a polyhedron $Sx^Z+Rx^Q \leq b$ contains a hypercube Fe(z) that is less than full dimensional:

Proposition 5 Let Fe(z) be a flat cube of dimension k and $Sx^Z+Rx^Q \leq b$ be a polyhedron. $Fe(z) \subseteq P^A_b$ if and only if $Sz^Z+Rz^Q \leq b'$, where $b'_i=b_i-e/2\|s_i\|_1$.

Since the hypercube Fe(z) only expands in the directions that correspond to integer variables, the inequality bounds b′ of the modified linear cube transformation are only influenced by the coefficients of the integer variables. Using Proposition 5, we can now modify our fast cube tests so that they work for mixed linear integer and rational arithmetic. For the largest cube test, we compute the center point of a largest cube Fe(z) that is flat in the directions that correspond to rational variables and fits into the polyhedron $Sx^Z+Rx^Q \leq b$. We determine the center z of this largest cube and the associated edge length e with the following LP:

$$\begin{aligned} &\text{maximize} \quad x_e \\ &\text{subject to} \quad Sx^{\mathbb{Z}} + Rx^{\mathbb{Q}} + s'\tfrac{x_e}{2} \leq b, \ \text{where } s'_i = \|s_i\|_1 \\ &\qquad\qquad\quad x_e \geq 0\,. \end{aligned}$$

From the resulting center point z we receive a candidate mixed integer rational solution by applying the rounding operator ⌈.⌋⌈.⌋ to the components z^Z of z that correspond to integer variables. For the unit cube test, we search for a cube $F_1(z)$ that is flat in the directions that correspond to rational variables, has edge length 1, and fits into the polyhedron $Sx^Z+Rx^Q \leq b$. A linear program that accomplishes this task is: $Sx^Z+Rx^Q \leq b'$, where $b'_i=b_i-1/2\|s_i\|_1$.

Again, 1 is the smallest value that we can fix as an edge length to guarantee a mixed rational integer solution for all cubes $F_1(z)$.

EXPERIMENTS

While our tests are useful for many types of polyhedra, the motivation for our tests stems from a special type of polyhedron, a so-called *infinite lattice width* polyhedron [20]. A polyhedron $Ax \leq b$ has *infinite lattice width* if for every objective $c \in Q^n \setminus \{0^n\}$ c, either its maximum or minimum objective value is unbounded:

$$\forall c \in \mathbb{Q}^n \setminus \{0^n\}.\ \sup\left\{c^T x : x \in P_b^A\right\} = \infty \text{ or } \inf\left\{c^T x : x \in P_b^A\right\} = -\infty .$$

Polyhedra with infinite lattice width seem trivial at first glance because their interior expands arbitrarily far in all directions (see Figure 2b). Therefore, a polyhedron with infinite lattice width contains an infinite number of integer solutions [20]. Nonetheless, many SMT solvers are inefficient on those polyhedra because they use a branch-and-bound approach with an underlying simplex solver [14]. Although such an approach terminates inside finite a priori bounds [25], it does not explore the infinite interior, but rather directs the search along the solutions suggested by the simplex solver: the vertices of the polyhedron. Thus, the SMT solvers concentrate their search on a bounded part of the polyhedron. This bounded part contains only a finite number of integer solutions, whereas the complete interior contains infinitely many integer solutions. The advantage of our cube tests is that they actually exploit the infinite interior because polyhedra with infinite lattice width contain cubes for every edge length (see Figure 2b). Our tests are, therefore, always successful on polyhedra with infinite lattice width and usually need only a small number of pivoting steps before finding a solution.

Lemma 4 ([8]) Let $Ax \leq b$ be a polyhedron. Let $a' \in Q^m$ be a vector such that its components are $a'_i = \|a_i\|_1$. Then $Ax \leq b$ contains a cube Ce(z) for every non-negative $e \in Q\delta$ if and only if $Ax \leq b$ has infinite lattice width.

We have found instances of polyhedra with the infinite lattice width property in some classes of the SMT-LIB benchmarks. These instances are 229 of the 233 *dillig* benchmarks designed by Dillig et al. [11], 503 of the 591 *CAV-2009* benchmarks also by Dillig et al. [11], 229 of the 233 *slacks* benchmarks which are the dillig benchmarks extended with slack variables [18], and 19 of the 37 *prime-cone* benchmarks, that is, "a group of crafted benchmarks encoding a tight *n*-dimensional cone around the point whose coordinates are the first *n* prime numbers" [18]. The remaining problems (4 from dillig, 88 from CAV-2009, 4 from slacks, and 18 from prime-cone) do not have infinite lattice width because they are either tightly bounded or unsatisfiable. For our experiments, we look only at the instances of those benchmark classes that actually fulfill the infinite lattice width property.

Using these benchmark instances, we have confirmed our theoretical assumptions (Lemma 4) in practice. We integrated the unit cube test into our own branch-and-bound solver *SPASS-IQ*(http://www.spass-prover.org/spass-iq) and ran it on the infinite lattice width instances; once with the unit cube test turned on (*SPASS-IQ-0.1+uc*) and once with the test turned off (*SPASS-IQ-0.1*). For every problem, SPASS-IQ-0.1+uc applies the unit cube test exactly once. This application happens before we start the branch-and-bound approach. We also compared our solver with state-of-the-art SMT solvers for linear integer arithmetic: *cvc4-1.4* [1], *mathsat5-3.13* [10], *yices2.5.1* [13], and *z3-4.4.1* [24]. All these solvers employ a branch-and-bound approach with an underlying dual simplex solver [14]. The only exception is mathsat5, which, subsequent to our first publication on the unit cube test [8], now also performs the unit cube test in advance. That is why we also test mathsat5 once with the unit cube test turned on (mathsat5-3.13+uc) and once with the test turned off (mathsat5-3.13).

The solvers had to solve each problem in under 10 min. For the experiments, we used a Debian Linux server with 32 Intel Xeon E5-4640 (2.4 GHz) processors and 512 GB RAM. Table 1 lists the results of the different solvers (column one) on the different benchmark classes (row one). Row two lists the number of benchmark instances we considered for our experiments. For each combination of benchmark class and solver, we have listed the number of instances the solver could solve in the given time as well as the total time (in seconds) of the instances solved (columns labelled with "solved" and "time", respectively).

Table 1: Experimental results

Benchmark name	**CAV-2009**		**DILLIG**		**PRIME-CONE**		**SLACKS**		**ROTATE**	
#Instances	**503**		**229**		**19**		**229**		**229**	
Solvers	**Solved**	**Time**	**Solved**	**Time**	**Solved**	**Time**	**Solved**	**Time**	**Solved**	**Time**
SPASS-IQ-0.1+uc	**503**	22	**229**	9	**19**	0.4	**229**	**26**	**229**	**9**
SPASS-IQ-0.1	**503**	713	**229**	218	**19**	0.4	197	95	**229**	214
ctrl-ergo	**503**	**12**	**229**	**5**	**19**	0.4	**229**	46	24	6760
cvc4-1.4	467	12903	206	4146	18	3	152	4061	208	6964
mathsat5-3.13+uc	**503**	42.37	**229**	18	**19**	0.4	**229**	39	**229**	21
mathsat5-3.13	500	4601	225	2315	**19**	3.5	181	4573	**229**	1507

yices-2.5.1	469	11403	213	2553	**19**	**0.1**	147	5725	180	10073
z3-4.4.1	466	682	213	475	**19**	**0.1**	158	371	213	473

The bold entries represent the best results for the given benchmark class

Our solver that employs the unit cube test solves all instances with the application of the unit cube test and is 25 times faster than our solver without the test. The SMT theory solvers in their standard setting were not able to solve all instances within the allotted time. Moreover, our unit cube test was over 100 times faster than any state-of-the-art SMT solver without the unit cube test. The results for mathsat5 further support the superiority of the test.

We also compared our test with the *ctrl-ergo* solver, which includes a subroutine that is essentially the dual to our largest cube test [4]. As expected, both approaches are comparable for infinite lattice width polyhedra. In order to also compare the two approaches on benchmarks without infinite lattice width, we created the *rotate* benchmarks by adding the same four inequalities to all infinite width instances of the dillig benchmarks. These four inequalities essentially describe a square bounding the variables x_0 and x_1 in an interval $[-u,u]$. For a large enough choice of u (e.g., $u=2^{10}$), the square is so large that the benchmarks are still satisfiable and not absolutely trivial for branch-and-bound solvers. To add a challenge, we rotated the square by a small factor $1 / r$, which resulted in the following four inequalities:

$$-b \cdot r \cdot r + r \leq b \cdot r \cdot x_0 - x_1 \leq b \cdot r \cdot r - r \text{, and}$$

$$-b \cdot r \cdot r + r \leq x_0 + b \cdot r \cdot x_1 \leq b \cdot r \cdot r - r .$$

These changes have nearly no influence on SPASS-IQ, and two SMT solvers even benefit from the proposed changes. For ctrl-ergo the *rotate* benchmarks are very hard because its subroutine detects only infinite lattice width. Without infinite lattice width, ctrl-ergo starts its search from the boundaries of the polyhedron instead of looking at the polyhedron's interior. We can even control the number of iterations (r^2) ctrl-ergo spends on the parts of the boundary without any integer solutions if we choose r accordingly (e.g., $r=2^{10}$). In contrast, we use our cube tests to also extract interior points for rounding. This difference makes our tests much more stable under small changes to the polyhedron.

There exist alternative methods for solving linear integer constraints that do not rely on a branch-and-bound approach [6, 18]. These have not yet matured enough to be competitive with our tests or state-of-the-art SMT theory solvers.

Most problems in the linear integer arithmetic SMT-LIB benchmarks with finite lattice width can be solved without using any actual integer arithmetic techniques. A standard simplex solver for the rationals typically finds a rational solution for such a problem that is also an integer solution. Applying the unit cube test on these trivial problem classes is a waste of time. In the worst case, it doubles the eventual solution time. For these examples it is beneficial to first compute a general rational solution and to check it for integer satisfiability before applying the unit cube test. This has the additional benefit that rational unsatisfiable problems are filtered out before applying the unit cube test. The unit cube test is also guaranteed to fail on problems containing boolean variables, i.e., variables that are either 0 or 1, unless they are absolutely trivial and describe a unit cube themselves. Whenever the problem contains a boolean variable, it is beneficial to skip the unit cube test. This is also the reason why we provide no experimental results for the theory of mixed linear integer and rational arithmetic, i.e., the few mixed benchmarks available in the SMT-LIB all contain boolean variables.

FROM CUBES TO EQUALITIES

If a polyhedron implies an equality, then it has only surface points and neither an interior nor a center. There is no way such a polyhedron contains a unit cube and a largest cube has edge length zero and is just a point in the original polyhedron. Equalities are, therefore, a challenge for the applicability of our cube tests.

There even exist systems of inequalities that imply infinitely many equalities. For instance, the system consisting of the inequalities $-2x_1 + x_2 \leq -2$, $x_1 + 3x_2 \leq 8$, and $x_1 - 2x_2 \leq -2$ has only one rational solution: the point $(x_1, x_2) = (2, 2)$. Therefore, it implies the equalities $-2x_1 + x_2 = -2$ and $x_1 + 3x_2 = 8$, and all linear combinations of those two equalities, i.e., $\lambda_1 \cdot (-2x_1 + x_2) + \lambda_2 \cdot (x_1 + 3x_2) = \lambda_1 \cdot (-2) + \lambda_2 \cdot 8$ for all $\lambda_1, \lambda_2 \in Q$. The above example also points us to another fact about equalities: there exists a finite representation of all equalities implied by a system of inequalities—even if the system implies infinitely many equalities.

One such finite representation is the equality basis for a satisfiable system of inequalities $Ax \leq b$. An equality basis is a system of equalities $D'x = c'$ such that all (explicit and implicit equalities) implied by $Ax \leq b$ are linear combinations of equalities from $D'x = c'$. We prefer to represent each equality basis $D'x = c'$ as an equivalent system of equalities $y - Dz = c$ such that $y = (y_1,..., y_n y)^T$ and $z = (z_1,...,z_n z)T$ are a partition of the variables in x, $D \in Q^{ny \times nz}$, and $c \in Q^{ny}$. The existence of such an equivalent system of equalities is guaranteed by Gaussian elimination. Moreover, each variable y_i appears exactly once in the system $y - Dz = c$, that is to say, yi appears only in the row $y_i - d^T{}_i z_i = c_i$. We choose to represent our equality bases in this manner because this form also correlates to a distinct substitution $\sigma^{D,c}_{y,z}$ that replaces variable y_i with $c_i + d^T{}_i z$:

$$\sigma^{D,c}_{y,z} := \{y_i \mapsto c_i + d_i^T z : i \in \{1, \ldots, n_y\}\}.$$

The substitution $\sigma^{D,c}_{y,z}$ is important because it allows us to eliminate all equalities from $Ax \leq b$. We simply apply the substitution $\sigma^{D,c}_{y,z}$ to $Ax \leq b$ and receive a new system $A'z \leq b'$ that neither contains the variables y nor implies any equalities.1 And the substitution $\sigma^{D,c}_{y,z}$ for the equality basis $y - Dz = c$ has even further applications. For instance, we can directly check whether an equality hT x = g is a linear combination of $y - Dz = c$ and, therefore, implied by both $Ax \leq b$ and $y - Dz = c$. We simply apply $\sigma^{D,c}_{y,z}$ to $h^T x = g$ and see if it simplifies to $0 = 0$. We even use $\sigma^{D,c}_{y,z}$ for the Nelson–Oppen style combination of theories (see Sect. 7).

Finding Equalities

The first step in computing an equality basis for a polyhedron $Ax \leq b$ is to detect whether the system contains any equalities. We have already stated a criterion that detects this:

Lemma 5 ([8]) Let $Ax \leq b$ be a polyhedron. Then exactly one of the following statements is true: (1) $Ax \leq b$ implies an equality $h^T x = g$ with $h \neq 0^n$, or (2) $Ax \leq b$ contains a cube with edge length $e > 0$.

A cube with positive edge length is enough to prove that there exists no implied equality. The actual edge length *e* of this cube is not relevant. Therefore, we can assume that the edge length *e* is arbitrarily small. We can even assume that our edge length is so small that we can ignore the different multiples ||ai|| 1 and any infinitesimals introduced by strict inequalities. We just have to turn all of our inequalities into strict inequalities.

Lemma 6 Let $Ax \le b$ be a polyhedron, where $a_i \neq 0^n$, $b_i=(p_i,q_i)$, $q_i \le 0$, and $b^\delta_i=(p_i,-1)$ be the strict versions of the bounds bi for all $i \in \{1,\ldots,m\}$. Then the following statements are equivalent: (1) $Ax \le b$ contains a cube with edge length $e>0$, and (2) $Ax \le b\delta$ is satisfiable.

Proof

(1) ⇒⇒ (2): If $Ax \le b$ contains a cube of edge length $e>0$, then $Ax \le b-a'$ is satisfiable, where $a'i=e/2\|a_i\|_1$. By Lemma 1, we know there exists a $\delta \in Q$ such that $Ax \le p+q\delta-a'$ Ax. Now, let $\delta'=\min\{a'_i-q_i\delta : i=1,\ldots,m\}$. Since $a'i-qi\delta \ge \delta'$, it holds that $Ax \le p-\delta' 1^m$. Since $q_i \le 0$ and $a'_i=\|a_i\|_1>0$, it also holds that $\delta'>0$. By Lemma 1, we deduce that $Ax<p$ and, therefore, $Ax \le b\delta$ holds.

(2) ⇒⇒ (1): If $Ax \le b\delta$ is satisfiable, then we know by Lemma 1 that there must exist a $\delta>0$such that $Ax \le p-\delta 1^m$ holds. Let $a_{max}=\max\{\|a_i\|_1 : i=1,\ldots,m\}$, $\delta'=\delta/2$, and $e=\delta a_{max}$. Then $p_i-\delta=p_i-\delta'-e/2a_{max} \le b_i-e/2\|a_i\|_1$. Thus, $Ax \le b$ contains a cube with edge length $e>0$.

In case $Ax \le b\delta$ is unsatisfiable, $Ax \le b$ contains no cube with positive edge length and, therefore by Lemma 5, an equality. In case $Ax \le b\delta$ is unsatisfiable, the algorithm returns an explanation, i.e., a minimal set C of unsatisfiable constraints $a^T_i x \le b^\delta_i$ from $Ax \le b\delta$. If $Ax \le b$ itself is satisfiable, we can extract equalities from this explanation: for every $a^T_i x \le b^\delta_i \in C$, $Ax \le b$ implies the equality $a^T_i x = b_i$.

Lemma 7 Let $Ax \le b$ be a satisfiable polyhedron, where $ai \neq 0^n$, $bi=(pi,qi)$, $qi \le 0$, and $b^\delta_i=(pi,-1)$ for all $i \in \{1,\ldots,m\}$. Let $Ax \le b\delta$ be unsatisfiable. Let *C* be a minimal set of unsatisfiable constraints $a^T_i x \le b^\delta_i$ from $Ax \le b\delta$. Then it holds for every $a^T_i x \le b^\delta_i \in C$ that $a^T_i x=bi$ is an equality implied by $Ax \le b$.

Proof

Because of transitivity of the subset and implies relationships, we can assume that $Ax \le b$ and $Ax \le b\delta$ contain only the inequalities associated with the explanation *C*. Therefore, $C=\{a^T_1 x \le b^\delta_1,\ldots,a^T_m x \le b^\delta_m\}$. By Lemma 2 and $Ax \le b\delta$ being unsatisfiable, we know that there exists a $y \in Q^m$ with $y \ge 0$, $y^TA=0^n$, and $y^T_b\delta<0$. By Lemma 2 and $Ax \le b$ being satisfiable, we know that $y^Tb \ge 0$ is also true. By Lemma 3, we know that $yk>0$ for every $k \in \{1,\ldots,m\}$.

Now, we use $y^Tb\delta<0$, $y^Tb \ge 0$, and the definitions of < and ≤≤ for $Q\delta$ to prove that $y^Tb=0$ and $b=p$. Since $y^T_b\delta<0$, we get that $y^Tp \le 0$. Since $y^Tb \ge 0$, we get that $y^T p \geq 0$. If we combine $y^T p \leq 0$ and $y^T p \geq 0$, we get that $y^T p=0$. From $y^T p=0$ and $y^T b \geq 0$, we get $y^T q \geq 0$. Since $y>0$ and $qi \leq 0$, we get that $y^T q=0$ and $qi=0$. Since $qi=0$, $b=p$.

Next, we multiply $y^T A = 0^n$ with an $x \in P A b$ to get $y^T Ax = 0$. Since $yk > 0$ for every $k \in \{1,..., m\}$, we can solve $y^T Ax = 0$ for every aT k x and get:

$$a_k^T x = - \sum_{i=1, i \neq k}^{m} \left(\frac{y_i}{y_k} a_i^T x \right).$$

Likewise, we solve $y^T b=0$ for every b_k to get: $b_k = - \sum_{i=1, i \neq k}^{m} \left(\frac{y_i}{y_k} b_i \right)$.

Since $x \in P^A_b$ satisfies all $a^T_i x \leq b_i$, we can deduce b_k as the lower bound of $a^T_k x$:

$$a_k^T x = - \sum_{i=1, i \neq k}^{m} \left(\frac{y_i}{y_k} a_i^T x \right) \geq - \sum_{i=1, i \neq k}^{m} \left(\frac{y_i}{y_k} b_i \right) = b_k ,$$

which proves that $Ax \leq b$ implies $a^T_k x = b_k$.

Lemma 7 justifies simplifications on $Ax \leq b\delta$. We can eliminate all inequalities in $Ax \leq b\delta$ that cannot appear in the explanation of unsatisfiability, i.e., all inequalities $a^T_i x \leq b^\delta_i$ that cannot form an equality $a^T_i x=b_i$ that is implied by $Ax \leq b$. For example, if we have an assignment $v \in Q^n_\delta$ such that $Av \leq b$ is true, then we can eliminate every inequality $a^T_i x \leq b^\delta_i$ for which $a^T_i v=b_i$ is false. According to this argument, we can also eliminate all inequalities $a^T_i x \leq b^\delta_i$ that were already strict inequalities in $Ax \leq b$.

Algorithm 1: EqBasis($Ax \leq b$)

Input : A satisfiable system of inequalities $Ax \leq b$, where $A \in \mathbb{Q}^{m \times n}$ and $b \in \mathbb{Q}^m_\delta$
Output : An equality basis $z - Dy = c$ for $Ax \leq b$
$l := 1$, $n_z := n$, $(z_1, \ldots, z_{n_z}) := (x_1, \ldots, x_n)$, $y := ()^T$, and $(y - Dz = c) := \emptyset$
Remove all rows $a_i^T z \leq b_i$ from $Az \leq b$ with $a_i = 0^n$ and $b_i = 0$
while $Az \leq b^\delta$ is unsatisfiable (i.e., $Az \leq b$ contains an equality) **do**
  Let C be an explanation for $Az \leq b^\delta$ being unsatisfiable
  Select $(a_i^T z \leq b_i^\delta) \in C$;   // by Lemma 7, $a_i^T z = b_i$ is implied by $Az \leq b$
  Select a variable z_k such that $a_{ik} \neq 0$.
  $\sigma' := \{z_k \mapsto \frac{b_i}{a_{ik}} - \sum_{j=1, j \neq k}^{n} \frac{a_{ij}}{a_{ik}} z_j\}$
  $z' := (z_1, \ldots, z_{k-1}, z_{k+1}, \ldots, z_n)^T$, $y' := (y_1, \ldots, y_l, z_k)^T$, $l := l + 1$
  $(A'z' \leq b') := (Az \leq b)\sigma'$
  $(y' - D'z' = c') := (y - Dz = c)\sigma' \cup \{z_k + \sum_{j=1, j \neq k}^{n} \frac{a_{ij}}{a_{,k}} z_j = \frac{b_i}{a_{ik}}\}$
  $z := z'$, $y := y'$, $(Az \leq b) := (A'z' \leq b')$, $(y - Dz = c) := (y' - D'z' = c')$
  Remove all rows $a_i^T z \leq b_i$ from $Az \leq b$ with $a_i = 0^n$ and $b_i = 0$
end
return $y - Dz = c$

Figure 3: EqBasis computes an equality basis.

Computing an Equality Basis

We now present the algorithm EqBasis(A'x ≤ b') (Fig. 3) that computes an equality basis for a polyhedron A'x ≤ b'. In a nutshell, EqBasis iteratively detects and removes equalities from our system of inequalities and collects them in a system of equalities until it has a complete equality basis. To this end, EqBasis computes in each iteration one system of inequalities Az ≤ b and one system of equalities y − Dz = c such that A'x ≤ b' is equivalent to (y − Dz = c) ∪ (Az ≤ b). While the variables z are completely defined by the inequalities Az ≤ b, the equalities y − Dz = c extend any assignment from the variables z to the variables y. Initially, z is just x, y − Dz = c is empty, and Az ≤ b is just A'x ≤ b'.

In every iteration l of the while loop, EqBasis eliminates one equality $a^T_i z$ = bi from Az ≤ b and adds it to y − Dz = c. EqBasis finds this equality based on the techniques we presented in the Lemmas 6 and 7 (line 3). If the current system of inequalities Az ≤ b implies no equality, then EqBasis is done and returns the current system of equalities y − Dz = c. Otherwise, EqBasis turns the found equality a^T_i z = bi into a substitution $\sigma' := \{z_k \mapsto \frac{b_i}{a_{ik}} - \sum_{j=1, j\neq k}^{n} \frac{a_{ij}}{a_{ik}} z_j\}$ (line 7) and applies it to Az ≤ b (line 9). This has the following effects: (1) the new system of inequalities A' z'≤ b' implies no longer the equality a^T_i z = bi ; and (2) it no longer contains the variable zk . Next, we apply σ' to our system of equalities (line 10) and concatenate the equality $z_k + \sum_{j=1, j\neq k}^{n} \frac{a_{ij}}{a_{ik}} z_j = \frac{b_i}{a_{ik}}$ to the end of (y−Dz = c)σ' . This has the following effects: (1) the new system of equalities y' −D' z' = c' implies a^T_i z = bi ; and (2) the variable zk appears exactly once in y' − D'z' = c' . This means that we can now re-partition our variables so that z := $(z_1,\ldots,z_{k-1},z_{k+1},\ldots,z_n)^T$ and yl := zk to get two new systems Az ≤ b and y − Dz = c that are equivalent to our original polyhedron (line 11). Finally, we remove all rows 0 ≤ 0 from Az ≤ b because those rows are trivially satisfied but would obstruct the detection of equalities with Lemma 6.

To prove the correctness of the algorithm, we first need to prove that moving the equality from our system of inequalities to our system of equalities preserves equivalence, i.e, the systems (Az≤b)∪(y−Dz=c) and (A'z'≤b')∪(y'−D'z'=c') are equivalent in line 10.

Lemma 8 Let Az ≤ b be a system of inequalities. Let y − Dz = c be a system of equalities. Let h^T z = g be an equality implied by Az ≤ b with h_k

$\neq 0$. Let $\sigma' := \{z_k \mapsto \frac{g}{h_k} - \sum_{j=1, j\neq k}^{n} \frac{h_j}{h_k} z_j\}$ be a substitution based on this equality. Let y' := $(y_1,\ldots, y_l, z_k)$T and z' := $(z_1,\ldots,z_{k-1},z_{k+1},\ldots,z_n)^T$. Let (A' z' ≤ b') := (Az ≤ b)σ' . Let (y' − D' z' = c') := (y − Dz = c)σ' ∪ $\sum_{j=1, j\neq k}^{n} \frac{h_j}{h_k} z_j = \frac{g}{h_k}\}$. Let u ∈ Qny δ , v ∈ Qnz δ , u' = $(u_1,\ldots, u_{ny}$, $v_k)^T$, and v' = $(v_1,\ldots,v_{k-1}, v_{k+1},\ldots,v_{nz})^T$. Then (Av ≤ b)∪(u−Dv = c) is true if and only if (A' v' ≤ b') ∪ (u' − D' v' = c') is true.

Proof

First, we create a new substitution $\sigma_v := \{z_k \mapsto \frac{g}{h_k} - \sum_{j=1, j\neq k}^{n} \frac{h_j}{h_k} v_j\}$ that is equivalent to σ′except that it directly assigns the variables zi to their values vi. Let us now assume that either (Av≤b)∪(u−Dv=c) or (A′v′≤b′)∪(u′−D′v′=c′) is true. This means that $h^T v=g$ is also true, either by definition of (Av≤b) or (u′−D′v′=c′). But $h^T v=g$ is true also implies that $v_k = \frac{g}{h_k} - \sum_{j=1, j\neq k}^{n} \frac{h_j}{h_k} v_j$ is true. Therefore, σv simplifies to the assignment zk↦vk. So (Av≤b)∪(u−Dv=c) and (A′v′≤b′)∪(u′−D′v′=c′) simplify to the same expressions and if one combined system is true, so is the other.

The algorithm EqBasis(A′x≤b′) decomposes the original system of inequalities A′x≤b′ into a reduced system Az≤b that implies no equalities, and an equality basis y−Dz=c. The algorithm is guaranteed to terminate because the variable vector *z* decreases by one variable in each iteration. Note that EqBasis(A′x≤b′) constructs y−Dz=c in such a way that the substitution $\sigma_{y,z}^{D,c}$ is the concatenation of all substitutions σ′ from every previous iteration. Therefore, we also know that $\sigma_{y,z}^{D,c}$ applied to A′x≤b′ results in the system of inequalities Az≤b that implies no equalities. We exploit this fact to prove the correctness of EqBasis(A′x≤b′), but first we need two more auxiliary lemmas.

Lemma 9 Let y−Dz=c be a satisfiable system of equalities. Let Ax≤b A∗x ≤ b∗ be two systems of inequalities, both implying the equalities in y − Dz = c. Let A' z ≤ b' := (Ax ≤ b) $\sigma_{y,z}^{D,c}$ and A∗∗z ≤ b∗∗ := (A∗x ≤ b∗) $\sigma_{y,z}^{D,c}$. Then A'z ≤ b' is equivalent to A∗∗z ≤ b∗∗ if Ax ≤ b is equivalent to A∗x ≤ b∗.

Proof

Let $Ax \leq b$ be equivalent to $A{*}x \leq b{*}$. Suppose to the contrary that $A'z \leq b'$ is not equivalent to $A{**}z \leq b{**}$. This means that there exists a $v \in Q^{nz}_{\delta}$ such that either $A'v \leq b'$ is true and $A{**}v \leq b{**}$ is false, or $A'v \leq b'$ is false and $A{**}v \leq b{**}$ is true. Without loss of generality we select the first case that $A'v \leq b'$ is true and $A{**}v \leq b{**}$ is false. We now extend this solution by $u \in Q^{ny}_{\delta}$, where $u_i := c_i + d^T_i v$, so $(A'v \leq b') \cup (u - Dv = c)$ is true. Based on the definition of $\sigma^{D,c}_{y,z}$ and ny recursive applications of Lemma 8, the four systems of constraints $Ax \leq b$, $A{*}x \leq b{*}$, $(A'z \leq b') \cup (y - Dz = c)$, and $(A{**}z \leq b{**}) \cup (y - Dz = c)$ are equivalent. Therefore, $(A{**}v \leq b{**}) \cup (u - Dv = c)$ is true, which means that $A{**}v \leq b{**}$ is also true. The latter contradicts our initial assumptions.

Now we can also prove what we have already explained at the beginning of this section. The equality $h^T x = g$ is implied by $Ax \leq b$ if and only if $y - Dz = c$ is an equality basis and $(hT\, x = g)\, \sigma^{D,c}_{y,z}$ simplifies to $0 = 0$. An equality basis is already defined as a set of equalities $y - Dz = c$ that implies exactly those equalities implied by $Ax \leq b$. So we only need to prove that $h^T x = g$ is implied by $y - Dz = c$ if $(h^T x = g)\, \sigma^{D,c}_{y,z}$ simplifies to $0 = 0$.

Lemma 10 Let $y - Dz = c$ be a satisfiable system of equalities. Let $h^T x = g$ be an equality. Then $y - Dz = c$ implies $h^T x = g$ iff $(h^T x = g)\, \sigma^{D,c}_{y,z}$ simplifies to $0 = 0$.

Proof

First, let us look at the case where $h^T x = g$ is an explicit equality $y_i - d^{T\,i} z = c_i$ in $y - Dz = c$. Then $(y_i - d^{T\,i} z = c_i)\sigma\, \sigma^{D,c}_{y,z}$ simplifies to $0 = 0$ because $\sigma^{D,c}_{y,z}$ maps y_i to $d^{T\,i} z + c_i$ and the variables z_j are not affected by $\sigma^{D,c}_{y,z}$.

Next, let us look at the case where $h^T x = g$ is an implicit equality in $y - Dz = c$. Since both $y - Dz = c$ and $(y - Dz = c) \cup (h^T z = g)$ imply $h^T z = g$ and the equalities in $y - Dz = c$, both $(y - Dz = c)\, \sigma^{D,c}_{y,z}$ and $((y - Dz = c) \cup (hT\, z = g))\sigma\, \sigma^{D,c}_{y,z}$ must be equivalent (see Lemma 9). As we stated at the beginning of this proof, $(y_i - d^T_i z = c_i)\, \sigma^{D,c}_{y,z}$ simplifies to $0 = 0$. An equality

$h'^T c = g'$ that simplifies to $0 = 0$ is true for all $v \in Qnz\ \delta$. Moreover, only equalities that simplify to $0 = 0$ are true for all $v \in Q^{nz}{}_{\delta}$. This means $(y - Dz = c)\ \sigma^{D,c}_{y,z}$ is satisfiable for all assignments and, therefore, $(h^T z = g)\ \sigma^{D,c}_{y,z}$ must simplify to $0 = 0$.

Finally, let us look at the case where $h^T x = g$ is not an equality implied by $y - Dz = c$. Suppose to the contrary that$((y-Dz = c) \cup (h^T z = g))\ \sigma^{D,c}_{y,z}$ is satisfiable for all assignments. We know based on Lemma 8 and transitivity of equivalence that $(y - Dz = c) \cup (h^T z = g)$ and $(y - Dz = c) \cup \emptyset$ are equivalent. Therefore, $h^T z = g$ is implied by $y - Dz = c$, which contradicts our initial assumption.

With Lemma 10, we have now all auxiliary lemmas needed to prove that the algorithm EqBasis correct:

Lemma 11 Let $A'x \leq b'$ be a satisfiable system of inequalities. Let $y-Dz=c$ be the output of EqBasis($A'x \leq b'$). Then $y-Dz=c$ is an equality basis of $A'x \leq b'$.

Proof

Let $Az \leq b$ be the result of applying $\sigma^{D,c}_{y,z}$ to $A'x \leq b'$. Since $y - Dz = c$ is the output of EqBasis($A'x \leq b'$), the condition in line 3 of EqBasis guarantees us that $Az \leq b$ implies no equalities. Let us now suppose to the contrary of our initial assumptions that $A'x \leq b'$ implies an equality $h'^T x = g'$ that $y - Dz = c$ does not imply. Since $h'^T x = g'$ is not implied by $y - Dz = c$, the output of $(h'^T x = g')\ \sigma^{D,c}_{y,z}$ is an equality $hT\ z = g$, where $h \text{ '}= 0nz$. This also implies that $(Az \leq b) \cup (h^T z = g)$ is the output of $((A'\ x \leq b') \cup (h'^T x = g'))\ \sigma^{D,c}_{y,z}$. By Lemma 9, $Az \leq b$ and $(Az \leq b) \cup (h^T z = g)$ are equivalent. Therefore, $Az \leq b$ implies the equality $hT\ z = g$, which contradicts the condition in line 3 of EqBasis and, therefore, our initial assumptions.

IMPLEMENTATION AND APPLICATION

It is not straight forward how to efficiently integrate our method that finds an equality basis into an SMT solver. Therefore, we now explain how to implement our method as an extension of Dutertre and de Moura's version [14] of the dual simplex algorithm [2, 22]. We choose to specialize this version of the dual simplex algorithm because it is implemented in most SMT solvers and has all properties necessary for an efficient theory solver: it produces minimal conflict explanations, handles backtracking efficiently, and is highly incremental. Whenever we refer to the simplex algorithm in this section, we refer to the specific version of the dual simplex algorithm presented by Dutertre and de Moura [14].

We defined the theory for the equality basis by representing our input constraints through inequalities $Ax \leq b$ because inequalities represent the set of solutions more intuitively. In the simplex algorithm, the input constraints are represented instead by a so-called tableau $Ax=0^m$ and two bounds $l_i \leq x_i \leq u_i$ for every variable x_i in the tableau. Therefore, it might seem difficult to efficiently integrate our method in the simplex algorithm. The truth, however, is that the tableau-and-bound representation grants us several advantages for the implementation of our equality basis method. For example, we do not have to explicitly eliminate variables via substitution, but we do so automatically via pivoting.

Later in this section, we also explain how the integration of our methods in the simplex algorithm can be used for the combination of theories with the Nelson–Oppen Method. For the Nelson–Oppen style combination of theories inside an SMT solver [9], each theory solver has to return all valid equations between variables in its theory. Linear arithmetic theory solvers sometimes guess these equations based on one satisfying assignment. Then the equations are transferred according to the Nelson–Oppen method without verification. This leads to a backtrack of the combination procedure in case the guess was wrong and eventually led to a conflict. With the availability of an equality basis, the guesses can be verified directly and efficiently. Therefore, the method helps the theory solver in avoiding any conflicts due to wrong guesses together with the overhead of backtracking. This comes at the price of computing the equality basis, which should be negligible because the integration we propose is incremental and includes justified simplifications.

The Dual Simplex Algorithm

The input of the simplex algorithm (Figure 4) is a set of equalities $Ax=0^m$ and a set of bounds for the variables $l_j \leq x_j \leq u_j$ (for $j=1,\ldots,n$). If there is no lower bound $l_j \in Q\delta$ for variable x_j, then we simply set $l_j=-\infty$. Similarly, if there is no upper bound $u_j \in Q\delta$ for variable x_j, then we simply set $u_j=\infty$.

Algorithm 2: pivot(x_i, x_j)

Input : A basic variable x_i and a non-basic variable x_j so that a_{ij} is non-zero
Effect : Transforms the tableau so x_i becomes non-basic and x_j basic

Let $x_i = \sum_{k \in \mathcal{N}} a_{ik} x_k$ be the row defining the basic variable x_i
We rewrite this row as $x_j = \frac{1}{a_{ij}} x_i - \sum_{k \in \mathcal{N} \setminus \{x_j\}} \frac{a_{ik}}{a_{ij}} x_k$ so it defines x_j instead
foreach $x_k \in \mathcal{N} \setminus \{x_j\}$ **do** $a_{jk} := -\frac{a_{ik}}{a_{ij}}$;
$a_{ji} := \frac{1}{a_{ij}}$
Substitute x_j in all other rows with $\frac{1}{a_{ij}} x_i - \sum_{k \in \mathcal{N} \setminus \{x_j\}} \frac{a_{ik}}{a_{ij}} x_k$
for $x_l \in \mathcal{B}$ **do**
  foreach $x_k \in \mathcal{N} \setminus \{x_j\}$ **do** $a_{lk} := a_{lk} + a_{lj} a_{jk}$;
  $a_{li} := a_{lj} a_{ji}$; $a_{lj} := 0$
end
$\mathcal{N} = (\mathcal{N} \cup \{x_i\}) \setminus \{x_j\}$; $\mathcal{B} = (\mathcal{B} \cup \{x_j\}) \setminus \{x_i\}$

Algorithm 3: update(x_j, v)

Input : A non-basic variable x_j and a value $v \in \mathbb{Q}_\delta$
Effect : Sets the value $\beta(x_j)$ of x_j to v and updates the values of all basic variables

foreach $x_i \in \mathcal{B}$ **do** $\beta(x_i) := \beta(x_i) + a_{ij}(v - \beta(x_j))$;
$\beta(x_j) := v$

Algorithm 4: pivotAndUpdate(x_i, x_j, v)

Input : A basic variable x_i, a non-basic variable x_j, and a value $v \in \mathbb{Q}_\delta$
Effect : Pivots variables x_i and x_j and updates the value $\beta(x_i)$ of x_i to v

$\theta := \frac{v - \beta(x_i)}{a_{ij}}$
$\beta(x_i) := v$; $\beta(x_j) := \beta(x_j) + \theta$
foreach $x_k \in \mathcal{B} \setminus \{x_i\}$ **do** $\beta(x_k) := \beta(x_k) + a_{kj}\theta$;
pivot(x_i, x_j)

Algorithm 5: Check()

Output : Returns *true* iff there exists a satisfiable assignment for the tableau and the bounds u and l; otherwise, it returns (*false*,x_i), where x_i is the conflicting basic variable

while *true* **do**
  select the smallest basic variable x_i such that $\beta(x_i) < l_i$ or $\beta(x_i) > u_i$
  if there is no such x_i **then return** *true*;
  if $\beta(x_i) < l_i$ **then**
    select the smallest non-basic variable x_j such that
    ($a_{ij} > 0$ and $\beta(x_j) < u_j$) or ($a_{ij} < 0$ and $\beta(x_j) > l_j$)
    if there is no such x_j **then return** (*false*,x_i) ;
    pivotAndUpdate(x_i, x_j, l_i)
  end
  if $\beta(x_i) > u_i$ **then**
    select the smallest non-basic variable x_j such that
    ($a_{ij} < 0$ and $\beta(x_j) < u_j$) or ($a_{ij} > 0$ and $\beta(x_j) > l_j$)
    if there is no such x_j **then return** (*false*,x_i) ;
    pivotAndUpdate(x_i, x_j, u_i)
  end
end

Figure 4: The functions of the dual simplex algorithm by Dutertre and de Moura [14].

We can easily transform a system of inequalities $Ax \leq b$ into the above format if we introduce a so-called slack variable si for every inequality in our system. Our system is then defined by the equalities $Ax - s = 0m$, and the bounds $-\infty \leq x\,j \leq \infty$ for every original variable x j and the bounds $-\infty \leq s_i \leq b_i$ for every slack variable introduced for the inequality $a^T_i\, x \leq bi$. We can even reduce the number of slack variables if we transform rows of the form $ai\,j \cdot x\,j \leq c\,j$ directly into bounds for x j . Moreover, we can use the same slack variable for multiple inequalities as long as the left side of the inequality is similar enough. For example, the inequalities $a^T_i\, x \leq bi$ and $-a^T_i\, x \leq ci$ can be transformed into the equality $a^T_i\, x - si = 0$ and the bounds $-ci \leq si \leq bi$. SMT solvers typically assign the slack variables during a preprocessing step with a normalization procedure based on a variable ordering. After the normalization, all terms are represented in one directed acyclic graph (DAG) so that all equivalent terms are assigned to the same node and, thereby, to the same slack variable. For more details on these simplifications we refer to [14].

The simplex algorithm also partitions the variables into two sets: the set of non-basic variables N and the set of basic variables B. Initially, our original variables are the non-basic variables and the slack variables are the basic variables. The non-basic variables N define the basic variables over a tableau derived from our system of equalities. Each row in this tableau represents one basic variable $xi \in B$: $xi = \sum_{x_j \in \mathcal{N}} a_{ij} x_j$. The simplex algorithm exchanges variables from $xi \in B$ and $x\,j \in N$ with the pivot algorithm. To do so, we also have to change the tableau via substitution. All tableaux constructed in this way are equivalent to the original system of equalities $Ax = 0^m$.

The goal of the simplex algorithm is to find an assignment β that maps every variable xi to a value $\beta(xi) \in Q\delta$ that satisfies our constraint system, i.e., $A(\beta(x)) = 0^m$ and $li \leq \beta(xi) \leq ui$ for every variable xi . The algorithm starts with an assignment β that fulfills $A(\beta(x)) = 0^m$ and $l\,j \leq \beta(x\,j) \leq u\,j$ for every non-basic variable $x\,j \in N$. Initially, we get such an assignment through our tableau. We simply choose a value $l\,j \leq \beta(x\,j) \leq u\,j$ for every non-basic variable $x\,j \in N$ and define the value of every basic variable $xi \in B$ over the tableau: $\beta(xi) := \sum_{x_j \in \mathcal{N}} a_{ij} \beta(x_j)$. As an invariant, the simplex algorithm continues to fulfill $A(\beta(x)) = 0^m$ and $l\,j \leq \beta(x\,j) \leq u\,j$ for every non-basic variable $x\,j \in N$ and every intermediate assignment β.

The simplex algorithm finds a satisfiable assignment or an explanation of unsatisfiability through the Check() algorithm. Since all non-basic variables fulfill their bounds and the tableau guarantees that Ax = 0^m, Check() only looks for a basic variable that violates one of its bounds. If all basic variables x_i satisfy their bounds, then β is a satisfiable assignment and Check() returns true. If Check() finds a basic variable xi that violates one of its bounds, then it looks for a non-basic variable x_j fulfilling the conditions in lines 6 or 12 of Check(). If it finds a non-basic variable x_j fulfilling the conditions, then we pivot x_i with x_j and update our β assignment so β(xi) is set to the previously violated bound value, which satisfies our invariant once more. If it finds no non-basic variable fulfilling the conditions, then the row of xi and all non-basic variables x_j with $a_{ij} \neq 0$ build an unresolvable conflict. Hence, Check() has found a row that explains the conflict and it can return unsatisfiable. The algorithm terminates due to a variable selection strategy called Bland's rule. Bland's rule is based on a predetermined variable order and always selects the smallest variables fulfilling the conditions for pivoting.

Implementation Details

In case of the tableau-and-bound representation, an equality basis simplifies to the tableau Ax = 0^m and a set oftightly bounded variables, i.e., a set of variables x_j such that $\beta(x_j) := l_j$ or $\beta(x_j) := u_j$ for all satisfiable assignments β. Therefore, one way of determining an equality basis is to find all tightly bounded variables.

```
Algorithm 6: Initialize()
  Effect : Removes all bounds l_k and u_k that cannot produce equalities; turns as
           many basic variables x_i with l_i = u_i into non-basic variables as is possible;
           the bounds for all variables x_k are turned into strict bounds if l_k < u_k
for x_k ∈ B ∪ N do
    l'_k := l_k; u'_k := u_k ;                        // Remember the original bounds
    if β(x_k) > l_k then l_k := −∞;
    if β(x_k) < u_k then u_k := +∞;
    if β(x_k) = p_k + q_k δ such that q_k ≠ 0 then l_k := −∞; u_k := ∞;
end
for x_i ∈ B do
    if l_i = u_i then
        select the smallest non-basic variable x_j such that a_ij is non-zero and l_j < u_j
        if there is such an x_j then pivot(x_i, x_j);
    end
end
for x_k ∈ B ∪ N do
    if l_k < u_k then
        if l_k ≠ −∞ then l_k := l_k + δ;
        if u_k ≠ +∞ then u_k := u_k − δ;
        if l_k ≠ −∞ and x_k ∈ N then update(x_k, l_k);
        if u_k ≠ +∞ and x_k ∈ N then update(x_k, u_k);
    end
end
```

Algorithm 7: FixEqs(x_i)

Input : A basic variable x_i that explains the conflict
Effect : Turns the bounds of all variables responsible for the conflict into equalities

for $x_j \in \mathcal{N}$ **do**
  if $l_j < u_j$ **then**
    if $\beta(x_i) < l_i$ and $a_{ij} > 0$ **then** $u_j := u'_j$; $l_j := u'_j$; update(x_j, u_j);
    if $\beta(x_i) < l_i$ and $a_{ij} < 0$ **then** $l_j := l'_j$; $u_j := l'_j$; update(x_j, l_j);
    if $\beta(x_i) > u_i$ and $a_{ij} > 0$ **then** $l_j := l'_j$; $u_j := l'_j$; update(x_j, l_j);
    if $\beta(x_i) > u_i$ and $a_{ij} < 0$ **then** $u_j := u'_j$; $l_j := u'_j$; update(x_j, u_j);
  end
end
if $\beta(x_i) > u_i$ **then** $u_i := u'_i$; $l_i := u'_i$;
if $\beta(x_i) < l_i$ **then** $l_i := l'_i$; $u_i := l'_i$;

Algorithm 8: FindTBnds()

Effect : Finds as many tightly bounded variables as possible
Output : *false* iff the system of linear arithmetic constraints is unsatisfiable

if Check() returns (*false*,x_i) **then return** *false*;
Initialize()
while Check() returns (*false*,x_i) **do**
  FixEqs(x_i)
end
/* For all variables x_k with $l_k < u_k$ recover their original bounds l'_k, u'_k */
for $x_k \in \mathcal{B} \cup \mathcal{N}$ **do**
  if $l_k < u_k$ **then** $l_k := l'_k$; $u_k := u'_k$;
end
return *true*

Figure 5: The functions used to turn our original tableau into a basis of equalities.

To find all tightly bounded variables, we present a new extension of the simplex algorithm called FindTBnds() (Fig. 5). This extension uses our Lemmas 6 and 7 to iteratively find all bounds $l_j \leq x_j$ ($x_j \leq u_j$) that hold tightly for all satisfiable assignments β, and then turns them into explicit equalities by setting $u_j := l_j$ ($l_j := u_j$). But first of all, FindTBnds() determines if our constraint system is actually satisfiable with a call of Check(). If the system is unsatisfiable, then it has no solutions and implies all equalities. In this case, FindTBnds() returns false.

Otherwise, we get a satisfiable assignment β from Check() and we use this assignment in Initialize() (Fig. 5) to eliminate all bounds that do not hold tightly under β (i.e., $\beta(x_i) > l_i$ or $\beta(x_i) < u_i$). We know that we can eliminate these bounds without losing any tightly bounded variables because we only need the bounds that can be part of an equality explanation, i.e., only bounds that hold tightly for all satisfiable assignments (see Lemma 7). For the same reason, Initialize() eliminates all originally strict bounds, i.e., bounds with a non-zero delta part.

Next, Initialize() tries to turn as many variables xi with $l_i = u_i$ into non-basic variables. We do so because xi is guaranteed to stay a non-basic variable if $l_i = u_i$ (see lines 6 and 12 of Check). Pivoting like this essentially eliminates the tightly bounded nonbasic variable xi and replaces it with the constant value l_i. There only exists one case when Initialize() cannot turn the variable xi with $l_i = u_i$ into a non-basic variable. This case occurs whenever all non-basic variables x_j with non-zero coefficient a_{ij} also have tight bounds $l_j = u_j$. In this case, the complete row xi = $\sum_{x_j \in \mathcal{N}} a_{ij} x_j$ simplifies to $x_i = l_i$, so it never produces a conflict and we can also ignore this row.

As its last action, Initialize() turns the bounds of all variables x_j with $l_j < u_j$ into strict bounds. Since Initialize() transformed these bounds into strict bounds, the condition of the while loop in line 3 of FindTBnds() checks whether the system contains another tightly bounded variable (see also Lemma 6). If Check returns (false, xi), then the row xi represents an equality explanation and all variables x j with a non-zero coefficient in the row hold tightly (see Lemma 7). FindTBnds() uses FixEqs(xi) (Fig. 5) to turn these tightly bounded variables x_j into explicit equalities by setting l j = u j . After FixEqs(xi) is done, we go back to the beginning of the loop in FindTBnds() and do another call to Check.

If Check returns true, then the original system of inequalities implies no further tightly bounded variables (Lemma 6). We exit the loop and revert the bounds of the remaining variables x_j with $l_j < u_j$ to their original values. As a result, we have also reverted to a linear system equivalent to our original constraint system. The only difference is that now all tightly bounded variables xi are explicit equalities because $l_i = u_i$. Moreover, the tableau $Ax = 0^m$ and the non-basic variables that are tightly bounded represent an equality basis for our original constraint system. The simplex algorithm even represents the current tableau and the tightly bounded non-basic variables in such a way that they also describe a substitution σ for the elimination of equalities: the rows of the tableau map each basic variable xi to their row definition $\sum_{x_j \in \mathcal{N}} a_{ij} x_j$ and the tightly bounded non-basic variables x_j, i.e., all variables x_j with j ∈ N and $l_j = u_j$, are mapped to their tight bound l_j.

After applying FindTBnds(), we can efficiently find all valid equations between variables as needed for the Nelson–Oppen style combination of theories. For each variable x_i, we use the substitution σ that we get from the tableau and the tightly bounded variables to get a normalized term that represents each variable. If the variable xi is non-basic and tightly bounded (i.e., $l_i = u_i$), then the normalized term is the constant value l_i . If the variable x_i is non-basic and not tightly bounded (i.e., $l_i \neq u_i$), then the normalized term is the variable x_i itself. If the variable x_i is basic, then the normalized term is $\left(\sum_{x_j \in \mathcal{N}, l_j \neq u_j} a_{ij} x_j\right) + \left(\sum_{x_j \in \mathcal{N}, l_j = u_j} a_{ij} l_j\right)$, where all basic mathematical operations between constant values are replaced by the results of those operations.

We know from Lemma 10 that $x_i\sigma = x_k\sigma$ simplifies to $0 = 0$ if σ is the substitution we get from an equality basis and $x_i = x_k$ is implied by our constraints. Therefore, both $x_i\sigma$ and $x_k\sigma$ must be represented by the same normalized term if x_i and x_k are equivalent. So the equality basis together with a normalization procedure has turned semantic equivalence into syntactic equivalence. It is very easy to find variables xi represented by the same normalized term if we store these terms in a DAG, which most SMT solvers already provide for assigning slack variables.

Incrementality, Explanations, and Justifications

Note that asserting additional bounds to our system can increase the number of tightly bounded variables. In this case, we have to apply FindTBnds() again to find all tightly bounded variables and to complete the new equality basis. We already mentioned that Check() never pivots a non-basic variable x j into a basic one if $l_j = u_j$ because of the conditions in the lines 6 and 12 of Check(). So even if the SMT solver asserts additional bounds for the variables and applies Check() again, the tightly bounded non-basic variables we have computed in the last call to FindTBnds() stay non-basic. Hence, our next application of FindTBnds() does not perform any computations for the tightly bounded variables that were detected by earlier applications of FindTBnds(). This means that our algorithm to compute the equality basis is highly incremental.

Another important feature of an efficient SMT theory solver is that it produces good—maybe even minimal—conflict explanations. In a typical SMT solver, a SAT solver based on CDCL (conflict-driven clause learning) selects and asserts a set of theory literals that satisfy the boolean model. Then the theory solvers verify that the asserted literals that belong to their theory are consistently satisfiable. If the theory solver finds a conflict between the asserted literals, then it returns a conflict explanation. The SAT solver uses the conflict explanation to start a conflict analysis that determines a good point for back jumping so it can select a new set of theory literals. Naturally, a good conflict explanation greatly enhances the conflict analysis and, therefore, the remaining search.

The literals asserted in our simplex based theory solver are bounds for our variables.[2] Our algorithm FindTBnds() asserts bounds independently of the SAT solver. This leads to problems in the conflict analysis because the conflict explanation is no longer comprehensible for the SAT solver. Hence, we have to extend FindTBnds()so it produces justifications (for the bounds it asserts in FixEqs(xi)) that the SAT solver can comprehend and reproduce.

We only need to justify bounds asserted by FixEqs(xi) because all other bounds asserted by FindTBnds() are reverted to their original bounds $x_k \geq 1\ k$ and $x_k \leq u'_k$. And even in FixEqs(x_i), we only have to justify the bounds $x_k \leq 1\ k$ ($x_k \geq u'_k$) that make the tight bounds $x_k \geq 1_k$ ($x_k \leq u'\,k$) explicit. We also see that the bounds asserted by FixEqs(xi) are just linear combinations of existing bounds if we look again at the proof of Lemma 7. The proof also shows that we can derive this linear combination from the conflict explanation C of the strict system. For instance, if the call to Check() from line 3 of FindTBnds() exits in line 7 with (false, xi), then the conflict explanation is

$$C = \{x_i > l'_i\} \cup \{x_j < u'_j : j \in \mathcal{N} \text{ and } a_{ij} > 0\} \cup \{x_j > l'_j : j \in \mathcal{N} \text{ and } a_{ij} < 0\}. \quad [14]$$

If the call to Check() exits instead in line 13 with (false, xi), then the conflict explanation is

$$C = \{x_i < u'_i\} \cup \{x_j > l'_j : j \in \mathcal{N} \text{ and } a_{ij} > 0\} \cup \{x_j < u'_j : j \in \mathcal{N} \text{ and } a_{ij} < 0\}. \quad [14]$$

We receive the set of tightly propagating bounds that we found with the last call to Check() if we turn all bounds in C into non-strict bounds:

$$C' = \{x_k \leq u'_k : (x_k < u'_k) \in C\} \cup \{x_k \geq l'_k : (x_k > l'_k) \in C\}.$$

FixEqs(x_i) asserts now for every bound ($x_k \geq l'_k$)∈Cthe bound $x_k \leq l'_k$. From the proof of Lemma 7, we see that the bound $x_k \leq l'_k$ is a linear combination of the bounds $C' \setminus \{x_k \geq l'_k\}$. Hence, $x_k \leq l'_k$ is implied by the bounds $C' \setminus \{x_k \geq l'_k\}$ and, therefore, the clause

$$\left(\bigvee_{(x_i \geq l'_i) \in C', x_i \neq x_k} x_i < l'_i \right) \vee \left(\bigvee_{(x_i \leq u'_i) \in C'} x_i > u'_i \right) \vee (x_k \leq l'_k)$$

justifies the asserted bound $x_k \leq l'_k$. Together with the slack variable definitions stored in the simplex tableau, this clause is a tautology and the SAT solver can learn it without restrictions. Moreover, all literals in this clause except for $x_k \leq l'_k$ are asserted as unsatisfiable in the current model of our SAT solver. Therefore, the SAT solver can assert the literal $x_k \leq l'_k$ on its own through unit propagation.

Symmetrically, FixEqs(x_i) asserts for every bound $(x_k \leq u'_k) \in C'$ the bound $x_k \geq u'_k$ and the justification for this bound is the clause:

$$\left(\bigvee_{(x_i \geq l'_i) \in C'} x_i < l'_i \right) \vee \left(\bigvee_{(x_i \leq u'_i) \in C', x_i \neq x_k} x_i > u'_i \right) \vee (x_k \geq u'_k) .$$

But FindTBnds() is not our only method that asserts literals independently of the SAT solver. If we use the equality basis computed by FindTBnds() for a Nelson–Oppen style combination of theories, then we also assert equalities $x_i = x_k$ for all pairs of equivalent variables x_i, x_k. Hence, we also have to justify these assertions to the SAT solver.

We get these justifications by looking at the normalized representations of the variables x_i and x_k that are equivalent. The current set of non-basic variables defines a basis and, therefore, already on its own a normalized representation for all variables. Since this normalized representation only depends on the current tableau $Ax = 0^m$, it is also independent of any of the asserted bounds. The normalized representation we use for the Nelson–Oppen style combination is only an extension of this representation by the tight bounds $x_j = c_j$ of all tightly bounded non-basic variables. Therefore, the equality $x_i = x_k$ is implied by those tight bounds $x_j = c_j$ that were actively used to compute this representation.

For instance, if x_i and x_k are both non-basic, both variables must be tightly bounded so that $x_i = x_k = v$. Otherwise, they cannot have the same normalized representation. Therefore, $x_i = v$ and $x_k = v$ imply $x_i = x_k$, or as a clause:

$$(x_i < v \vee x_i > v) \vee (x_k < v \vee x_k > v) \vee (x_i = x_k) .$$

Next, we look at the case where two basic variables x_i and x_k are equivalent. But before we give the complete formal justification, let us look at an example. Let the variables x_1, x_2, x_3, x_4, x_5 be non-basic and the variables x_6 and x_7 be basic. In this example, the basic variables are defined by the non-basic variables as follows: $x_6 = 2x_1 - x_2 + 3x_4$ and $x_7 = 2x_1 - x_2 + x_5$.

Moreover, let the variables x_2, x_3, x_4, and x_5 be tightly bounded such that $x_2 = 1$, $x_3 = 0$, $x_4 = 1$, and $x_5 = 3$. If we now replace the tightly non-basic variables, in the definitions of x6 and x7 we get that both of their normalized representations are $2x_1$ and we have actively used the tight bounds $x_2 = 1$, $x_4 = 1$, $x_5 = 3$ to compute this normalization. Hence, $x_6 = x_7$ is implied by the tight bounds $x_2 = 1$, $x_4 = 1$, and $x_5 = 3$. The variables x_6 and x_7 are also equivalent if we had not asserted that $x_2 = 1$ because the normalized representation of both variables without $x_2 = 1$ is $2x_1 - x_2$. Hence, $x_2 = 1$ is redundant for the justification and $x_6 = x_7$ is also implied by just the tight bounds $x_4 = 1$ and $x_5 = 3$.

To find which tightly bounded variables are redundant, we can just look at the coefficients. If a_{ij} and a_{kj} are the same, then any tight bound $x_j = c_j$ is redundant in the justification. This gives us the following clause as a general justification:

$$\left(\bigvee_{j \in \mathcal{N}, l_j = u_j, a_{ij} \neq a_{kj}, (a_{kj}, a_{ij}) \neq (0,0)} x_j < l_j \vee x_j > u_j \right) \vee x_i = x_k$$

(1)

From this clause, we also get the justification for the mixed case, i.e., the case where xi is basic and x_k non-basic. We simply treat x_k as if it were defined as a basic variable by itself ($x_k = 1 \cdot x_k$), so $a_{kk} = 1$ and all other $a_{kj} = 0$. If we simplify these restrictions in the clause justification (1) for the case with two basic variables, then we receive the following general justification for the mixed case:

$$\left(\bigvee_{j \in \mathcal{N}, l_j = u_j, a_{ij} \neq 0, (j, a_{ij}) \neq (k,1)} x_j < l_j \vee x_j > u_j \right) \vee x_i = x_k$$

All literals in the above clauses except for $x_i = x_k$ are asserted as unsatisfiable in the current model of our SAT solver. This holds because these literals contain only tightly bounded variables. Hence, the SAT solver can assert the literal $x_i = x_k$ on its own through unit propagation. Note also that all justifications we defined are in some sense minimal: each of the above clauses is a tautology and, if we remove one literal from the clause, then it is no longer a tautology. This fact is another property that enhances any potential conflict analysis.

CONCLUSIONS

We have presented the linear cube transformation (Proposition 3), which allows us to efficiently determine whether a polyhedron contains a cube of a given edge length. Based on this transformation we have created two tests for linear integer arithmetic: the largest cube test and the unit cube test. Our tests can be integrated into SMT theory solvers without sacrificing the advantages that SMT solvers gain from the incremental structure of subsequent subproblems. Furthermore, our experiments have shown that these tests increase efficiency on certain polyhedra such that previously hard sets of constraints become trivial.

One obstacle for our cube tests are equalities. Resolving these obstacles led to an additional application of the linear cube transformation: investigating equalities. Through Lemmas 6 and 7, we have presented a method that efficiently checks whether a system of linear arithmetic constraints implies an equality at all. We use this method in the algorithm EqBasis($Ax \leq b$) to compute an equality basis $y - Dz = c$, which is a finite representation of all equalities implied by the inequalities $Ax \leq b$.

We also presented various applications for the equality basis $y - Dz = c$. (1) We can use the equality basis to eliminate all equalities from $Ax \leq b$. It is, therefore, useful as a preprocessing step for our cube tests. (2) We can use the equality basis to directly check whether an equality $h^T x = g$ is implied by $Ax \leq b$. (3) In Sect. 7, we also use the equality basis to efficiently compute all pairs of equivalent variables in $Ax \leq b$. These pairs are necessary for a backjump-free *Nelson–Oppen* style combination of theories.

The results presented in this paper have further applications. For instance, our methods for detecting implied equalities are also useful for quantifier elimination. In general, a *quantifier elimination* (QE) procedure takes a formula $\exists y.\phi(y)$, where $\phi(y)$ itself is quantifier-free but may contain extra variables x called *parameters*, and returns an equivalent formula ϕ' that is quantifier-free. *Linear virtual substitution* is a complete QE procedure for the theory of linear rational arithmetic [23]. It eliminates the variable y by creating a case distinction exploiting the following fact: a linear real arithmetic formula $\phi(y)$ is satisfiable if and only if $\phi(l)$ is satisfiable, where l is the strictest lower bound (or upper bound) of y, i.e., the smallest value for y in any solution to the problem. This value is either represented by one of the inequalities in $\phi(y)$ containing y or $-\infty$ ($+\infty$). There are only finitely many inequalities in $\phi(y)$, so by a case distinction over all inequalities containing y satisfiability can be preserved:

$$\exists y.\phi(y) \equiv \phi(-\infty) \vee \bigvee_{a_{iy}y + a_i^T x \leq b_i \text{ in } \phi(y) \text{ with } a_{iy} < 0} \phi\left(-\frac{b_i}{a_{iy}} + \frac{a_i^T x}{a_{iy}}\right).$$

This case distinction is the source of the worst-case doubly exponential complexity of the procedure in case of quantifier alternations. At the same time, there are also instances that we can resolve without case distinctions. For instance, if the formula $\phi(y)$ implies an equality $h_y \cdot y + h^T x = g$ where $h_y \neq 0$, then we already know one guaranteed definition for the strictest lower bound of *y*:

$$\frac{g}{h_y} - \frac{h^T x}{h_y}.$$

A quantifier-free formula that is equivalent to the original one is simply:

$$\exists y.\phi(y) \equiv \phi\left(\frac{g}{h_y} - \frac{h^T x}{h_y}\right).$$

This technique is well-known and integrated in many QE implementations [12, 23, 30]. Even so, we are unaware of any implementation that makes use of non-explicit equalities for this purpose. This is where our methods that find implicit equalities come into play. Our methods are applicable because QE procedures typically keep ϕ in a disjunctive form and the respective disjuncts contain often only conjuncts of inequalities. This allows us to efficiently search for an equality.

For future research, we plan to implement the methods around the equality basis and investigate their performance for the above mentioned applications. Moreover, we want to work out even more applications for the linear cube transformation.

Footnotes

1. 1If we combine the equality basis with a *diophantine equation handler* [16], then we even receive a substitution σ' that eliminates the equalities in such a way that we can reconstruct an integer solution from them. The result is a new system of inequalities that implies no equalities and has an integer solution if and only if $Ax \leq b$ has one.
2. Actually, the literals we assert are full inequalities $a_i^T x \leq bi$. Due to slacking, the left side of those constraints is abstracted to a slack

variable *s* such that $s = a^T_i x$. The definition of the slack variable $s = a^T_i x$ is directly stored in the simplex solver and only a bound $s \leq bi$ remains as the literal for the SMT solver.

ACKNOWLEDGEMENTS

Open access funding provided by Max Planck Society. The authors would like to thank the anonymous reviewers of FMSD, IJCAR 2016, and SMT 2016 for their valuable comments, suggestions, and for directing us to related work. Special thanks are also due to Bruno Dutertre, Tim King, and Andrew Reynolds for drawing our attention to additional applications.

REFERENCES

1. Barrett C, Conway C, Deters M, Hadarean L, Jovanovi´c D, King T, Reynolds A, Tinelli C (2011) CVC4. In: CAV, LNCS, vol 6806, pp 171–177
2. Beale EML (1954) An alternative method for linear programming, vol 50, issue 4, pp 513–523
3. Bjørner N (1999) Integrating decision procedures for temporal verification. Ph.D. thesis, Stanford, CA, USA
4. Bobot F, Conchon S, Contejean E, Iguernelala M, Mahboubi A, Mebsout A, Melquiond G (2012) A simplex-based extension of Fourier–Motzkin for solving linear integer arithmetic. In: IJCAR 2012, LNCS, vol 7364, pp 67–81
5. Boyd S, Vandenberghe L (2004) Convex optimization. Cambridge University Press, Cambridge
6. Bromberger M, Sturm T, Weidenbach C (2015) Linear integer arithmetic revisited. In: CADE-25, LNCS, vol 9195, pp 623–637
7. Bromberger M, Weidenbach C (2016) Computing a complete basis for equalities implied by a system of LRA constraints. In: SMT 2016, pp 15–30
8. Bromberger M, Weidenbach C (2016) Fast cube tests for lia constraint solving. In: IJCAR 2016, LNCS, vol 9706
9. Bruttomesso R, Cimatti A, Franzen A, Griggio A, Sebastiani R (2009) Delayed theory combination vs. Nelson–Oppen for satisfiability modulo theories: a comparative analysis. AMAI 55(1):63–99
10. Cimatti A, Griggio A, Schaafsma B, Sebastiani R (2013) The MathSAT5 SMT Solver. In: TACAS, LNCS, vol 7795
11. Dillig I, Dillig T, Aiken A (2009) Cuts from proofs: a complete and practical technique for solving linear inequalities over integers. In: CAV, LNCS, vol 5643, pp 233–247
12. Dolzmann A, Sturm T, Weispfenning V (1999) Real quantifier elimination in practice. In: Algorithmic algebra and number theory. Springer, pp 221–247
13. Dutertre B (2014) Yices 2.2. In: CAV 2014, LNCS, vol 8559
14. Dutertre B, de Moura, L (2006) A fast linear-arithmetic solver for DPLL(T). In: CAV, LNCS, vol. 4144, pp. 81–94 (2006). Extended version: Integrating simplex with DPLL(T). Technical report, CSL,

SRI International

15. Faure G, Nieuwenhuis R, Oliveras A, Rodríguez-Carbonell E (2008) Sat modulo the theory of linear arithmetic: exact, inexact and commercial solvers. In: SAT 2008, LNCS, vol 4996, pp 77–90
16. Griggio A (2012) A practical approach to satisfiability modulo linear integer arithmetic. JSAT 8(1/2):1–27
17. Hillier FS (1969) Efficient heuristic procedures for integer linear programming with an interior. Oper Res 17(4):600–637
18. Jovanovi´c D, de Moura L (2013) Cutting to the chase. JAR 51(1):79–108
19. Jünger M, Liebling TM, Naddef D, Nemhauser GL, Pulleyblank WR, Reinelt G, Rinaldi G, Wolsey LA (eds) (2010) 50 years of integer programming 1958–2008
20. Kannan R, Lovász L (1986) Covering minima and lattice point free convex bodies. FSTTCS, LNCS, vol 241, pp 193–213
21. Karmarkar N (1984) A new polynomial-time algorithm for linear programming. Combinatorica 4(4):373– 396
22. Lemke CE (1954) The dual method of solving the linear programming problem. Nav Res Logist Quart 1(1):36–47
23. Loos R, Weispfenning V (1993) Applying linear quantifier elimination. Comput J 36(5):450–462
24. de Moura L, Bjørner N (2008) Z3: an efficient SMT solver. In: Tools and algorithms for the construction and analysis of systems, LNCS, vol 4963, pp 337–340
25. Papadimitriou CH (1981) On the complexity of integer programming. J ACM 28(4):765–768
26. Pugh W (1991) The omega test: a fast and practical integer programming algorithm for dependence analysis. In: Supercomputing 1991, Supercomputing '91. ACM, pp 4–13
27. Refalo P (1998) Approaches to the incremental detection of implicit equalities with the revised simplex method. In: PLILP 1998, LNCS, vol 1490, pp 481–496
28. Rueß H, Shankar N (2004) Solving linear arithmetic constraints. Technical report, SRI International, Computer Science Laboratory
29. Schrijver A (1986) Theory of linear and integer programming. Wiley, New York

30. Sturm T (1996) Real quadratic quantifier elimination in risa/asir. Technical report, ISIS-RM-5E, Fujitsu Laboratories Ltd
31. Telgen J (1983) Identifying redundant constraints and implicit equalities in systems of linear constraints. Manag Sci 29(10):1209–1222
32. Van Hentenryck P, Graf T (1992) Standard forms for rational linear arithmetic in constraint logic programming. AMAI 5(2):303–319

INDEX

D

E

F

U

V

W